AF598011

Methods in Molecular Biology

Series Editor
John M. Walker
School of Life Sciences
University of Hertfordshire
Hatfield, Hertfordshire, AL10 9AB, UK

For further volumes:
http://www.springer.com/series/7651

Protocols in In Vitro Hepatocyte Research

Edited by

Mathieu Vinken

Department of In Vitro Toxicology and Dermato-Cosmetology, Vrije Universiteit Brussel, Belgium

Vera Rogiers

Department of In Vitro Toxicology and Dermato-Cosmetology, Vrije Universiteit Brussel, Belgium

Editors
Mathieu Vinken
Department of In Vitro Toxicology
and Dermato-Cosmetology
Vrije Universiteit Brussel
Belgium

Vera Rogiers
Department of In Vitro Toxicology
and Dermato-Cosmetology
Vrije Universiteit Brussel
Belgium

ISSN 1064-3745 ISSN 1940-6029 (electronic)
ISBN 978-1-4939-2073-0 ISBN 978-1-4939-2074-7 (eBook)
DOI 10.1007/978-1-4939-2074-7
Springer New York Heidelberg Dordrecht London

Library of Congress Control Number: 2014954297

© Springer Science+Business Media New York 2015
This work is subject to copyright. All rights are reserved by the Publisher, whether the whole or part of the material is concerned, specifically the rights of translation, reprinting, reuse of illustrations, recitation, broadcasting, reproduction on microfilms or in any other physical way, and transmission or information storage and retrieval, electronic adaptation, computer software, or by similar or dissimilar methodology now known or hereafter developed. Exempted from this legal reservation are brief excerpts in connection with reviews or scholarly analysis or material supplied specifically for the purpose of being entered and executed on a computer system, for exclusive use by the purchaser of the work. Duplication of this publication or parts thereof is permitted only under the provisions of the Copyright Law of the Publisher's location, in its current version, and permission for use must always be obtained from Springer. Permissions for use may be obtained through RightsLink at the Copyright Clearance Center. Violations are liable to prosecution under the respective Copyright Law.
The use of general descriptive names, registered names, trademarks, service marks, etc. in this publication does not imply, even in the absence of a specific statement, that such names are exempt from the relevant protective laws and regulations and therefore free for general use.
While the advice and information in this book are believed to be true and accurate at the date of publication, neither the authors nor the editors nor the publisher can accept any legal responsibility for any errors or omissions that may be made. The publisher makes no warranty, express or implied, with respect to the material contained herein.

Printed on acid-free paper

Humana Press is a brand of Springer
Springer is part of Springer Science+Business Media (www.springer.com)

Preface

Legislation in the area of safety testing of chemicals has drastically changed during the last decade. Thus, adopted in 2007, the European chemicals policy, commonly known as REACH (i.e., *r*egistration, *e*valuation, *a*uthorization and restriction of *ch*emicals), demands the safety assessment of thousands of chemical substances [1]. Clearly, the large-scale testing exercise imposed by the REACH regulation inherently requires high amounts of laboratory animals, not only raising serious ethical questions but also having a substantial economic impact. On the other hand, in compliance with the European Regulation 1223/2009, which became fully valid in 2013, animals can no longer be used for the toxicity testing of cosmetic products and their ingredients [2]. These clear-cut regulatory developments collectively illustrate that there is a ubiquitous need for alternative methods for the safety assessment of chemicals, such as in vitro testing platforms, which is in line with the 3R concept of Russell and Burch, calling for *r*efinement, *r*eduction, and *r*eplacement of animal experimentation [3]. In fact, the tendency for adopting nonanimal methods for toxicity testing is not only expected to affect other chemical areas in Europe in the near future, including the plant protection product and pharmaceutical fields, but also gains increasing importance in other parts of the world (e.g., the USA, Brazil, Canada, and Korea).

The liver fulfills a number of vital functions in the body, such as the control of lipid and carbohydrate homeostasis, the synthesis of various proteins, the storage of vitamins, the processing of bile, immunological defense, and the biotransformation of endogenous molecules and xenobiotics. Most of these functions are mediated by the hepatocytes, which are parenchymal cells that constitute as much as 80 % of the total liver mass [4]. Inherent to these critical tasks, in particular their biotransformation activity, hepatocytes form major targets for disease and systemic toxicity. For this reason, most attention has been traditionally paid, and is still being paid, to liver-based in vitro models in the area of 3R alternative methods for toxicity testing. A plethora of hepatic in vitro systems is currently available, ranging from hepatocyte subcellular fractions to whole isolated perfused livers [5, 6]. Among those, cultures of primary hepatocytes, especially of human origin, are generally considered as the gold standard in the field of liver-based in vitro modeling [6, 7]. Primary hepatocytes are typically isolated from freshly removed livers by means of a two-step collagenase perfusion technique [8]. As they are scarce, isolated human hepatocytes may be cryopreserved prior to cultivation and use [9]. Primary hepatocytes provide a good reflection of the hepatic in vivo situation for merely a couple days when properly cultured. Hence, their conventional cultures can only be used for short-term purposes. Indeed, long-term cultivation of primary hepatocytes is largely impeded by the progressive loss of the hepatocyte-specific phenotype both at the morphological and at the functional level. This so-called dedifferentiation process, which is initiated during the hepatocyte isolation procedure, can be counteracted, at least in part, by a number of classical cultivation strategies that try to restore the natural hepatocyte microenvironment in vitro as well as by a number of novel (epi)genetic approaches [5, 7, 10, 11]. Such optimized primary hepatocyte cultures are of great value, not only for long-term toxicity testing schemes but also for studying basic hepatocyte (patho)physiology. In this context, insight into the biological development of hepatocytes has greatly increased in the last few years through in vitro differentiation of different types of stem cells into hepatocyte-like cells using a number of experimental approaches [12–15]. Along the same line, proliferative events can be mimicked in vitro either by cultivating primary hepatocytes in very specific conditions [16] or by using pathologically altered or artificially modified hepatocytes [17]. In addition, primary hepatocyte cultures have been proven ideal tools to study the biological counterpart of liver cell growth, namely hepatocyte cell death [18].

Functional hallmarks of cultured primary hepatocytes, such as metabolic competence, are frequently evaluated by monitoring the transcriptional and translational levels of liver-specific gene expression. Expression studies using transcriptomics and proteomics methodologies can indeed shed light onto the comprehensive suite of events at the most upstream regulatory level of liver-specific functionality. More recently, other disciplines have also entered the "-omics" arena, including metabonomics and epigenomics [6]. Such cutting-edge technologies can be equally applied for evaluating the response of cultured primary hepatocytes to toxic substances. Among those are several compounds which may trigger different types of drug-induced liver injury. Drug-induced liver injury is of high clinical concern, as it is the leading cause of acute liver failure. Furthermore, drug-induced liver injury is also of clear relevance to pharmaceutical industry, since it underlies the withdrawal of a considerable number of drugs during premarketing and postmarketing phases of drug development [19]. These various types of hepatotoxicity all have their own very specific morphological and functional characteristics that can be investigated in liver-based in vitro models, including primary hepatocyte cultures [6, 20].

The present book provides a state-of-the-art compilation of protocols and reviews to practically set up and apply primary hepatocyte cultures for research and screening purposes. In a first part, methods for isolating hepatocytes from liver tissue (*see* Chapter 1) and their cryopreservation (*see* Chapter 2) as well strategies for studying the different aspects of the hepatocyte life cycle in vitro are outlined. In particular, hepatocyte proliferation can be induced in vitro by using specific mitogens as cell culture medium additives for primary hepatocytes (*see* Chapter 3) or by using immortalized (*see* Chapter 4) or tumor-derived (*see* Chapter 5) hepatic cell lines. Following is a protocol on how to study and evaluate apoptotic cell death in cultured primary hepatocytes (*see* Chapter 6). Thereafter, the in vitro generation of hepatocyte-like cells from embryonic stem cells (*see* Chapter 7), adult stem cells (*see* Chapter 8), and induced pluripotent stem cells (*see* Chapter 9) is described. Considerable attention is subsequently paid to approaches to maintain the adult differentiated status in cultured hepatocytes, including the use of differentiation-promoting cell culture medium additives (*see* Chapter 10), the re-establishment of heterologous cell–cell contacts (*see* Chapter 11), the re-introduction of an extracellular matrix backbone (*see* Chapter 12), the setup of advanced liver bioreactors (*see* Chapter 13), chromatin remodeling (*see* Chapter 14), and the overexpression of genes that promote the liver-specific phenotype (*see* Chapter 15).

In the second part of the book, a number of routinely used and recently introduced in vitro techniques to monitor hepatocyte functionality and toxicity are presented. Global profiling at the transcriptomic (*see* Chapter 16), epigenomic (*see* Chapter 17), proteomic (*see* Chapter 18), and metabonomic (*see* Chapter 19) level can provide detailed mechanistic information on modifications in cellular pathways in primary hepatocyte cultures upon exposure to xenobiotics, whether or not of deleterious nature. The latter may affect critical liver-specific functions, which can be probed in vitro, namely biotransformation capacity (*see* Chapter 20), drug transporter activity (*see* Chapter 21), albumin secretion (*see* Chapter 22), synthesis of blood coagulation factors (*see* Chapter 23), ammonia detoxification (*see* Chapter 24), and bile secretion (*see* Chapter 25). In the final section, protocols to evaluate toxic responses in liver-based in vitro models are presented, both general cytotoxicity (*see* Chapter 26) and cell death (*see* Chapter 27) as well as specific types of drug-induced liver injury, including cholestasis (*see* Chapter 28), steatosis (*see* Chapter 29), and hepatic fibrosis (*see* Chapter 30).

The current book is intended for basic and applied researchers, ranging from the undergraduate to the postdoctoral and professional level, in the area of pharmacology and toxicology, both in academic and industrial settings. It can be used by investigators familiar and unfamiliar with the field of liver-based in vitro modeling and testing.

At the start of this book, the editors would like to express their deepest gratitude to all chapter contributors. Furthermore, the editors greatly acknowledge the Springer team and, in particular, series editor John M. Walker for his continuous assistance during the preparation of this book.

Brussels, Belgium

Mathieu Vinken
Vera Rogiers

References

1. Regulation (EC) No. 1907/2006 of the European Parliament, of the Council of 18 December 2006 concerning the Registration, Evaluation, Authorisation, Restriction of Chemicals (REACH), establishing a European Chemicals Agency, amending Directive 1999/45/EC, repealing Council Regulation (EEC) No 793/93, Commission Regulation (EC) No 1488/94 as well as Council Directive 76/769/EEC, Commission Directives 91/155/EEC, 93/67/EEC, 93/105/EC, 2000/21/EC. Off J Eur Union L396:1–849
2. Regulation (EC) No. 1223/2009 of the European Parliament, of the Council of 30 November 2009 on cosmetic products (recast). Off J Eur Union L342:59–209
3. Russell WMS, Burch RL (1959) The principles of humane experimental technique. Methuen, Co Ltd., UK
4. Arias IM, Alter HJ, Boyer JL, Cohen DE, Fausto N, Shafritz DA, Wolkoff AW (2009) The liver: biology, pathobiology. 5th edn. Wiley-Blackwell, Oxford
5. Vinken M, Vanhaecke T, Rogiers V (2012) Primary hepatocyte cultures as in vitro tools for toxicity testing: *quo vadis*? Toxicol In Vitro 26:541–544
6. Godoy P, Hewitt NJ, Albrecht U et al (2013) Recent advances in 2D, 3D in vitro systems using primary hepatocytes, alternative hepatocyte sources, non-parenchymal liver cells, their use in investigating mechanisms of hepatotoxicity, cell signaling, ADME. Arch Toxicol 87:1315–1530
7. Fraczek J, Bolleyn J, Vanhaecke T et al (2013) Primary hepatocyte cultures for pharmaco-toxicological studies: at the busy crossroad of various anti-dedifferentiation strategies. Arch Toxicol 87:577–610
8. Papeleu P, Vanhaecke T, Henkens T et al (2006) Isolation of rat hepatocytes. Methods Mol Biol 320:229–237
9. Hewitt NJ (2010) Optimisation of the cryopreservation of primary hepatocytes. Methods Mol Biol 640:83–105
10. Vinken M, Papeleu P, Snykers S et al (2006) Involvement of cell junctions in hepatocyte culture functionality. Crit Rev Toxicol 36:299–318
11. Elaut G, Henkens T, Papeleu P et al (2006) Molecular mechanisms underlying the dedifferentiation process of isolated hepatocytes, their cultures. Curr Drug Metab 7:629–660
12. Snykers S, De Kock J, Vanhaecke T et al (2011) Hepatic differentiation of mesenchymal stem cells: in vitro strategies. Methods Mol Biol 698:305–314
13. Snykers S, De Kock J, Rogiers V et al (2009) In vitro differentiation of embryonic, adult stem cells into hepatocytes: state of the art. Stem Cells 27:577–605
14. Al Battah F, De Kock J, Vanhaecke T et al (2011) Current status of human adipose-derived stem cells: differentiation into hepatocyte-like cells. Sci World J 11:1568–1581
15. Szkolnicka D, Zhou W, Lucendo-Villarin B et al (2013) Pluripotent stem cell-derived hepatocytes: potential, challenges in pharmacology. Annu Rev Pharmacol Toxicol 53:147–159

16. Papeleu P, Loyer P, Vanhaecke T et al (2004) Proliferation of epidermal growth factor-stimulated hepatocytes in a hormonally defined serum-free medium. Altern Lab Anim 32:57–64
17. Donato MT, Lahoz A, Castell JV et al (2008) Cell lines: a tool for in vitro drug metabolism studies. Curr Drug Metab 9:1–11
18. Vinken M, Maes M, Oliveira AG et al (2014) Primary hepatocytes, their cultures in liver apoptosis research. Arch Toxicol 88:199–212
19. Vinken M, Maes M, Vanhaecke T et al (2013) Drug-induced liver injury: mechanisms, types, biomarkers. Curr Med Chem 20:3011–3021
20. Gomez-Lechon MJ, Lahoz A, Gombau L et al (2010) In vitro evaluation of potential hepatotoxicity induced by drugs. Curr Pharm Des 16:1963–1977

Contents

Contributors

PAULA M. ALVES • *iBET, Instituto de Biologia Experimental e Tecnológica, Oeiras, Portugal*
PIETER ANNAERT • *Drug Delivery, Disposition, Department of Pharmaceutical, Pharmacological Sciences, Katholieke Universiteit Leuven, Leuven, Belgium*
MARTA BENET • *Instituto de Investigación Sanitaria La Fe, Unidad de Hepatología Experimental, Valencia, Spain*
JENNIFER BOLLEYN • *Department of In Vitro Toxicology, Dermato-Cosmetology, Vrije Universiteit Brussel, Brussels, Belgium*
ROQUE BORT • *Experimental Hepatology Unit-CIBERehd, IIS Hospital La Fe, Valencia, Spain*
JORIS BRASPENNING • *Medicyte GmbH, Heidelberg, Germany*
CATARINA BRITO • *iBET, Instituto de Biologia Experimental e Tecnológica, Oeiras, Portugal*
KAROLIEN BUYL • *Department of In Vitro Toxicology, Dermato-Cosmetology, Vrije Universiteit Brussel, Brussels, Belgium*
KATE CAMERON • *MRC Centre for Regenerative Medicine, University of Edinburgh, Edinburgh, UK*
SANDRA COECKE • *EURL ECVAM, Systems Toxicology Unit, JRC, Institute for Health, Consumer Protection, European Commission, Joint Research Center, Ispra, Italy*
BRUNO COGLIATI • *Department of Pathology, School of Veterinary Medicine, Animal Science, University of São Paulo, São Paulo, Brazil*
MERAV COHEN • *Department of Cell, Developmental Biology, Silberman Institute of Life Sciences, The Hebrew University of Jerusalem, Jerusalem, Israel*
MAR COLL • *Laboratory of Liver Fibrosis, Institut d'Investigacions Biomèdiques August Pi i Sunyer, Barcelona, Spain*
ANNE CORLU • *Inserm, UMR 991, Liver, Metabolisms, Cancer, University of Rennes 1, Rennes Cedex, France*
RITA COSTA • *iBET, Instituto de Biologia Experimental e Tecnológica, Oeiras, Portugal*
CLAIRE DENIZOT • *Centre de Pharmacocinétique, Technologie Servier, Orléans, France*
LISA DIETZ • *Leibniz-Institut für Analytische Wissenschaften-ISAS-e.V., Dortmund, Germany*
TATYANA Y. DOKTOROVA • *Department of In Vitro Toxicology, Dermato-Cosmetology, Vrije Universiteit Brussel, Brussels, Belgium*
MARÍA TERESA DONATO • *Unidad de Hepatología Experimental, Instituto de Investigación Sanitaria La Fe, Valencia, Spain; Bioquímica y Biología Molecular, Facultad de Medicina, Universidad de Valencia, Valencia, Spain; CIBERehd, FIS, Barcelona, Spain*
MINA EDWARDS • *Institute of Structural, Molecular Biology, University College London, London, UK*
EWA C.S. ELLIS • *Unit of Transplantation Surgery, Liver Cell Laboratory, Department of Clinical Investigation, Science, Technology, Karolinska University Hospital, Karolinska Institute, Stockholm, Sweden*

ALMUDENA ESPÍN-PÉREZ • *Department of Toxicogenomics, Maastricht University, Maastricht, The Netherlands*
OLIVIER FARDEL • *Institut de Recherches en Santé, Environnement et Travail, Université de Rennes 1, Rennes, France*
JOANNA FRACZEK • *Department of In Vitro Toxicology, Dermato-Cosmetology, Vrije Universiteit Brussel, Brussels, Belgium*
RICHARD L. GIESECK III • *Wellcome Trust-Medical Research Council Stem Cell Institute, Anne McLaren Laboratory for Regenerative Medicine, Department of Surgery, University of Cambridge, Cambridge, UK*
MARÍA JOSÉ GÓMEZ-LECHÓN • *Unidad de Hepatología Experimental, Instituto de Investigación Sanitaria La Fe, Valencia, Spain; CIBERehd, FIS, Barcelona, Spain*
VARVARA GOULIARMOU • *EURL ECVAM, Systems Toxicology Unit, JRC, Institute for Health, Consumer Protection, European Commission, Joint Research Center, Ispra, Italy*
NICHOLAS R.F. HANNAN • *Wellcome Trust-Medical Research Council Stem Cell Institute, Anne McLaren Laboratory for Regenerative Medicine, Department of Surgery, University of Cambridge, Cambridge, UK*
DAVID C. HAY • *MRC Centre for Regenerative Medicine, University of Edinburgh, Edinburgh, UK*
STEFAN HEINZ • *Medicyte GmbH, Heidelberg, Germany*
JAN HENGSTLER • *Leibniz Research Centre for Working Environment, Human Factors, Technical University of Dortmund, Dortmund, Germany*
JÜRGEN HESCHELER • *Center of Physiology, Pathophysiology, Institute of Neurophysiology, University of Cologne, Cologne, Germany*
NICOLA J. HEWITT • *SWS, Erzhausen, Germany*
LYNDSEY HOUSEMAN • *Institute of Structural, Molecular Biology, University College London, London, UK*
HARTMUT JAESCHKE • *Department of Pharmacology, Toxicology, Therapeutics, University of Kansas Medical Center, Kansas City, KS, USA*
HELENE JOHANSSON • *Unit of Transplantation Surgery, Liver Cell Laboratory, Department of Clinical Investigation, Science, Technology, Karolinska University Hospital, Karolinska Institute, Stockholm, Sweden*
ELODIE JOUAN • *Institut de Recherches en Santé, Environnement et Travail, Université de Rennes 1, Rennes, France*
RAMIRO JOVER • *Instituto de Investigación Sanitaria La Fe, Unidad de Hepatología Experimental, Valencia, Spain*
JANNEKE KEEMINK • *Drug Delivery, Disposition, Department of Pharmaceutical, Pharmacological Sciences, Katholieke Universiteit Leuven, Leuven, Belgium*
JOS C. KLEINJANS • *Department of Toxicogenomics, Maastricht University, Maastricht, The Netherlands*
JOERY DE KOCK • *Department of In Vitro Toxicology, Dermato-Cosmetology, Vrije Universiteit Brussel, Brussels, Belgium*
THEO M. DE KOK • *Department of Toxicogenomics, Maastricht University, Maastricht, The Netherlands*
MAREK KOŘÍNEK • *Apigenex s.r.o., Prague, Czech Republic*
JULIAN KRAUSKOPF • *Department of Toxicogenomics, Maastricht University, Maastricht, The Netherlands*
GAHL LEVY • *Alexander Grass Center for Bioengineering, Benin School of Computer Science, Engineering, The Hebrew University of Jerusalem, Jerusalem, Israel*

Albert P. Li • *In Vitro ADMET Laboratories LLC, Columbia, MD, USA*
Pascal Loyer • *Inserm, UMR 991, Liver, Metabolisms, Cancer, University of Rennes 1, Rennes Cedex, France*
Baltasar Lucendo-Villarin • *MRC Centre for Regenerative Medicine, University of Edinburgh, Edinburgh, UK*
Michaël Maes • *Department of In Vitro Toxicology, Dermato-Cosmetology, Vrije Universiteit Brussel, Brussels, Belgium*
Yaakov Nahmias • *Department of Cell, Developmental Biology, Silberman Institute of Life Sciences, The Hebrew University of Jerusalem, Jerusalem, Israel*
Harshal Nemade • *Center of Physiology, Pathophysiology, Institute of Neurophysiology, University of Cologne, Cologne, Germany*
Marlies Oorts • *Drug Delivery, Disposition, Department of Pharmaceutical, Pharmacological Sciences, Katholieke Universiteit Leuven, Leuven, Belgium*
Yannick Parmentier • *Centre de Pharmacocinétique, Technologie Servier, Orléans, France*
Olavi Pelkonen • *Department of Pharmacology, Toxicology, Institute of Biomedicine, Medical Research Center Oulu, University of Oulu, Oulu, Finland*
Luis Perea • *Laboratory of Liver Fibrosis, Institut d'Investigacions Biomèdiques August Pi i Sunyer, Barcelona, Spain*
Ian R. Phillips • *Institute of Structural, Molecular Biology, University College London, London, UK*
Eva Ramboer • *Department of In Vitro Toxicology and Dermato-Cosmetology, Vrije Universiteit Brussel, Belgium*
Sofia P. Rebelo • *iBET, Instituto de Biologia Experimental e Tecnológica, Oeiras, Portugal*
Robim M. Rodrigues • *Department of In Vitro Toxicology, Dermato-Cosmetology, Vrije Universiteit Brussel, Belgium*
Vera Rogiers • *Department of In Vitro Toxicology and Dermato-Cosmetology, Vrije Universiteit Brussel, Brussels, Belgium*
Agapios Sachinidis • *Center of Physiology, Pathophysiology, Institute of Neurophysiology, University of Cologne, Cologne, Germany*
Pau Sancho-Bru • *Liver Unit, Hospital Clínic, Institut d'Investigacions Biomèdiques August Pi i Sunyer, Barcelona, Spain*
Irena Selicharová • *Institute of Organic Chemistry, Biochemistry, Academy of Sciences of the Czech Republic, v.v.i., Prague, Czech Republic*
Elizabeth A. Shephard • *Institute of Structural, Molecular Biology, University College London, London, UK*
Vaibhav Shinde • *Center of Physiology, Pathophysiology, Institute of Neurophysiology, University of Cologne, Cologne, Germany*
Albert Sickmann • *Leibniz-Institut für Analytische Wissenschaften-ISAS-e.V., Dortmund, Germany*
Isaia Sotiriadou • *Center of Physiology, Pathophysiology, Institute of Neurophysiology, University of Cologne, Cologne, Germany*
Marcos F.Q. Sousa • *iBET, Instituto de Biologia Experimental e Tecnológica, Oeiras, Portugal*
Regina Stöber • *Leibniz Research Centre for Working Environment, Human Factors, Technical University of Dortmund, Dortmund, Germany*
Dagmara Szkolnicka • *MRC Centre for Regenerative Medicine, University of Edinburgh, Edinburgh, UK*

LAIA TOLOSA • *Unidad de Hepatología Experimental, Instituto de Investigación Sanitaria La Fe, Valencia, Spain*

LUDOVIC VALLIER • *Wellcome Trust-Medical Research Council Stem Cell Institute, Anne McLaren Laboratory for Regenerative Medicine, Department of Surgery, University of Cambridge, Cambridge, UK*

TAMARA VANHAECKE • *Department of In Vitro Toxicology, Dermato-Cosmetology, Vrije Universiteit Brussel, Brussels, Belgium*

MARC LE VÉE • *Institut de Recherches en Santé, Environnement et Travail, Université de Rennes 1, Rennes, France*

MATHIEU VINKEN • *Department of In Vitro Toxicology and Dermato-Cosmetology, Vrije Universiteit Brussel, Belgium*

JOOST WILLEBRORDS • *Department of In Vitro Toxicology, Dermato-Cosmetology, Vrije Universiteit Brussel, Brussels, Belgium*

BENJAMIN L. WOOLBRIGHT • *Department of Pharmacology, Toxicology, Therapeutics, University of Kansas Medical Center, Kansas City, KS, USA*

SARA CRESPO YANGUAS • *Department of In Vitro Toxicology, Dermato-Cosmetology, Vrije Universiteit Brussel, Brussels, Belgium*

Part I

In Vitro Models of the Hepatocyte Life Cycle

Chapter 1

Isolation and Culture of Mouse Hepatocytes: Gender-Specific Gene Expression Responses to Chemical Treatments

Lyndsey Houseman, Mina Edwards, Ian R. Phillips, and Elizabeth A. Shephard

Abstract

In this chapter, the isolation of primary mouse hepatocytes and their response to chemical treatment are described. We show that it is important to consider, in the experimental design, the sex of the animals to be used. We demonstrate this by measuring the effect of sex hormones or xenobiotics on the expression of flavin-containing monooxygenase 5 in cultures of primary hepatocytes isolated from male and female mice.

Key words Mouse hepatocyte isolation, Mouse hepatocyte viability, Hormones

1 Introduction

Primary cultures of hepatocytes are a useful tool to study cellular processes, such as changes in gene expression and endogenous and xenobiotic responses. Primary hepatocytes are not wholly representative of a normal liver, given that the cells are removed from their situation in vivo and forced to grow in vitro. Nevertheless, the functional capabilities of freshly isolated hepatocytes more closely mirror the capacities of a normal liver in vivo than do liver-derived cell lines [1]. Previously, we have reported methods for the isolation of mouse hepatocytes [2] and rat hepatocytes and their transfection [3]. Here, we extend the protocol for the isolation of mouse hepatocytes to include considerations of gender-specific responses to chemical treatments.

Flavin-containing monooxygenase 5 (FMO5) is highly expressed in the liver of male and female mice, with the expression being highest in males [4]. We use the expression of FMO5 to illustrate how results obtained following chemical treatment can differ depending on whether hepatocytes are isolated from male or female mice. FMO5 expression is induced to a similar extent in

Mathieu Vinken and Vera Rogiers (eds.), *Protocols in In Vitro Hepatocyte Research*, Methods in Molecular Biology, vol. 1250, DOI 10.1007/978-1-4939-2074-7_1, © Springer Science+Business Media New York 2015

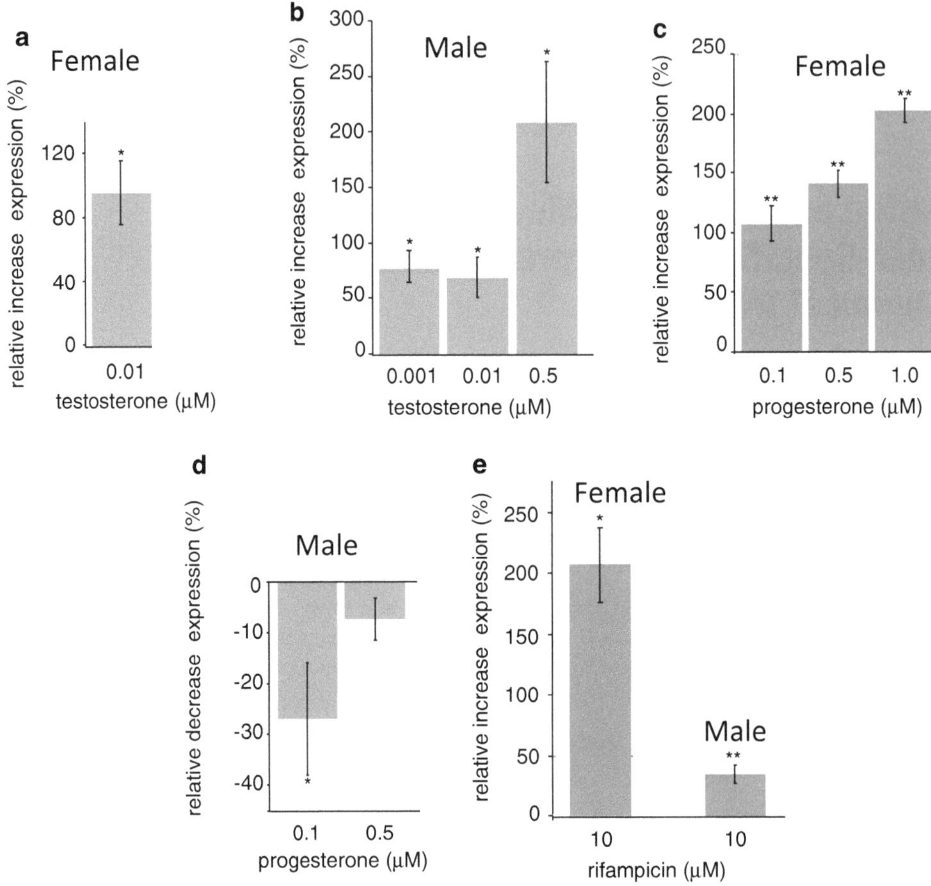

Fig. 1 Effects of hormones and antibiotics on FMO5 expression in freshly isolated mouse hepatocytes. Hepatocytes were isolated from female (**a**, **c**, **e**) or male (**b**, **d**, **e**) mice and were incubated with selected chemicals for 24 h. Following RNA extraction, FMO5 mRNA levels were measured by quantitative real-time reverse transcriptase polymerase chain reaction analysis. In the graphs, 0 % represents the expression of FMO5 mRNA when cells are treated with the vehicle alone. Significance is calculated with respect to vehicle alone (*$p < 0.05$ and **$p < 0.005$). Error bars are standard error of the mean

male and female mouse hepatocyte cultures treated with testosterone (Fig. 1). However, progesterone treatment increases FMO5 expression only in female hepatocyte cultures (Fig. 1). The difference in response to progesterone by the male and female hepatocytes could be due to differential expression of progesterone receptor isoforms (i.e., A or B) in males and females [5, 6]. For instance, FMO5 was induced up to tenfold in a breast cancer cell line stably expressing progesterone B receptor, but not in cells stably expressing progesterone receptor A [7].

FMO5 expression is known to be responsive to chemical exposure to the antibiotic rifampicin [8]. In hepatocytes from female mice treated with rifampicin, a pregnane X receptor ligand, the expression of FMO5 is much higher than that observed in treated male hepatocytes (Fig. 1). These results with different chemical treatments of male and female hepatocytes demonstrate differences in degree of response of the sexes. This shows that the response to a particular chemical treatment of female and male hepatocytes can be the same (i.e., induction in both sexes), different (i.e., induction in only one sex), or can differ in extent (i.e., levels of expression differ markedly between the two sexes). Sex differences in drug-metabolizing enzymes are known in vivo [9]. In this chapter, we demonstrate that the liver, even when disrupted by collagenase treatment and subjected to culture conditions, exhibits a memory of gender.

2 Materials

1. 5.7 mg/mL ketamine in deionized water. Stable for 6 months at −80 °C.
2. 2.6 mg/mL xylamine hydrochloride in deionized water. Prepare just before use.
3. Anesthetizing combined solution. Mix 1.5 mL of the ketamine solution and 1.5 mL of the xylamine hydrochloride solution. Mix well and sterilize by passing through a 0.22 μm filter.
4. 50 IU/mL sodium heparin. Sterilize by filtration through a 0.22 μm filter and store at 4 °C. Stable for months at 4 °C.
5. 0.4 % trypan blue in 0.85 % NaCl.
6. 100 mM sodium pyruvate in distilled water. Filter sterilize. Stable up to 24 months at 4 °C.
7. 50 mM ethylene glycol tetraacetic acid in water, pH 7.4. Prepare 100 mL. Filter sterilize. Stable at room temperature.
8. 25 % D-glucose in water. Prepare 50 mL. Sterilize by autoclaving and allow to cool to room temperature. Store at 4 °C.
9. 100 mM $CaCl_2$ in water. Filter sterilize. Can be stored at 4 °C for 1–2 weeks.
10. Collagenase type H (*Clostridium histolyticum*). Warm the bottle to room temperature before use and weigh the required amount directly into perfusion buffer 2.
11. Perfusion buffer 1, pH 7.4. Calcium- and magnesium-free Hanks Balanced Salt Solution with 5.33 mM KCl, 0.441 mM KH_2PO_4, 4.17 mM $NaHCO_3$, 137.93 mM NaCl, 0.338 mM Na_2HPO_4, 4.75 mg/mL D-glucose, 26.6 μM phenol red, 0.5 mM ethylene glycol tetraacetic acid. Sterilize by passing through a 0.22 μm filter. Stable for 6 months at 4 °C (*see* **Note 1**).
12. Perfusion buffer 2, pH 7.4. Calcium- and magnesium-containing Hanks Balanced Salt Solution with 1.26 mM $CaCl_2$, 0.493 mM $MgCl_2$, 0.407 mM $MgSO_4$, 5.33 mM KCl,

0.441 mM KH_2PO_4, 4.17 mM $NaHCO_3$, 137.93 mM NaCl, 0.338 mM Na_2HPO_4, 4.75 mg/mL D-glucose, 26.6 μM phenol red, 0.72 % bovine serum albumin fraction IV (*see* **Note 2**).

13. Perfusion buffer 2, pH 7.4, supplemented with 3 mM $CaCl_2$. Prepare just before use (*see* **Note 2**).
14. Perfusion buffer 2, pH 7.4, supplemented with 3 mM $CaCl_2$ and collagenase (*see* **Note 2**). Add collagenase H to a final concentration of 0.08 U/mL to perfusion buffer 2 supplemented with 3 mM $CaCl_2$ immediately before use and mix well.
15. 100× concentrated antibiotic/antimycotic solution. 10,000 U/mL penicillin, 10 mg/mL streptomycin, 25 μg/mL amphotericin B. Store at −20 °C.
16. 100× concentrated insulin–transferrin–selenium liquid supplement. 1.0 g/L insulin, 0.67 mg/L sodium selenite, 0.55 g/L transferrin, 11.0 g/L sodium pyruvate. Store at 4 °C.
17. Standard William's Medium E medium. William's Medium E containing 2 mg/mL D-glucose and 50 mg/mL L-glutamic acid. Store at 4 °C.
18. 20 μg/mL dexamethasone solution (*see* **Note 3**).
19. Matrigel basement membrane matrix: 8.4 mg/mL Matrigel (BD Biosciences, United Kingdom) in phenol red-free Dulbecco's Modified Eagle's Medium supplemented with 10 μg/mL gentamycin. Store Matrigel at −20 °C. Stable for 3 months (*see* **Note 4**).
20. Matrigel-coated culture plates (*see* **Note 5**).
21. Hepatocyte culture medium. Standard William's Medium E supplemented with 7 % dialyzed fetal bovine serum, 10 mL/L insulin–transferrin–selenium supplements, 10 mL/L antibiotic/antimycotic solution, 30 mM sodium pyruvate, and 5 nM dexamethasone.
22. Hepatocyte washing medium. Hepatocyte culture medium without dexamethasone.
23. Sterile nylon 70 μm cell strainers.
24. 24 G gavage needle (i.e., cannula) (Popper and Son, United Kingdom). This should be autoclaved.
25. Hemocytometer.
26. Perfusion apparatus (*see* **Note 6**).
27. Laminar flow cabinets (i.e., category II safety cabinets), one for perfusion apparatus and one for washing and plating hepatocytes.
28. Inverse-phase light microscope.

29. Perfusion pump and appropriate plastic tubing.
30. Thermocirculator.
31. Cylinder of medical gas mixture (i.e., 95 % O_2 and 5 % CO_2).
32. CO_2 incubator.
33. Daylight magnifying lens illuminator.
34. Thermostated water bath.
35. Dissection mat.
36. Surgical instruments: Deschamps ligature needle, Mayo surgical scissors with blunt tips and curved blades, medium and small sharp scissors with flat blades, clamps (Lawton, United Kingdom), home-made vein-lifter (*see* **Note 7**).
37. Syringes and 29 G needle micro-fine 0.33 mm × 12.7 mm 0.5 mL U100 insulin syringe.
38. Thread: braided silk, sterile, nonabsorbable suture, size 3.
39. Selected chemicals for treatment of hepatocytes.

3 Methods

3.1 Sterilization of the Perfusion Apparatus

1. Install the perfusion apparatus in the laminar flow hood.
2. The day before the hepatocyte isolation, fill the apparatus with 70 % alcohol and leave it completely full for 24 h.
3. The following day, circulate the 70 % alcohol for 15 min.
4. Turn on the connected thermocirculator to 37 °C to heat the water of the thermocirculator's heating jacket.
5. Rinse the apparatus six times with sterile double distilled water. Each rinse should circulate through the apparatus for 10 min.

3.2 Coating the Culture Plates with Matrigel

These steps are carried out before the isolation procedure (*see* **Notes 4** and **5**).

1. Precool 12-well culture plates and keep on ice.
2. Dilute ice-cold Matrigel with ice-cold William's E medium in a ratio 1:9. Keep the solution on ice at all times.
3. Use a cold pipette tip to dispense 200 μL of diluted Matrigel into a well. Use a cold standard sterile cell scraper that fits the well to help spread the Matrigel evenly.
4. Place the coated plate in an incubator set at 37 °C with 5 % CO_2 for 1 h.
5. When the hepatocytes are almost ready to be plated, rinse the plates once with William's E medium to remove any traces of nongelled Matrigel.

3.3 Cannulation and Perfusion of Liver (See **Note 8***)*

1. Fill the perfusion apparatus reservoir with 50 mL of perfusion buffer 1. Circulate for 5–10 min and ensure there are no air bubbles. The reservoir for perfusion buffer 1 must be full before connection of the cannula.
2. Set the temperature on the thermocirculator to 42 °C.
3. Spray and clean the dissection mat with 70 % alcohol. Place the surgical box containing the sterile instruments next to the mat.
4. Aseptically, place the gavage needle, serving as the cannula, into a sterile 6 cm diameter cell culture plate containing 8 mL of heparin solution. Fill the gavage needle with heparin using an insulin syringe and needle (*see* **Note 9**).
5. Weigh the mouse by placing it in a beaker on a bench balance.
6. Anesthetize the animal by injecting intraperitoneally 0.1 mL/20 g body weight of the anesthetizing combined solution. When the animal fails to respond to stimuli, swab the abdomen thoroughly with 70 % alcohol. Place the animal on its back on the dissection mat and secure in place.
7. Maintain sterile conditions and open the abdominal cavity by making a U-shaped incision. Move the intestines to the left of the animal's torso to reveal and expose the *vena cava inferior*.
8. Thread a ligature needle with thread. Loosely tie a ligature around the *vena cava inferior* with the help of a vein lifter (*see* **Note 7**) just below the junction to the kidney (i.e., the *vena subrenalis*).
9. Make a slight incision with a small sharp pair of flat scissors parallel to the wall of the *vena cava inferior* below the ligature and large enough to insert the cannula (i.e., gavage needle). The cannula has previously been filled with heparin (*see* **Note 9**).
10. Check to ensure that there are no air bubbles in the cannula (*see* **Note 9**).
11. Insert the cannula at the lower end of the *vena cava inferior*. Do not push too far into the vessel.
12. After the insertion of the cannula into the *vena cava inferior*, tie the loose ligature with a double knot around the cannula.
13. Place a second ligature around the *vena cava inferior* just below the heart and tie with a double knot. This prevents the flow of perfusion buffers throughout the body.
14. Check the perfusion apparatus and make sure there are no air bubbles anywhere in the system. Check also that the thermocirculator has reached 42 °C. This is to obtain a 37 °C temperature in the liver.
15. Carefully transfer the mouse from the dissecting mat and place gently on the platform of the perfusion apparatus. Take care to ensure that the cannula remains inserted in the correct position.

16. Connect the cannula to the perfusion apparatus tubing, which must be free of air bubbles. From here on, work very quickly. **Steps 17–20** must take no longer than 1–2 min in total.
17. Start the perfusion immediately. Perfuse perfusion buffer 1 through the liver at a rate of 2 mL/min. If the perfusion is successful, the liver will bleach to a light beige color within 10 s and have a shiny and soft surface (*see* **Note 10**).
18. Immediately make a small incision in the portal vein to allow the perfusion buffers to flow freely out of the liver.
19. Increase the flow rate to 7 mL/min.
20. Cut the diaphragm and clamp the *vena cava inferior* just below the heart.
21. Massage the liver gently between the thumb and the forefinger to assist the digestion of the organ and to prevent clot formation.
22. Rinse the abdominal cavity of the animal with distilled sterile water to clear out the blood.
23. Continue perfusion with perfusion buffer 1 for about 7 min.
24. Change to perfusion buffer 2 (*see* **Note 2**) and continue the perfusion for about 8–10 min.

3.4 Isolation of Hepatocytes

1. Carefully remove the liver. Do not cut through the intestines, as this could cause contamination. Transfer the liver to a 10 cm diameter cell culture plate containing 10 mL of ice-cold hepatocyte washing medium.
2. Using a pair of round-ended forceps, carefully peel away the Glisson's capsule enclosing the liver in order to disperse the hepatocytes.
3. Disperse the cells further through a large-bore pipette (e.g., 25 mL pipette).
4. Place a 70 μm nylon cell strainer on top of a 50 mL sterile tube. Filter the cell suspension through the strainer and leave the cells to settle for 10 min.
5. Carefully remove the supernatant, which contains mostly dead cells.
6. Dilute the settled cells in 20 mL of ice-cold hepatocyte washing medium.
7. Centrifuge at $50 \times g$ and 4 °C for 2 min.
8. Decant the supernatant and resuspend the cells very gently in 20 mL of ice-cold hepatocyte washing medium.
9. Repeat **steps 6–9** twice.
10. Remove the supernatant carefully with a sterile pipette and resuspend the cell pellet very gently in 20 mL of ice-cold hepatocyte culture medium.

11. Keep the cell suspension on ice.
12. Count the cells with a hemocytometer to determine the number and percentage of viable cells using the trypan blue method (*see* **Note 11**).
13. Dispense 1 mL (5×10^6 cells/mL) of viable hepatocytes into each well of a Matrigel-coated 12-well culture plate (*see* **Note 12**).
14. Incubate the cells at 37 °C with 5 % CO_2 for 2 to 3 h.
15. Check cell attachment under the microscope. Most of the cells should be attached.
16. Using a pipette, remove the medium and dead or unattached cells. Add fresh hepatocyte culture medium with or without fetal bovine serum depending on how the hepatocytes are to be used in further experimental work.

3.5 Treatment of Hepatocytes with Chemicals of Interest

1. After 2–3 h, replace the medium with fresh culturing medium without fetal bovine serum, but containing the chemical of interest or vehicle alone (*see* **Note 13**).
2. Incubate the plates at 37 °C with 5 % CO_2 for 24 h.
3. Carefully remove the medium and add chosen cell lysis solution (*see* **Note 14**).

4 Notes

1. Before perfusion, prepare 100 mL of perfusion buffer 1 and saturate for 15 min with 95 % O_2 and 5 % CO_2. Check the pH and readjust to pH 7.4 with either 1 M HCl or 0.5 M NaOH. Sterilize by filtration through a 0.22 μm filter unit. Prewarm perfusion buffer 1 and keep at 37 °C until ready to use. Before starting the dissection of the mouse, circulate 50 mL of the buffer through the perfusion apparatus to ensure an air bubble-free unit.
2. Before perfusion, prepare 100 mL of perfusion buffer 2 and supplement with $CaCl_2$ to a final concentration of 3 mM. Check the pH and readjust to pH 7.4 with either 1 M HCl or 0.5 M NaOH. Sterilize by filtration through a 0.22 μm filter unit. Prewarm the buffer and keep at 37 °C until ready to use. Collagenase H is added to perfusion buffer 2 just before use to a final concentration of 0.08 U/mL. The solution is mixed well and used immediately.
3. Store the powder at 2–8 °C. Prepare a 20 μg/mL stock solution by first dissolving 1 mg of dexamethasone powder in 1 mL of absolute ethanol. Swirl gently to dissolve and then add 49 mL of sterile hepatocyte culture medium. Prepare working

aliquots and store at −20 °C. Avoid repeated freezing and thawing.

4. Matrigel is supplied in phenol red-free Dulbecco's Modified Eagle's Medium. We dilute in standard William's E medium. Matrigel solidifies rapidly if warmed. Thaw the frozen matrix at 4 °C overnight on ice. Keep the thawed Matrigel on ice at all times and mix well in an ice-waterbath to homogeneity. Dispense appropriate volumes aseptically into precooled tubes, using precooled pipettes and tissue culture plastic ware. We found that the most reliable way to precool plastic ware is to store it in a −20 °C freezer and remove as required.
5. Before and during the coating with Matrigel, the plates and other plastic ware must be kept ice-cold.
6. We use a home-made perfusion apparatus [2].
7. We use an L-shaped metal instrument (i.e., 3 cm long) for ease of handling.
8. A successful liver perfusion requires speed and coordination. It is essential to practice inserting the cannula, tying the ligatures, and transferring the mouse to the perfusion platform, before embarking on a full-scale isolation procedure. On the day of the perfusion, check your list and make sure that everything one needs is at hand.
9. Any air bubbles in the cannula (i.e., gavage needle) will seriously impair the quality of the perfusion. Heparin is used in the cannula to help prevent blood coagulation in the liver lobes. Coagulation can hinder complete perfusion of the organ.
10. If, on starting the perfusion with perfusion buffer 1, the liver does not bleach to a light beige color, stop the perfusion and discard the liver.
11. Nonviable cells will take up the trypan blue stain into their nuclei. A very low viability generally gives poor results and suggests a possible problem with the procedure or the reagents used. The viability obtained using this test with mouse hepatocytes is a lot lower than that observed for isolated rat hepatocytes. We found that a viability of 65 % produces good cultures of mouse hepatocytes. The key criteria for viability are speed and how quickly the liver bleaches.
12. Mouse hepatocytes do not respond well when plated at too high or too low a density. Under our conditions, we found 5×10^6 cells/well of a 12-well plate to be optimum.
13. Control cultures must be treated with vehicle alone. In the examples shown in Fig. 1, we used ethanol as the vehicle to prepare solutions of the chemicals used to treat hepatocytes. Ethanol is known to induce gene expression in cultured cells [10]. FMO5 expression was moderately increased in response

to vehicle-alone treatment. In the results presented in Fig. 1, the vehicle-alone expression level is set to 0 and levels of expression obtained for treated cells are expressed as percentage increase or decrease.

14. We have found the Tri reagent (Sigma Chemical Co., United Kingdom) suitable for RNA extraction from hepatocytes. This solution has the advantage that protein, RNA, and DNA can be isolated from the same cells.

Acknowledgements

We thank Dr. Rick Moore (NIEHS, North Carolina, United States of America) for advice. Lyndsey Houseman was a recipient of a Drummond Scholarship.

References

1. Wilkening S, Stahl F, Bader A (2003) Comparison of primary human hepatocytes and hepatoma cell line HepG2 with regard to their biotransformation properties. Drug Metab Dispos 31:1035–1042
2. Edwards M, Houseman L, Phillips IR et al (2013) Isolation of mouse hepatocytes. Methods Mol Biol 987:283–293
3. Edwards M, Wong SC, Chotpadiwetkul R et al (2006) Transfection of primary cultures of rat hepatocytes. Methods Mol Biol 320:273–282
4. Janmohamed A, Hernandez D, Shephard EA et al (2004) Cell-, tissue-, sex- and developmental stage-specific expression of mouse flavin-containing monooxygenases (Fmos). Biochem Pharmacol 68:73–83
5. Guerra-Araiza C, Cerbon MA, Morimoto S et al (2000) Progesterone receptor isoforms expression pattern in the rat brain during the estrous cycle. Life Sci 66:1743–1752
6. Guerra-Araiza C, Reyna-Neyra A, Salazar AM et al (2001) Progesterone receptor isoforms expression in the prepuberal and adult male rat brain. Brain Res Bull 54:13–17
7. Miller MM, James RA, Richer JK et al (1997) Progesterone regulated expression of flavin-containing monooxygenase 5 by the B-isoform of progesterone receptors: implications for tamoxifen carcinogenicity. J Clin Endocrinol Metab 82:2956–2961
8. Rae JM, Johnson MD, Lippman ME et al (2001) Rifampin is a selective, pleiotropic inducer of drug metabolism genes in human hepatocytes: studies with cDNA and oligonucleotide expression arrays. J Pharmacol Exp Ther 299:849–857
9. Dannan GA, Guengerich FP, Waxman DJ (1986) Hormonal regulation of rat liver microsomal enzymes: role of gonadal steroids in programming, maintenance, and suppression of delta 4-steroid 5 alpha-reductase, flavin-containing monooxygenase, and sex-specific cytochromes P-450. J Biol Chem 261:10728–10735
10. DeBast G, Coecke S, Akrawi M et al (1995) Effect of ethanol on glutathione S-transferase expression in co-cultured rat hepatocytes. Toxicol In Vitro 6:467–471

Chapter 2

Cryopreservation of Hepatocytes

Nicola J. Hewitt and Albert P. Li

Abstract

The use of cryopreserved hepatocytes has increased in the last decade due to the improvement of the freezing and thawing methods, and has even achieved acceptance by the US Food and Drug Administration for use in drug metabolizing enzyme induction studies. This chapter provides an overview of the theories behind the process of cryopreservation as well as practical advice on methods to cryopreserve hepatocytes, which retain functions similar to fresh cells after thawing. Parameters, such as cell density, cryoprotectants, freezing media, storage conditions, and thawing techniques, should be critically considered. Special emphasis is put on human hepatocytes, but information for the cryopreservation of animal hepatocytes is also described.

Key words Cryopreservation, Human, Animal, Hepatocytes, Purification, Plateability, Thawing, Recovery

1 Introduction

As a result of the improvement of the cryopreservation of human hepatocytes and their increased plateability after thawing, their application to different assays has been extended dramatically. Early cryopreservation methods only allowed for short-term incubations (i.e., less than 6 h) with thawed cell suspensions. Improved methods have resulted in plateable cryopreserved hepatocytes that can be used in long-term assays, such as drug metabolizing enzyme induction studies, time-dependent inhibition studies, long-term drug metabolism, hepatotoxicity, and bile transporter function assays [1–3]. The quality of thawed plateable hepatocytes from different species is apparent as highly confluent cultures with morphologies similar to that of fresh cells (Fig. 1). The confidence in cryopreservation methods has reached a level such that regulatory agencies accept fresh and cryopreserved cells data interchangeably [4].

The method of cryopreservation is based on preventing cellular damage due to ice crystal formation and chemical changes in cells as they cool and eventually freeze. If cells are frozen using a

Mathieu Vinken and Vera Rogiers (eds.), *Protocols in In Vitro Hepatocyte Research*, Methods in Molecular Biology, vol. 1250, DOI 10.1007/978-1-4939-2074-7_2, © Springer Science+Business Media New York 2015

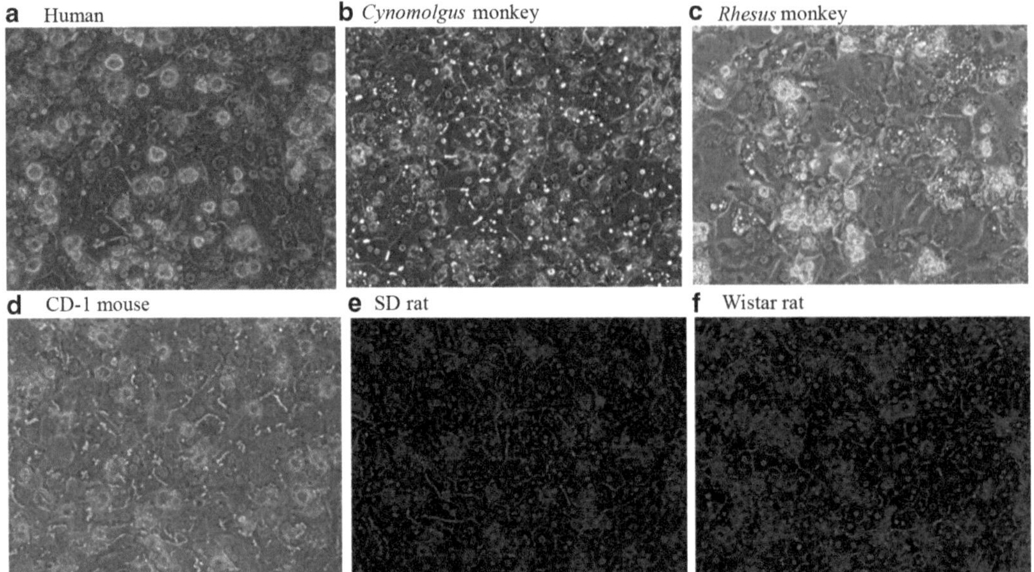

Fig. 1 Morphology of cultures of primary hepatocytes from different species. The morphology and confluency of thawed cryopreserved (**a**) human, (**b**) *Cynomolgus* monkey, (**c**) *Rhesus* monkey, (**d**) CD-1 mouse, (**e**) SD rat and (**f**) Wister rat hepatocyte cultures indicates optimal health of the hepatocytes. Hepatocytes that are well cryopreserved would retain the typical polygonal cell shape with defined plasma membranes, clear (multi-) nuclei and high confluency (10× magnification)

constant slow decline in temperature, regardless of the rate, the cells lose viability and functions [5]. This is because water in the medium starts to freeze before intracellular water, resulting in a higher osmolarity in the medium. Water leaves the cells by osmosis, causing them to shrink. Although a certain amount of shrinkage is not detrimental, if the temperature continues to decrease at a slow rate, too much water leaves the cells and they start to collapse on themselves (i.e., plasmolysis). Excessive loss of intracellular water causes precipitation of solutes, changes in pH, and denaturation of proteins, which causes them to die [6]. In contrast to slow freezing, rapid freezing simply freezes all water and does not allow intracellular water to leave the cells by osmosis. The intracellular water forms ice crystals, which disrupt membranes, leading to the demise of the cells. As a result of the understanding of the effects of freezing, methods for preventing cryo-injury due to ice crystals or biochemical imbalances due to dehydration have been developed [7]. These have been shown to produce cryopreserved hepatocytes with the highest and reproducible viability (Table 1). In this chapter, the entire process of cryopreservation, from selection of the best fresh hepatocytes to their thawing, is described. The procedures described represent the most current practice based on available published information rather than procedures from a specific laboratory. The procedures may be further optimized for best results by investigators interested in cryopreservation of their hepatocytes of interest.

Table 1
Yield and viability of primary human hepatocytes from different donors

Human donor number	Experiment number	Age (years)	Gender	Race	Yield (million)	Viability (%)
1	1	35	M	H	7.9	95
2	1	26	F	C	6.0	98
3	1	47	F	C	5.3	95
4	1	25	M	C	5.8	79
5	1	64	F	H	7.0	93
6	1	15	M	C	6.0	75
7	1	17	M	C	4.4	97
8	1	21	F	C	6.4	89
9	1	59	F	C	8.3	97
10	1	56	M	H	7.5	97
11	1	51	M	C	5.8	89
12	1	40	F	C	5.8	92
13	1	17	F	C	4.4	93
14	1	45	M	C	5.1	98
15	1	55	M	C	8.0	98
	2				7.8	92
	3				8.8	94
	4				7.0	92
16	1	42	M	H	6.9	97
	2				6.0	93
	3				7.7	90
	4				7.8	91

The cryopreserved human hepatocytes were recovered using Universal Cryopreservation Recovery Medium. Viability was determined based on trypan blue exclusion. The reproducibility of the yield and viability for each donor can be seen in the results for human donor numbers 15 and 16 where results of four independent evaluations, each with a different vial, are shown. Yield is the number of viable cells recovered after thawing (C, Caucasian; F, female; H, Hispanic; M, male)

2 Materials

2.1 Preincubation Medium

1. Complete William's Medium E. Add 5.5 mL 2 mg/mL insulin solution dissolved in water, 2.5 mL 10 mg/mL gentamycin solution and/or 5 mL 100× concentrated penicillin/streptomycin solution to 500 mL William's E Medium GlutaMAX (Life Technologies, Germany).
2. Fructose or alpha lipoic acid. If the performance of the freshly isolated cells is suboptimal (e.g., the viability is below 70 %, the

morphology is poor or there is low attachment when seeded), the preincubation medium can be supplemented with fructose or alpha lipoic acid. To this end, add fructose or alpha lipoic acid at final concentrations of 200 mM and 5 mM, respectively, to complete William's Medium E (*see* **Note 1**).

3. Glucose and insulin. If the performance of the cells after isolation is suboptimal, the preincubation medium can be modified to include glucose and insulin. For this purpose, add glucose and insulin at final concentrations of 5 mM and 1 nM, respectively, to complete William's Medium E (*see* **Note 2**).
4. *N*-Acetyl-L-cysteine. If the performance of the cells after isolation is suboptimal, the preculture medium and the culture medium after thawing can be supplemented with a final concentration of 1 mM *N*-acetyl-L-cysteine. Alternatively, a medium which already has L-cysteine as a basal supplement (e.g., William's Medium E) can be used (*see* **Note 3**).

2.2 Freezing Medium

1. Basal freezing medium. Standard basal media can be used, such as William's Medium E, Leibovitz, Dulbecco's Modified Essential Medium, Minimal Essential Medium, or Modified Earle's Medium (*see* **Note 4**). Animal component-free medium could also be considered (*see* **Note 5**).
2. Bovine or human serum. The concentration of serum can be between 10 and 90 % [8, 9] (*see* **Note 6**).
3. Dimethylsulfoxide (DMSO). The most common and effective cryoprotectant is DMSO [10]. It should be slowly added to the diluted cell suspension in freezing medium to a final concentration of 10 %. The concentration of DMSO used for cryopreserving hepatocytes from any species should be noncytotoxic and give rise to maximal postthaw recovery. This can be tested by checking the viability 30 min after adding and *prior* to freezing (*see* **Note 7**).
4. Trehalose. Use at a final concentration of 0.2 M in the freezing medium (*see* **Note 8**).

2.3 Thawing Medium

There are a number of options for thawing media, including basic and commercial types. The best option is to test each and compare the quality of the cells after thawing.

1. Standard thawing medium. Add 5 mL fetal bovine serum to 45 mL complete William's Medium E or other appropriate culture medium. This is sufficient for 1–5 cryovials.
2. 90 % Percoll solution. Add one part 10× concentrated Hank's Balanced Salt Solution (Life Technologies, Germany) to nine parts Percoll with density 1.124 g/mL. Mix thoroughly and make sure the pH is 7.4. If it is not, add 0.5 M HCl until the pH is 7.4. It is very important that there are no crystals in the

Percoll solution (*see* **Note 9**). If there are, remake the solution, but adjust the pH of the 10× concentrated HBSS before adding to the Percoll. The limits of pH are 7.2–7.6. Anything outside this range will be toxic to the cells.

3. Percoll thawing medium. Add 25 mL thawing medium to a 50 mL centrifuge tube. Then add 90 % Percoll to the 50 mL centrifuge tube according to the hepatocyte species, namely, 18 mL for rat, dog, monkey, mouse and 16 mL for human.
4. Recovery medium. Two commercial media are recommended for recovery of thawed cryopreserved human or animal hepatocytes, namely Cryopreserved Hepatocytes Recovery Medium (Life Technologies, Germany) and Universal Cryopreservation Recovery Medium (In Vitro ADMET Laboratories, United States of America).

2.4 Freezing and Storage

1. Programmable freezer. A programmable freezer allows for a controlled and precise rate of freezing. Most freezing regimens follow the freezing profile example shown in Table 2. The freezing program takes into account the increase in temperature of the cell suspension when the latent heat of fusion is released (i.e., at about −9.5 °C when the concentration of DMSO is 10 %) by counteracting it with a small shock freeze (Table 2). The cells are maintained at 10 °C for 10 min and then gradually frozen down to −30 °C, which includes the shock freeze to maintain the continued gradual decrease in temperature in the cell suspension itself. This slow freezing

Table 2
Programmable freezer temperature profile

Time (min)	Temperature (°C)
0	20
10	0
20	0
27	-9.5
27.01	-28
27.5	-25
30	-25
31	-15
41	-30
51	-100

step allows for the loss of some intracellular water, but is not enough to cause changes in solute concentrations. After this, the temperature is dropped by 70 to −100 °C in 10 min, which effectively freezes all water remaining in the cells and that prevents further loss of water.

2. Mr. Frosty freezing container (Thermo Scientific, Germany). This freezing container has 12–18 places for cryovials. It requires isopropyl alcohol and a −70 °C freezer.

3 Methods

3.1 Hepatocyte Quality Check

Cryopreserved hepatocytes are more likely to attach after thawing if the fresh cells frozen are of the highest quality (*see* **Note 10**). If the viability is lower than 80 % and the morphology of the cells is compromised (e.g., with blebs and vacuoles), the isolation procedure itself should be optimized (*see* **Note 11**) or the viability should be increased by Percoll purification.

3.2 Percoll Purification

1. Resuspend human hepatocytes in 36 mL basal culture medium and add 16 mL 90 % Percoll solution, pH 7.4 (*see* **Note 12**).
2. Mix the cells and Percoll solution thoroughly but gently by inverting the tubes 3–4 times.
3. Centrifuge the cells at 168 × *g* and room temperature for 20 min.
4. Remove the upper layer of dead cells and all of the supernatant, leaving only the cell pellet at the bottom of the tube.
5. Gently dislodge the cells in the pellet by agitating the tubes.
6. Add 5 mL basal freezing medium to the cells and gently mix the cells to result in an even single cell suspension.
7. Count the viability and density of the cells using either trypan blue exclusion or a cell counter.

3.3 Preincubation

If the cell viability after isolation is below 70 % or the morphology is poor, the performance of the cells and their recovery from cryopreservation may be improved by preincubating them with medium supplemented with ingredients known to have a beneficial effect on the cells (*see* **Notes 1–3**). This should be carried out *prior* to cryopreservation to allow the hepatocytes to recover from the effects of isolation and washing [11, 12]. The addition of fructose, alpha lipoic acid, glucose, insulin, and *N*-acetyl-L-cysteine may improve the recovery of the cells after cryopreservation, but this is dependent on the quality of the cells. If they are already highly viable (i.e., over 95 % viable) with a good morphology (i.e., clear, spherical cells with no blebs), then the improvement in viability *prior* to cryopreservation is likely to be minimal, but recovery after cryopreservation may be improved. Low-quality cell suspensions (i.e., 50 % viability with poor morphology and cell debris) are unlikely to survive cryopreservation. In these cases, either the cells can be purified using the Percoll method to increase the initial cell viability for cryopreservation or can be used only as fresh cells.

1. The preincubation medium can be supplemented with fructose, alpha lipoic acid, glucose, insulin, and *N*-acetyl-L-cysteine to increase the success of the cryopreservation.
2. Dilute the hepatocytes to 5 to 10 million cells/mL in the preculture medium and add 30 mL to a T175 culture flask (*see* **Note 13**).
3. Place the flask in an incubator at 37 °C and 5 % CO_2 for 30 min.
4. After the preincubation, transfer the cells to a 50 mL centrifuge tube and centrifuge at 60 × *g* and room temperature for 5 min.
5. Resuspend the cells in an appropriate volume of freezing medium.

3.4 Preparation of Cells for Cryopreservation

1. Dilute human hepatocytes to 11 million cells/mL in freezing medium (*see* **Note 14**).
2. Place the hepatocyte suspension on ice-water and keep them on ice until they are cryopreserved (*see* **Note 15**).
3. Slowly add DMSO over at least 10 min to allow for its equilibration between intracellular matrix and the medium [13]. If DMSO is added too quickly, then it creates an osmotic imbalance and the cells are compromised [3].
4. Aliquot 1 mL of the cell suspension into each cryovial.

3.5 Cryopreservation with a Programmable Freezer

1. Prepare the programmable freezer for use by ensuring it has sufficient liquid nitrogen for the entire run.
2. Precool the programmable freezer to 10 °C.
3. Place the vials in a programmable freezer and start the program.
4. Once finished, transfer the cryovials to a suitable storage tank.

3.6 Cryopreservation Without a Programmable Freezer

This method has shown to be successful in the cryopreservation of hepatocytes from a number of species [13–16].

3.6.1 Method 1

1. Place the cryovials into a noninsulated rack (i.e., not a polystyrene holder with a tight fit).
2. Place the cryovials into a −20 °C freezer for 12 min.
3. Transferred to a −80 °C freezer for 1 h.
4. Plunge the cryovials into liquid nitrogen.

3.6.2 Method 2

1. Place the cryovials into cooling boxes containing isopropanol.
2. Place the Mr. Frosty box into a −80 °C freezer for 8–24 h [8, 11]. The cells will be frozen at a constant rate of 1 °C/min (*see* **Note 16**).

3.7 Storage of Cells

1. The vials of cells should be transferred to a storage container as quickly as possible to avoid cell damage by slow warming.
2. The storage container should be maintained at less than −130 °C, since some chemical reactions (i.e., proteases) are still possible at temperatures warmer than this, and they may compromise cell viability and recovery (i.e., the number of viable cells surviving cryopreservation).
3. Ideally, cells can be stored at −150 °C in a freezer (*see* **Note 17**), −160 °C in the vapor phase of nitrogen, or −196 °C in liquid nitrogen (*see* **Note 18**). Storage at −80 °C for longer than 2 h leads to lower initial postthaw viabilities. Thus, if stored correctly, the recovery (e.g., postthaw viability and attachment) can be maintained for years [13].

3.8 Thawing and Handling of Cryopreserved Hepatocytes

The success of cryopreservation depends on both the freezing process and the method of thawing and handling of the cryopreserved cells (*see* **Note 19**). Thawing can include a Percoll purification step if required. The correct Percoll concentration and centrifuge speeds will result in maximal recovery of viable cells whilst effectively removing dead cells (*see* **Notes 9** and **20**).

1. Prepare the standard or Percoll thawing medium and warm it to 37 °C and keep it in the water bath until the cells are thawed.
2. Transfer the cryovial from the storage tank to a water bath maintained at 37 °C. Transfer of the vial from the liquid nitrogen tank to the water bath should be completed with in 10 s (*see* **Note 21**).
3. Thaw the cells until all the ice is melted. It is better to slightly overthaw them, than to leave ice crystals in because these will damage the cells. Thawing should take about 90–120 s, but this will vary.
4. Transfer to the vial and the warmed thawing medium to a sterile flow hood.
5. Transfer the cells into the tube containing the thawing medium by pouring or gently pipetting.
6. Wash all the cells from the cryovial by gently adding 1 mL thawing medium and then pouring it back into the centrifuge tube.
7. Centrifuge cells in standard thawing medium at room temperature for 5 min at an appropriate speed (i.e., 60 × *g* for rat, dog, monkey, and human, and 30 × *g* for mouse hepatocytes).

8. For cells thawed in Percoll containing medium, make the volume up to 50 mL with extra thawing medium, but not with Percoll.
9. Close the lid, invert the tube slowly about 1–2 times to resuspend the cells, and mix the Percoll to result in an even concentration.
10. Centrifuge the cells in Percoll thawing medium at room temperature for 20 min at an appropriate speed (i.e., $168 \times g$ for rat, dog, monkey, and human, and $60 \times g$ for mouse hepatocytes).
11. Pour off the supernatant, loosen the cell pellet before adding medium, and resuspend the cells in appropriate culture medium (*see* **Note 22**).

4 Notes

1. Attachment of rat and human cryopreserved hepatocytes may be improved by preincubating the fresh hepatocytes with fructose or alpha lipoic acid *prior* to freezing [17–19].
2. These supplements increase the glycogen content of hepatocytes, which is broken down to glucose-6-phosphate after thawing, and thus provides an energy source which is reported to be lacking in these cells [20, 21]. Preincubating fresh cells in glucose-containing Krebs–Henseleit buffer for 30 min in a 95 % air and 5 % CO_2 atmosphere improves the attachment efficiency of both rat and human cryopreserved hepatocytes [22].
3. *N*-Acetyl-L-cysteine is an amino acid and is a precursor of reduced glutathione, a cytoprotective tripeptide found in high concentrations (i.e., 5 mM) in hepatocytes [23]. The isolation procedure and cryopreservation cause hepatocytes to lose glutathione [24, 25]. Glutathione protects against the effects of reactive oxygen species, lipid peroxidation, and prevents apoptosis [26]. Therefore, the survival from cryo-injury is increased by increasing glutathione levels. The addition of glutathione per se to the culture medium will not increase intracellular glutathione because it will not cross the plasma membrane. Therefore, hepatocytes could be incubated with *N*-acetyl L-cysteine, a precursor of glutathione, which is actively taken up into the cells by transpeptidase and then converted to L-cysteine in hepatocytes.
4. The basal medium used for freezing cells varies between laboratories, but there is no *consensus* as to which is optimal for cryopreservation. Basal media with strong buffering capacities, such as phosphate-based media, may offer an advantage in maintaining pH.

5. Animal component-free media has gained in use, either due to transport requirements or clinical reasons, such as a lack of tolerance to animal derived products [27].
6. Positive attributes of serum include quenching of proteolytic enzymes released from dead cells, cytoprotection from free oxygen radicals [28], and slowing down cell sedimentation during freezing due to its viscosity.
7. DMSO has been shown to cause toxicity if incubated with hepatocytes at high concentrations (i.e., 10–20 %) for longer than 30 min *prior* to cryopreservation, especially if maintained at 37 °C [7, 13]. This could explain why cryopreservation of large quantities of cells (i.e., more than 300 vials) may result in poorer postthaw viabilities. They are compromised because the time between addition of DMSO and the start of freezing is too long. The key is to allow the DMSO to equilibrate in the cell suspension, but not to allow the DMSO to cause toxicity. Other cryoprotectants have been included in freezing medium, such as polyvinylpyrrolidone and polyethyleneglycol, but these are used in combination with DMSO, rather than as single supplements. A drawback of these compounds is that they may interfere with the liquid chromatography-mass spectrometry analysis of test compounds after the cells have been thawed and used in a metabolite profiling assay. Washing thawed cells a number of times may reduce this effect, but because polyethyleneglycol interacts strongly with the plasma membrane, it is unlikely to be completely removed.
8. This disaccharide protects proteins and cellular membranes from inactivation or denaturation due to different stress conditions. The basis for the idea came from the fact that lower organisms, such as yeasts, bacteria, fungi, and insects, all survive freezing or drying and all have concentrated levels of disaccharides, especially trehalose [29]. Membrane stabilization may be a result of trehalose interacting with the plasma membrane to counteract the changes in membrane fluidity. An additional mechanism of cryoprotection may be due to scavenging of free radicals, which may be released during oxidative stress [30].
9. A drawback of Percoll is that it may form large crystals if it is repeatedly heated and cooled and these will prevent hepatocytes from attaching. If this occurs, wash the cells until no Percoll crystals are visible. Another key factor in using Percoll is that it is nonphysiological (i.e., osmotically equivalent to water) and requires dilution in tenfold concentrated buffer, such as Hank's Balanced Salt Solution, before use. If Percoll is diluted in water, this will result in the demise of the hepatocytes due to water entering the cells through osmosis, causing them to swell and burst.

10. Attempting to improve the quality of poor cell preparations by adding supplements or preincubating cells will not increase the postthaw recovery [17].
11. The quality of hepatocytes can be improved by considering factors, such as the transport medium used to transfer the liver to the laboratory, the time between organ procurement and the start of collagenase digestion, the condition of the liver, and the isolation time itself.
12. It has been reported that purifying rat and dog hepatocytes using Percoll *prior* to cryopreservation results in higher cell recoveries after thawing [31]. However, Percoll is known to cause a substantial loss of viable cells [31] unless the method is optimized [32]. The use of Percoll does not have to result in a significant loss of viable cells. In general, the larger the number of cells which are purified, the higher the concentration of Percoll that is needed. Although in theory, all centrifuges should spin at the same speed, this is not always true in reality. Therefore, the speed at which the cells are centrifuged may need to be adjusted according to the specific centrifuge. Hence, it is recommended that the concentrations of Percoll or centrifuge speed is adjusted for hepatocytes of different species due to differences in their sizes [33].
13. If the hepatocyte isolation or subsequent wash steps do not include oxygenation of the buffers, hepatocytes will lose adenosine triphosphate. Incubation of freshly isolated hepatocytes under an atmosphere of 95 % air and 5 % CO_2 for 30 min at 37 °C increases the adenosine triphosphate content of hepatocytes and may increase their likelihood of survival [34]. The increased energy status of hepatocytes may make them more able to undergo energy-consuming processes such as urea synthesis [35] and gluconeogenesis [5].
14. This may not be optimal for hepatocytes from other species which may have different cell volumes. Mouse hepatocytes are much larger than rat or human hepatocytes (i.e., approximately threefold [33]) and this affects the optimal freezing density. The optimal freezing density of mouse hepatocytes is between 1.2 and 1.5 million cells/mL, whereas rat hepatocytes are optimally frozen at 4–10 million cells/mL.
15. Keeping cells at 4 °C reduces the toxicity of DMSO [21] and maintains metabolizing enzyme activities [35]. Repeated warming and cooling of hepatocytes leads to activation of proteolytic enzymes *via* alterations in the cellular iron homeostasis and apoptosis [36]. Thus, repetition of the warm and cool cycles should be avoided at all times.
16. This method of freezing human and pig hepatocytes was found to be equally effective as a programmable freezer with respect

to initial cell viabilities, attachment efficiencies, and some drug metabolizing enzyme activities [8, 11, 20].

17. An electrically powered freezer has the advantage of being low maintenance, but make sure it is connected to emergency electricity in case of power failure.
18. A disadvantage of liquid nitrogen is that it sometimes leaks into the vials. This will create pressure in the vial when it is removed from the storage tank and warmed. The obvious consequence of this is that the tube explodes. Many researchers store cryopreserved hepatocytes in vapor phase nitrogen, which decreases the likelihood of exploding vials and maintains them at a temperature well below the critical threshold of −130 °C.
19. If cells are warmed slowly, for instance, if the vials are removed from liquid nitrogen and placed on water ice, intracellular ice crystals will coalesce and form larger crystals, which subsequently disrupt cell membranes. Even cells placed on dry ice are warming up (i.e., from −196 to −80 °C) and this may also cause membrane damage due to ice crystals growing. Rapid thawing in a 37 °C water bath causes the intracellular ice crystals to instantly thaw, preventing any damage due to growing ice crystals.
20. If the concentration of Percoll is too high, viable cells are lost as well as dead cells [37]. Conversely, if the concentration is too low, the viability may even decrease because the viable and dead cells are rapidly sedimented into a tight pellet, which is hard to resuspend.
21. The most convenient transfer of vials to the water bath is when the storage tank is in the cell culture laboratory. However, this is not always possible and vials may well be kept in the basement of the building. Even if cells are stored in a different room, the method of transfer needs to be quick and safe. The best way to transfer cells from the storage tank to the water bath is in a container of liquid nitrogen, with enough liquid nitrogen to ensure the vials are immersed for the entire transfer time. It is not sufficient to pour a small amount of liquid nitrogen over vials in a polystyrene box and hope that it is still there by the time they arrive at the water bath. Those who do not have good access to liquid nitrogen can opt for transferring vials in dry ice pellets. Do not use ice-water and do not allow the vials to warm above −70 °C at any time during transfer.
22. After centrifugation, pour off the supernatant and loosen the cell pellet before adding any media. If you add media first and then try to resuspend the cells, they will form clumps. Loosen the pellet by gently shaking the bottom of the tube.

References

1. Kaneko A, Kato M, Endo C et al (2010) Prediction of clinical CYP3A4 induction using cryopreserved human hepatocytes. Xenobiotica 40:791–799
2. Bi YA, Kazolias D, Duignan DB (2006) Use of cryopreserved human hepatocytes in sandwich culture to measure hepatobiliary transport. Drug Metab Dispos 34:1658–1665
3. Li AP (2007) Human hepatocytes: isolation, cryopreservation and applications in drug development. Chem Biol Interact 168:16–29
4. US Food and Drug Administration (2012) Guidance for industry: drug interaction studies: study design, data analysis, implications for dosing, and labeling recommendations. pp 1–79
5. Gomez-Lechon MJ, Lopez P, Castell JV (1984) Biochemical functionality and recovery of hepatocytes after deep freezing storage. In Vitro 20:826–832
6. Meryman HT (1961) Freezing of living cells: biophysical considerations. Natl Cancer Inst Monogr 7:7–15
7. Chesne C, Guillouzo A (1988) Cryopreservation of isolated rat hepatocytes: a critical evaluation of freezing and thawing conditions. Cryobiology 25:323–330
8. Gomez-Lechon MJ, Lahoz A, Jimenez N et al (2006) Cryopreservation of rat, dog and human hepatocytes: influence of preculture and cryoprotectants on recovery, cytochrome P450 activities and induction upon thawing. Xenobiotica 36:457–472
9. Hengstler JG, Ringel M, Biefang K et al (2000) Cultures with cryopreserved hepatocytes: applicability for studies of enzyme induction. Chem Biol Interact 125:51–73
10. Loretz LJ, Li AP, Flye MW et al (1989) Optimization of cryopreservation procedures for rat and human hepatocytes. Xenobiotica 19:489–498
11. Alexandre E, Viollon-Abadie C, David P et al (2002) Cryopreservation of adult human hepatocytes obtained from resected liver biopsies. Cryobiology 44:103–113
12. Li AP, Lu C, Brent JA et al (1999) Cryopreserved human hepatocytes: characterization of drug-metabolizing enzyme activities and applications in higher throughput screening assays for hepatotoxicity, metabolic stability, and drug-drug interaction potential. Chem Biol Interact 121:17–35
13. Chesne C, Guyomard C, Fautrel A et al (1993) Viability and function in primary culture of adult hepatocytes from various animal species and human beings after cryopreservation. Hepatology 18:406–414
14. Swales NJ, Luong C, Caldwell J (1996) Cryopreservation of rat and mouse hepatocytes I: comparative viability studies. Drug Metab Dispos 24:1218–1223
15. Swales NJ, Johnson T, Caldwell J (1996) Cryopreservation of rat and mouse hepatocytes II: assessment of metabolic capacity using testosterone metabolism. Drug Metab Dispos 24:1224–1230
16. Price JA, Caldwell J, Hewitt NJ (2006) The effect of EGF and the comitogen, norepinephrine, on the proliferative responses of fresh and cryopreserved rat and mouse hepatocytes. Cryobiology 53:182–193
17. Terry C, Dhawan A, Mitry RR et al (2006) Preincubation of rat and human hepatocytes with cytoprotectants *prior* to cryopreservation can improve viability and function upon thawing. Liver Transpl 12:165–177
18. Terry C, Dhawan A, Mitry RR et al (2010) Optimization of the cryopreservation and thawing protocol for human hepatocytes for use in cell transplantation. Liver Transpl 16:229–237
19. Aghdai MH, Jamshidzadeh A, Nematizadeh M et al (2013) Evaluating the effects of dithiothreitol and fructose on cell viability and function of cryopreserved primary rat hepatocytes and HepG2 cell line. Hepat Mon 13:e7824
20. Loven AD, Olsen AK, Friis C et al (2005) Phase I and II metabolism and carbohydrate metabolism in cultured cryopreserved porcine hepatocytes. Chem Biol Interact 155:21–30
21. Guillouzo A, Rialland L, Fautrel A et al (1999) Survival and function of isolated hepatocytes after cryopreservation. Chem Biol Interact 121:7–16
22. Silva JM, Day SH, Nicoll-Griffith DA (1999) Induction of cytochrome P450 in cryopreserved rat and human hepatocytes. Chem Biol Interact 121:49–63
23. Orrenius S, Moldeus P (1984) The multiple roles of glutathione in drug metabolism. Trends Pharmacol Sci 5:432–435
24. Swales NJ (1993) Chapters 2 and 3. In: Caldwell J (Supervisor) Mechanistic studies of the cytotoxicity of trans-cinnamaldehyde. PhD thesis, St. Mary's Hospital Medical School, University of London, pp 1–228
25. Stevenson DJ, Morgan C, McLellan LI et al (2007) Reduced glutathione levels and expression of the enzymes of glutathione synthesis in cryopreserved hepatocyte monolayer cultures. Toxicol In Vitro 21:527–532
26. Vairetti M, Griffini P, Pietrocola G et al (2001) Cold-induced apoptosis in isolated rat hepatocytes: protective role of glutathione. Free Radic Biol Med 31:954–961

27. Selvaggi TA, Walker RE, Fleisher TA (1997) Development of antibodies to fetal calf serum with arthus-like reactions in human immunodeficiency virus-infected patients given syngeneic lymphocyte infusions. Blood 89:776–779
28. Gutteridge JM, Quinlan GJ (1993) Antioxidant protection against organic and inorganic oxygen radicals by normal human plasma: the important primary role for iron-binding and iron-oxidising proteins. Biochim Biophys Acta 1156:1444–1450
29. Katenz E, Vondran FW, Schwartlander R et al (2007) Cryopreservation of primary human hepatocytes: the benefit of trehalose as an additional cryoprotective agent. Liver Transpl 13:38–45
30. Leekumjorn S, Wu Y, Sum AK et al (2008) Experimental and computational studies investigating trehalose protection of HepG2 cells from palmitate-induced toxicity. Biophys J 94:2869–2883
31. Swales NJ, Utesch D (1998) Metabolic activity of fresh and cryopreserved dog hepatocyte suspensions. Xenobiotica 28:937–948
32. Hewitt NJ, Utesch D (2004) Cryopreserved rat, dog and monkey hepatocytes: measurement of drug metabolizing enzymes in suspensions and cultures. Hum Exp Toxicol 23: 307–316
33. Swales NJ, Caldwell J (1997) Phase 1 and 2 metabolism in freshly isolated hepatocytes and subcellular fractions from rat, mouse, chicken and ox livers. Pestic Sci 49:291–299
34. Zaleski J, Richburg J, Kauffman FC (1993) Preservation of the rate and profile of xenobiotics metabolism in rat hepatocytes stored in liquid nitrogen. Biochem Pharmacol 46: 111–116
35. Poullain MG, Fautrel A, Guyomard C et al (1992) Viability and primary culture of rat hepatocytes after hypothermic preservation: the superiority of the Leibovitz medium over the University of Wisconsin solution for cold storage. Hepatology 15:97–106
36. Doeppner TR, Grune T, de Groot H et al (2003) Cold-induced apoptosis of rat liver endothelial cells: involvement of the proteasome. Transplantation 75:1946–1953
37. Powis G, Santone KS, Melder DC et al (1987) Cryopreservation of rat and dog hepatocytes for studies of xenobiotic metabolism and activation. Drug Metab Dispos 15:826–832

Chapter 3

Culture Conditions Promoting Hepatocyte Proliferation and Cell Cycle Synchronization

Anne Corlu and Pascal Loyer

Abstract

The liver overcomes damages induced by harmful substances or viral infections and allows the use of extended resection in human therapy through its remarkable ability to regenerate. The regeneration process relies on the massive proliferation of differentiated hepatocytes that exit quiescence and undergo a limited number of cell cycles to restore the hepatic mass. Many discoveries on the regulation of hepatocyte proliferation have benefited from the use of in vitro models of cultures of primary hepatocytes as well as hepatoma cells as opposed to data obtained from in vivo models of liver regeneration, such as following partial hepatectomy in rodents. In this chapter, the most pertinent in vitro models used to promote the proliferation of hepatocytes and technical procedures to synchronize their progression throughout the cell cycle are presented with the goal to investigate the regulation of the hepatocyte cell cycle and the molecular pathways regulating liver regeneration.

Key words Liver regeneration, Hepatocytes, Hepatoma cells, Cell cycle, Growth factors

1 Introduction

The unique feature of the liver regeneration process is the massive proliferation in the remnant parenchyma of differentiated hepatocytes that exit quiescence and undergo a limited number of cell cycles to restore the hepatic mass [1, 2]. Hepatic progenitor cells are recruited only when the proliferation of mature cells is impaired [3]. Immediately following partial hepatectomy in rodents [4], hepatocytes exit quiescence and enter in the G1 phase [5]. In rats, the peak of DNA synthesis is observed at 22–24 h posthepatectomy followed by a peak of mitosis at 28–30 h [6, 7], while in mouse, the progression in the G1 phase of the cell cycle is slower and the peaks of DNA synthesis and mitosis are delayed by about 24 h [8]. Numerous studies have been performed to identify exogenous factors triggering liver regeneration and to define the mechanisms regulating hepatocyte cell cycle progression during the first synchronous wave of proliferation [9]. The proinflammatory

Mathieu Vinken and Vera Rogiers (eds.), *Protocols in In Vitro Hepatocyte Research*, Methods in Molecular Biology, vol. 1250, DOI 10.1007/978-1-4939-2074-7_3, © Springer Science+Business Media New York 2015

cytokines tumor necrosis factor alpha (TNFα) and interleukin 6 (IL6) are the early stimuli allowing the exit of hepatocytes from quiescence [10, 11]. Within 30 min posthepatectomy, their release induces a rapid activation of the transcription factors nuclear factor kappa light chain enhancer of activated B cells (NF-κB) and signal transducer and activator of transcription 3 (STAT3) which participate in the induction of the immediate early genes, including *c*-fos and *c*-jun [5]. This initiation phase results in the G0/G1 transition and early G1 progression allowing hepatocytes to respond to growth factors, such as hepatocyte growth factor (HGF), which renders the cells competent for commitment to DNA replication.

The isolation procedure of rodent and human hepatocytes [12–14] has allowed the establishment of in vitro models of primary hepatocytes, which have been used to characterize the mitogenic factors promoting their proliferation [9]. Our laboratory has reported that during isolation, hepatocytes sequentially express mRNAs of immediate early proto-oncogenes, like *c*-jun and *c*-fos (Fig. 1), followed by *c*-myc even *prior* to their plating in culture [15], demonstrating that quiescent hepatocytes undergo the G0/G1 transition during collagenase perfusion [16]. Furthermore, in the absence of growth factors, hepatocytes in primary culture are able to progress up to the mid-late G1 phase, as evidenced by the sequential overexpression of *c*-fos (Fig. 1), *c*-jun, jun B, *c*-myc, and *c*-Ki-ras (Fig. 2). Thus, hepatocytes maintained in pure culture in the absence of growth factors are arrested in the mid-late G1 phase and are no longer quiescent in a G0/G1 state like the cells in vivo in normal liver. In contrast, hepatocytes stimulated by HGF, epidermal growth factor (EGF), or transforming growth factor alpha (TGFα) [17–22] replicate their DNA (Fig. 2) and complete the cell cycle by a G2/M transition (Fig. 2). In these cell models, hepatocytes also exhibit partially synchronous progression throughout the cell cycle, which has been used to investigate the expression of cell cycle regulators, such as the cyclin D proteins and cyclin-dependent kinases (CDKs) (Fig. 2). The progression beyond the restriction point is strictly dependent on the stimulation by growth factors [16] and correlates with the induction of cyclin D1 immediately following EGF (Fig. 2) or HGF [23] stimulation.

More complex primary culture systems have been developed to improve survival and differentiation, such as by the addition of dimethylsulfoxide (DMSO) to the cell culture medium [24], cocultures associating hepatocytes and nonparenchymal liver epithelial cells [25], and extracellular matrix coating [26, 27]. In all these models, hepatocytes are arrested in the G1 phase of the cell cycle and do not replicate DNA upon stimulation by growth factors alone [28–31]. In our laboratory, we used a coculture model consisting of rat hepatocytes and rat liver epithelial cells (RLECs), also called biliary epithelial cells, in which heterotypic cell–cell contacts are restored and the abundant deposition of extracellular matrix is observed [32].

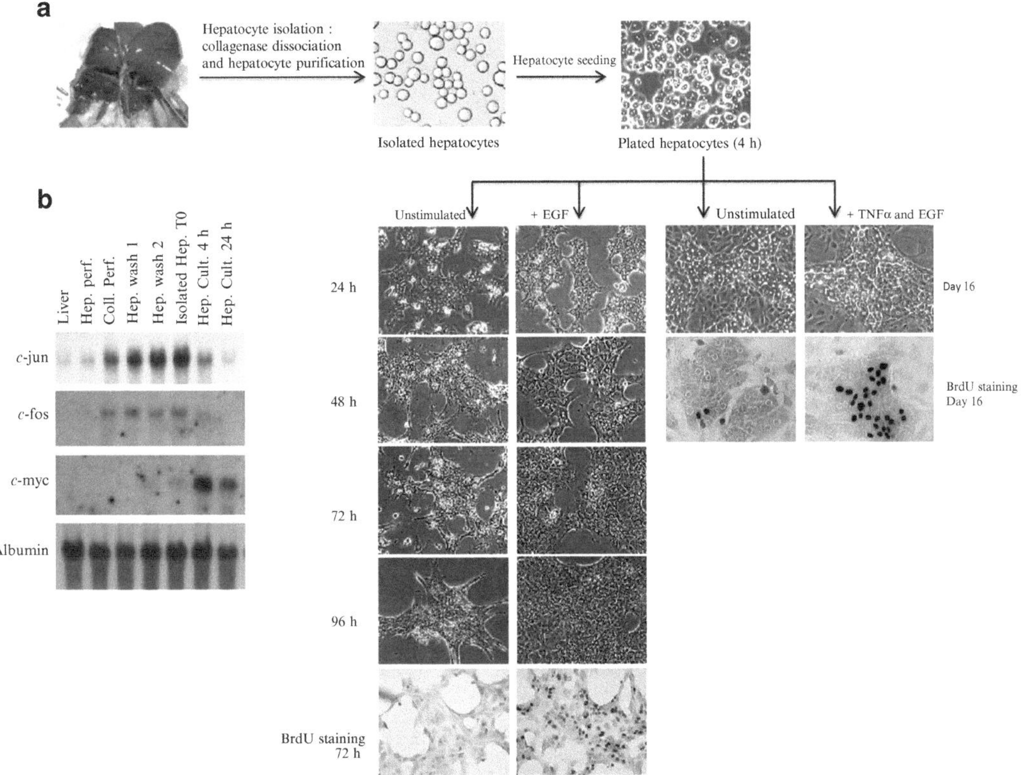

Fig. 1 Isolation and primary culture models of rodent hepatocytes. (**a**) Hepatocytes are isolated by enzymatic dissociation of the liver via in situ perfusion through the portal vein, purified by decantation and centrifugation. On the *lower left panel,* hepatocytes have been plated in pure culture and maintained in culture medium either without growth factors (i.e., unstimulated) or in presence of EGF (+EGF) that stimulates the DNA replication visualized by the BrdU staining at 72 h, while on the *lower right panel*, hepatocytes have been cocultured with RLECs without growth factors (i.e., unstimulated) or in presence of both TNFα and EGF promoting DNA replication (i.e., BrdU staining at day 16 of coculture during the second period of stimulation). (**b**) During liver dissociation, the purification of the hepatocytes (i.e., *Hep. Perf.* Hepes perfusion, *Coll. Perf. Collagenase perfusion, Hep. Wash.* Hepes washes) and the early culture time points (*Hep. Cult.* culture hepatocytes), a sequential induction of the immediate early proto-oncogenes *c*-jun, *c*-fos, and *c*-myc mRNA takes place characterizing the exit from quiescence and the entry into the cell cycle of the hepatocytes

As indicated, cocultured hepatocytes (Fig. 1) are unable to proliferate under EGF or HGF stimulation alone as in liver tissue [32], but they proliferate under costimulation with TNFα and EGF or HGF. In addition, they undergo several rounds of division when cocultures are alternatively maintained in the presence and absence of cytokines and growth factors (Figs. 1 and 3) [33]. TNFα alone does not act as a complete mitogen, since it induces DNA synthesis in less than 3 % of hepatocytes as observed in vivo [11], but promotes the degradation of the extracellular matrix located around the hepatocyte colonies. This extracellular matrix remodeling is required for the G1/S transition [33]. Analysis of cell cycle proteins has revealed that growth factors alone, such as

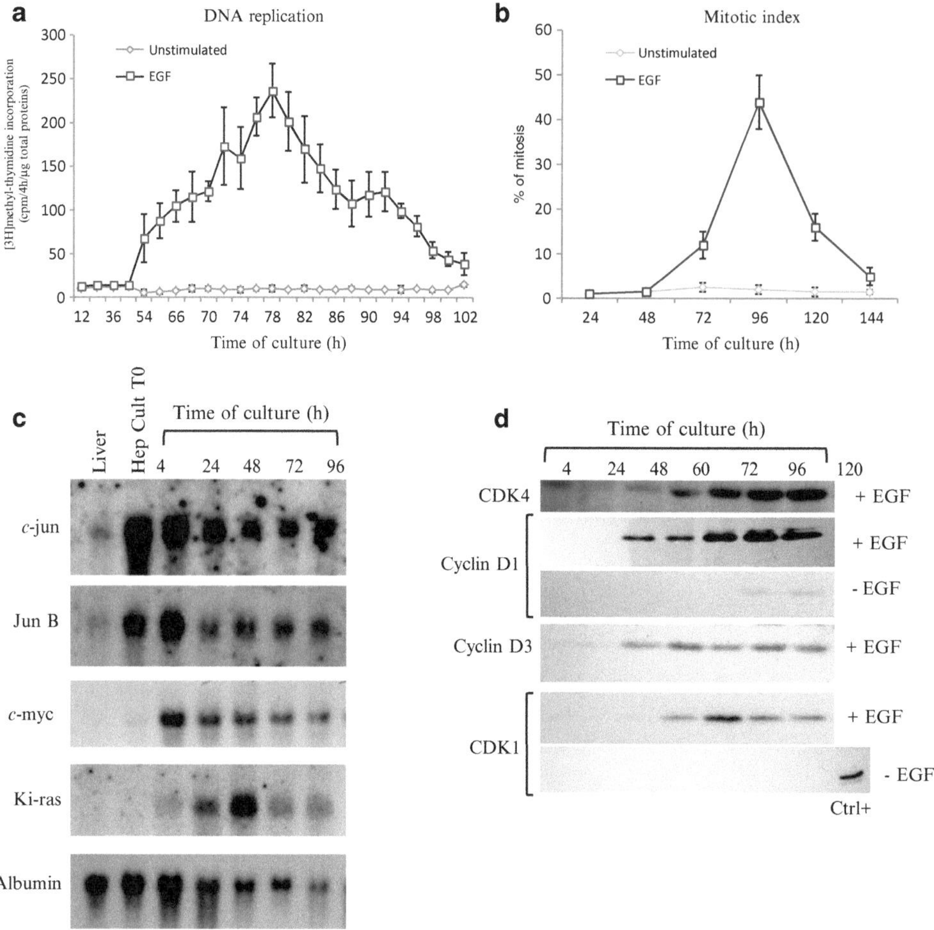

Fig. 2 Time course of DNA replication, mitosis, and expression of cell cycle regulators in pure culture of rat hepatocytes. (**a**) DNA replication measured by incorporation of [methyl ^{3}H]thymidine in genomic DNA. (**b**) Mitotic index established after 24 h of treatment with colcemid in cultures of rat hepatocytes in absence (i.e., nonstimulated) or presence of EGF (+EGF). (**c**) Northern blotting of the *c*-jun, jun B, *c*-myc, and Ki-ras mRNA proto-oncogene expression in normal liver, freshly isolated hepatocytes (Hep Cult T0), and in hepatocytes cultured during 4 days of culture. (**d**) Immunoblotting of CDK4 and CDK1, cyclins D1 and D3 in cultured hepatocytes in absence (–EGF) or presence (+EGF) of EGF

EGF, induce cyclin D1, CDK2, and CDK4, but not CDK1 (Fig. 3), yet hepatocytes fail to reach the G1/S boundary. In contrast, when TNFα is combined with EGF, hepatocytes express both CDK1 and CDK2, and progression into the S phase is observed.

Besides primary cultures of hepatocytes, hepatoma cell lines can also be used to investigate the mechanisms regulating the liver cell cycle. Human hepatoma HepaRG cells are human bipotent hepatic progenitors with high proliferation potential that are capable to differentiate into biliary and hepatocyte-like cells (Fig. 4) [34–36]. Confluent progenitor or differentiated HepaRG cells are quiescent, but following detachment and plating at low density, cells reenter the cell cycle in a partially synchronous manner (Fig. 4).

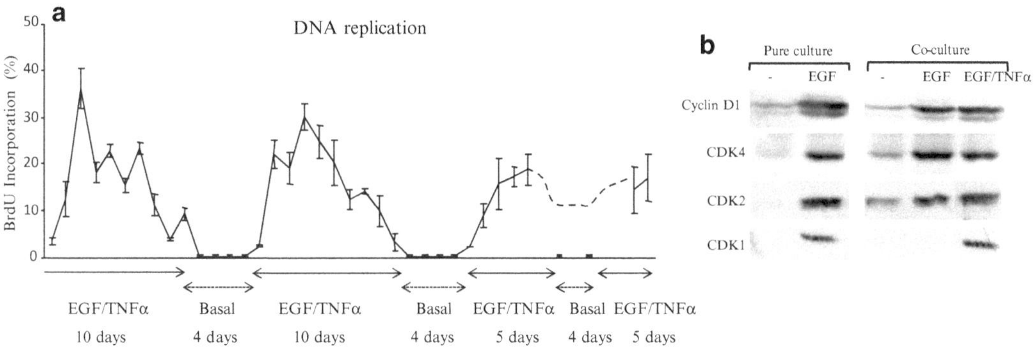

Fig. 3 Proliferation of hepatocytes in cocultures stimulated with TNFα and EGF. (**a**) Time course of DNA replication measured by BrdU incorporation in hepatocytes cocultured with RLECs for 42 days. (**b**) Immunoblotting of cyclin D1 and CDK1, 2, and 4 in hepatocytes maintained in pure culture or coculture whether or not stimulated with EGF alone or EGF and TNFα

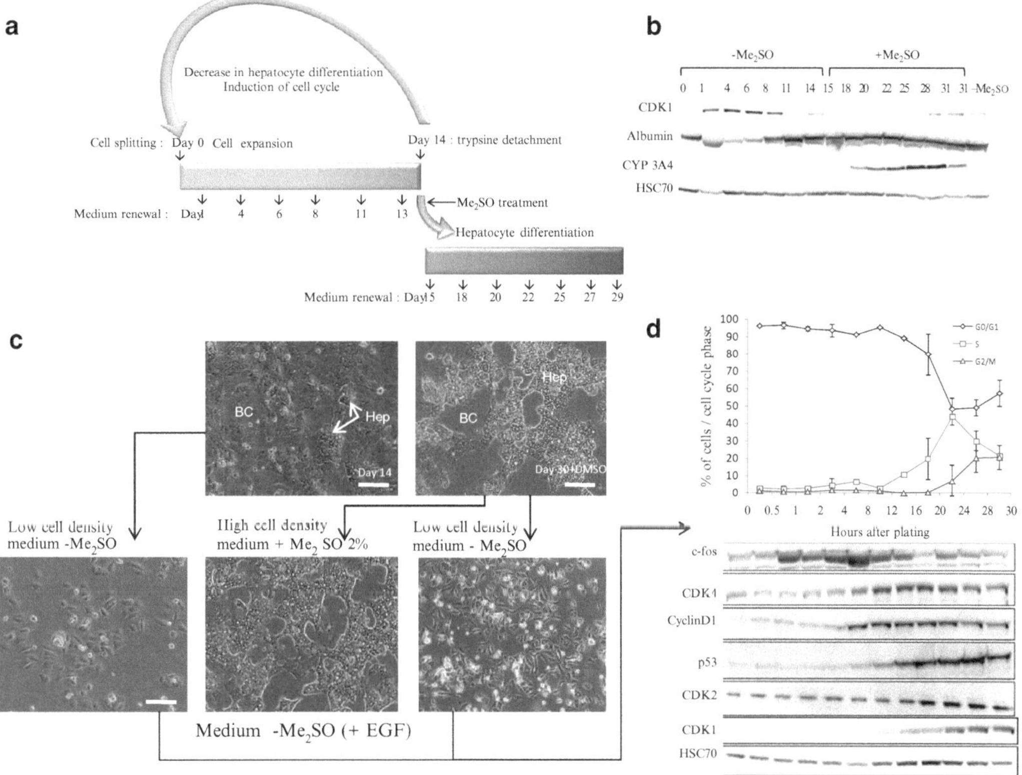

Fig. 4 Culture conditions and protocols for HepaRG cell cycle synchronization. (**a**) HepaRG cells are routinely cultured following a 30 day procedure. During the first 2 weeks following trypsinization, cells actively divide and slowly commit to either hepatocyte or cholangiocyte lineage. At day 14, cells can undergo a new passage for further expansion or be switched to a culture medium containing 2 % DMSO for enhancing the expression of the liver-specific functions in hepatocyte-like cells. (**b**) Immunoblotting of CDK1, albumin, cytochrome P450-3A4, and heat shock cognate protein 70 (HSC70) as loading control in HepaRG cells at different time points during proliferation and differentiation over 31 days postplating. (**c**) From cultures at day 14 or 31, cells can be plated to obtain a partially synchronized first cell cycle during the 30 h postplating. The graph represents the DNA content (i.e., % of cells in G0/G1, S, and G2/M phases measured by flow cytometry) in HepaRG cells at different times after plating of hepatocyte-like cells stimulated for 2 weeks with DMSO. (**d**) Immunoblotting of *c*-fos, CDK4, cyclin D1, p53, CDK2, and CDK1 protein levels at the different time points during the 30 h postplating

The aim of this chapter is to present the most pertinent in vitro models and commonly used mitogenic factors that promote proliferation of hepatocytes. Also described are culture conditions for primary hepatocytes, HepaRG, and rat hepatoma FAO cells proliferating in a synchronous manner with the goal to investigate the regulation of the hepatocyte cell cycle and the molecular pathways regulating liver regeneration. This chapter focuses on the canonical growth factors that stimulate hepatocytes, such as EGF, HGF, and TGFα, but of course many other regulators of hepatocytes have been identified either in vivo or in vitro [1, 2, 37]. These include potent mitogenic factors, such as the heparin binding-EGF-like factor and strong inhibitors of proliferation, including transforming growth factor beta (Table 1).

2 Materials

2.1 Culture Reagents: Solutions and Media (See Note 1*)*

1. Phosphate-buffered saline (PBS). 137 mM NaCl, 2.7 mM KCl, 10 mM $Na_2HPO_4 \cdot 2H_2O$, 1.8 mM KH_2PO_4 in deionized water. Adjust to pH 7.4 and sterilize by passing through a 0.22 μm filter.
2. 0.05 % trypsin-5.3 mM ethylenediaminetetraacetic acid solution (Life Technologies, France).
3. 100 mM L-glutamine solution (Life Technologies, France).
4. 50 IU/mL-50 μg/mL penicillin–streptomycin mix (Life Technologies, France) (*see* **Note 2**).
5. Fetal calf serum (Thermo Scientific, France).
6. Sodium pyruvate (Sigma-Aldrich, France). Prepare an 8 M stock solution in sterile water. Sodium pyruvate acts as a comitogen [22] and antioxidative compound [38].
7. Sodium butyrate (Sigma-Aldrich, France). Prepare a 500 mM stock solution in sterile water.
8. Primary hepatocyte culture media. William's E medium [39] or a mix of 1/3 Minimal Essential Medium and 2/3 Medium 199 supplemented with 2 mM L-glutamine, 1 % penicillin–streptomycin, 100 μg/mL fraction V bovine serum albumin (*see* **Note 3**), 5 μg/mL recombinant human or porcine insulin (*see* **Note 4**), and 5×10^{-7} M hydrocortisone sodium hemisuccinate (*see* **Note 5**).
9. RLEC medium [39, 40]. William's E medium supplemented with 2 mM L-glutamine, 1 % penicillin–streptomycin, and 10 % fetal calf serum (*see* **Note 6**).

Table 1
List of signals regulating the hepatocyte proliferation

Compound	Role	Reference
EGF	Potent growth factor stimulating the G1/S transition	[13]
TGFα	Potent growth factor stimulating the G1/S transition	[46]
HGF	Potent growth factor stimulating the G1/S transition	[19]
IL6	Proinflammatory cytokine promoting entry and progression in the early G1 phase	[10]
TNFα	Proinflammatory cytokine promoting entry and progression in early G1 phase	[11]
Insulin/glucagon	Potent survival factors and comitogens	[47]
IFNγ	Inhibitor of MMP9 activation and extracellular remodeling during hepatocyte proliferation	[33]
TGFβ	Inhibitor of hepatocyte proliferation	[48]
Amphiregulin	Potent mitogen for cultured hepatocytes	[49]
HB-EGF	Potent mitogen for cultured hepatocytes	[50]
Glucocorticoids	Survival factor with moderate activity as inhibitor of hepatocyte proliferation	[42]
Fetal calf serum	Factor of attachment and stimulator of hepatoma cell proliferation	[13]
Denaturated collagen gel	Coating factor enabling rapid attachment and spreading and favoring hepatocyte proliferation	[30]
Native collagen matrigel	Survival factor favoring the differentiation and inhibiting the hepatocyte proliferation	[31]
DMSO	Inhibitor of hepatocyte proliferation	[28]
Norepinephrine	Potentiator of mitogenic effects of growth factors	[51]
Serotonin	Putative growth factor	[52]
IL4	Facilitator of liver regeneration	[53]

DMSO dimethyl sulfoxide, *EGF* epidermal growth factor, *HB-EGF* heparin binding EGF-like growth factor, *HGF* hepatocyte growth factor, *IFNγ* interferon gamma, *IL4/6* interleukin 4/6, *MMP9* matrix metallopeptidase 9, *TGFα/β* transforming growth factor alpha/beta, *TNFα* tumor necrosis factor alpha

10. HepaRG medium. William's E medium supplemented with 2 mM L-glutamine, 1 % penicillin–streptomycin, 5 μg/mL of porcine or recombinant human insulin, 5×10^{-5} M hydrocortisone sodium hemisuccinate, and 10 % fetal calf serum (*see* **Note 7**). For full hepatocytic differentiation, the same medium is supplemented with 2 % DMSO.

11. FAO medium. Mix of NCTC135 and HAMF12 media (Gibco, France) supplemented with 2 mM L-glutamine, 1 % penicillin–streptomycin, and 10 % fetal calf serum. The FAO cell line has been subcloned from the rat hepatoma cell line H4IIEC3 isolated from the Reuber H35 hepatoma [41].

2.2 Growth Factors and Cytokines: Storage, Concentrations, and Combined Stimulations

Lyophilized growth factors must be reconstituted according to the manufacturer's instructions (*see* **Note 8**). Stock solutions must be aliquoted, frozen at −80 °C for long-term storage, and kept at 4 °C for rapid use to avoid multiple freezing and thawing cycles. For the most commonly used growth factors, prepare stock solutions as explained below.

1. 100 μg/mL recombinant human (rHu)-EGF (Promega, France) solution in sterile water.
2. 100 μg/mL TGFα (R&D systems Europe, France) solution in sterile water.
3. 5 μg/mL rHu-HGF (Sigma-Aldrich, France) solution in sterile water.
4. 10 μg/100 μL rHu-TNFα (Life Technologies, France) solution in sterile water.
5. 10 μg/100 μL rHu-IL6 (Miltenyi, France) solution in PBS supplemented with 0.1 % bovine serum albumin.

2.3 Materials and Reagents for Detection of DNA Replication and DNA Content in Hepatocytes

1. [Methyl ^{3}H]thymidine with specific activity at 325 GBecquerel (GBq)/mmole (Perkin Elmer, France).
2. 10 mg/mL 5-bromo-2′-deoxyuridine (BrdU) stock solution (BD Pharmigen, France). Prepare a working solution by diluting the stock solution 1/30 in cell culture medium.
3. 5 % acetic acid, 5 % water, and 90 % ethanol fixation solution.
4. 37 % HCl (Merck, France).
5. 0.1 M borate buffer, pH 8.5. Dissolve 6.18 g boric acid, 4.38 g NaCl, and 9.54 g sodium tetraborate in 900 mL of ultrapure water. Use a stir bar and low heat to ensure that the boric acid fully dissolves. Adjust the pH to 8.5 with NaOH and sterilize by vacuum filtration with a 0.22 μm filter.
6. PBS containing 0.05 % Tween 20 (Sigma-Aldrich, France).
7. Normal nonimmune donkey serum (Jackson ImmunoResearch Laboratories, United States of America).
8. Anti-BrdU antibody IIB5 (Abcam, United Kingdom).
9. Secondary anti-mouse antibody labeled with Dylight 488 (Eurobio, France).
10. 30 % trichloroacetic acid.
11. 99 % formic acid.

12. Methanol-free formaldehyde solution (Thermo Scientific, France). Prepare a 4 % working solution by diluting the 16 % commercial solution with PBS.
13. KaryoMax® Colcemid™ solution (Life Technologies, France).
14. Scintillation counters for the detection of radioactively labeled thymidine. The protocol in this chapter uses the TriCard device Perkin Elmer (France).
15. Inverted microscope equipped with phase-contrast and fluorescence modules.
16. Cycle test plus DNA reagent kit (Becton Dickinson, France) containing the permeabilization and RNase solution A, neutralization solution B, and the DNA staining agent propidium iodide solution C.
17. Flow cytometer FACScalibur equipped an argon/blue laser at 488 nm (Becton Dickinson, France).
18. Array Scan Cellomics imager system (Thermo Scientific, France).

2.4 Materials and Reagents for RNA and Protein Extraction and Analyses of Cell Cycle Regulator Expression

1. Lysis buffer for protein extraction. 50 mM 4-(2-hydroxyethyl)-1-piperazineethanesulfonic acid pH 7.5, 150 mM NaCl, 1 mM ethylenediaminetetraacetic acid pH 8, 2.5 mM ethyleneglycol tetraacetic acid pH 7.4, 0.1 % Tween 20, 10 % glycerol supplemented with 0.1 mM sodium orthovanadate, 1 mM sodium fluoride, 10 mM beta-glycerophosphate, 0.1 mM phenylmethanesulfonyl fluoride, and complete ethylenediaminetetraacetic acid-free (Roche, France) (*see* **Note 9**).
2. Bio-Rad protein assay dye reagent (Bio-Rad, France).
3. 4× concentrated denaturing loading buffer for protein denaturation and loading on polyacrylamide gels. 150 mM Tris–HCl pH 6.8, 0.2 % bromophenol blue, 5 % beta-mercaptoethanol, 20 % glycerol, and 8 % SDS.
4. RNA extraction kits Nucleospin RNA (Macherey-Nagel, France) or RNeasy (Qiagen, France).
5. Retrotranscription and quantitative polymerase chain reaction assays using kits, such as the high capacity cDNA reverse transcription kit (Life Technologies, France) and the power SYBRgreen PCR master mix (Applied Biosystems, France).
6. 20× concentrated Novex NuPage 3-(*N*-morpholino)propanesulfonic acid (MOPS) SDS running buffer (Life Technologies, France).
7. RNA sample buffer. 10 mM sodium phosphate pH 7.4 containing 2.2 M formaldehyde, 50 % deionized formamide, 0.1 % SDS, 5 % glycerol, and 5 mM bromophenol blue.

8. [α-^{32}P]deoxynucleotide triphosphates (dNTP) at 10 mCi/mL with specific activity about 3,000 Ci/mmol.
9. Klenow fragment of *Escherichia Coli* DNA polymerase I (New England BioLabs, United States of America).
10. Prehybridization and hybridization solution containing 0.5 M $Na_2HPO_4 \cdot 7H_2O$, 4 mL/L 85 % phosphoric acid, 7 % SDS, and 1 mM ethylenediaminetetraacetic acid pH 7.
11. 3MM chromatography paper (Whatman, France).
12. Molecular biology grade agarose, such as QA-Agarose (Q-Biogen, France).
13. 37 % formaldehyde solution (Sigma-Aldrich, France).
14. 20× concentrated saline sodium citrate buffer (Sigma-Aldrich, France).
15. Cassette with intensifying screen for X-ray film exposure (Kodak, France).
16. 1–2 mL sephadex G-50 spun chromatography column compatible with centrifugation at about 1,500 × *g*.
17. Novex bis/tris polyacrylamide gels (Life Technologies, France).
18. Ultrasonic liquid processor equipped with a small size probe adapted to small volumes of buffers and cells.
19. Vertical electrophoresis system and a protein transfer apparatus, such as the iBlot dry blotting system (Invitrogen, France).
20. Chemiluminescence imaging system, such as Imager ECL (Vilbert Lourmat, France).
21. Chemiluminescent horseradish peroxidase substrate, such as Super Signal West Dura extended duration substrate (Thermo Scientific, France).
22. Tris-buffered saline (TBS). 25 mM Tris–HCl pH 7.4, 137 mM NaCl, 2.7 mM KCl, 0.2 % Tween 20 containing grade V bovine albumin or 2 % defat free milk.
23. PBS or TBS with 0.2 % Tween 20.
24. SpeedVac (SC110, Savant-France).
25. Nylon membrane (Amersham GE Health Care, France).
26. Ultraviolet irradiator.
27. Spectrophotometer for small volume samples, such as the NanoDrop ND-1000 device (LabTech, France).
28. Spectrophotometer for 96-well plates with filter at 595 nm.
29. Water bath.
30. Horizontal electrophoresis apparatus.
31. Ultraturrax (IKA, France).

3 Methods

3.1 Primary Culture of Hepatocytes

3.1.1 Pure Culture of Rodent Hepatocytes

1. Following isolation from the liver, plate hepatocytes in medium supplemented with 10 % fetal calf serum at different cell densities and maintained in various culture medium conditions (Fig. 1) according to the aims of the designed experiments (*see* **Note 10**). Plate at densities of about 5×10^4 hepatocytes/cm^2 in order to obtain robust proliferation activity. Alternatively, seeding can be performed by plating cells in collagen-coated dishes even in the absence of fetal calf serum. Attachment should occur within 4 h after seeding.
2. 4–6 h after plating, carefully discard the medium to eliminate damaged cells and gently replace with fresh medium containing a growth factor, such as 25 and 50 ng/mL EGF, 2–10 ng/mL HGF, or 20 ng/mL TGFα. Alternatively, addition of growth factors can be done while seeding cells. A similar procedure can be applied to human primary hepatocytes (*see* **Note 11**).
3. Renew the medium with growth factors daily.

3.1.2 Coculture of Hepatocytes with RLECs

1. Seed rodent hepatocytes at a density of 5×10^4 hepatocytes/cm^2.
2. Detach RLECs using trypsin. Resuspend the cells in William's E medium containing 5 μg/mL insulin, 5×10^{-7} M hydrocortisone sodium hemisuccinate, and 10 % fetal calf serum. Plate the RLECs onto the hepatocytes 4 h after hepatocyte plating at a density of 10^5 RLEC/cm^2. RLECs attach and spread between hepatocyte colonies and proliferate until the coculture is totally confluent (Fig. 1).
3. At day 1, renew culture medium with 5×10^{-5} M hydrocortisone sodium hemisuccinate.
4. Maintain the culture condition until reaching a confluent monolayer. The coculture is considered functional when the monolayer is totally confluent and neocanaliculi are well visible (i.e., usually at day 2–4 postplating).
5. Stimulate proliferation with a combination of proinflammatory cytokines and growth factors 4–6 days after plating hepatocytes and RLECs. Specifically, a proinflammatory cytokine, either 10 and 100 ng/mL TNFα or 20 ng/mL IL6, and a growth factor, such as EGF, TNFα, or HGF, should be used (*see* **Note 12**).
6. Renew the medium daily with fresh cytokines and growth factors.

3.2 Hepatoma Cell Lines

3.2.1 Regular Expansion of HepaRG Cells

1. Expand HepaRG progenitor cells in William's E medium supplemented with 5 μg/mL insulin (*see* **Note 4**), 5×10^{-5} M hydrocortisone sodium hemisuccinate, and 10 % fetal calf serum. Detach confluent HepaRG cells 2 weeks after the previous passage (Fig. 4). For a 75 cm² flask, discard the culture medium, rinse the cell monolayer with 10 mL PBS, add 5 mL trypsin solution, allow the trypsin solution to cover the entire surface, and discard 3 mL. Incubate the culture flask in an incubator at 37 °C and 5 % CO_2 for 5 min. Resuspend the cells in 10 mL medium. Disrupt cell clumps by pipetting up and down.
2. Plate the cells at a density of 2.6×10^4 cells/cm², which is about 2×10^6 cells/75 cm² flask.
3. Renew the medium the day after seeding and then every 2 or 3 days (Fig. 4).
4. Two weeks later, either expand the progenitor cells following the procedure described above or maintain the cells for another 2 weeks in the same medium supplemented with 2 % DMSO and renew every 2 days (*see* **Note 13**).

3.2.2 Synchronous Proliferation of Progenitor HepaRG Cells

1. Detach confluent cells at day 14 following the previous passage by trypsin treatment described above for a regular passage.
2. Resuspend the cells in William's E medium supplemented with 20 % fetal calf serum and plate the cells at density of about 2.5×10^4 cells/cm².
3. To further improve cell proliferation, 25 ng/mL EGF may be added to the medium.
4. Follow cell cycle progression over a 36 h time course (Fig. 4) (*see* **Note 14**).

3.2.3 Synchronous Proliferation of Hepatocyte-Like HepaRG Cells

1. Detach hepatocyte-like cells at a purity of at least 80 % by a simple trypsin treatment. Use only 75 cm² flasks.
2. Rinse cell cultures twice with 10 mL PBS. Add 2.5 mL PBS and 2.5 mL trypsin solution.
3. Close tightly the flask and mix gently. Incubate at room temperature for 5 min and monitor the morphological change of the hepatocyte-like cells using a microscope equipped with phase contrast. When hepatocyte-like cells have rounded up, shake the flask containing the PBS–trypsin solution laterally to allow complete detachment of the cells.
4. Collect and add the 5 mL diluted trypsin solution containing the cells to 5 mL of William's E medium.
5. Spin the cells at $50 \times g$ and room temperature for 2 min, resuspend the pellet in 5 mL of William's E medium and count the hepatocyte-like cells. Typically, $3–5 \times 10^6$ hepatocyte-like cells can be purified from a 75 cm² flask.

6. Plate the cells at a density of 2.5×10^4 cells/cm^2 in William's E medium without DMSO for studying hepatocyte proliferation.
7. Alternatively, hepatocyte-like cells cultured in DMSO-containing medium are selectively detached as described above, but are plated at a density of 5×10^4 cells/cm^2 in William's E medium supplemented with 2 % DMSO for maintaining high expression of hepatocyte-specific functions and to keep cells quiescent. At day 2 or 3 postplating, discard the culture medium containing DMSO, wash cells with PBS, and stimulate proliferation with William's E medium deprived in DMSO. To further improve cell proliferation, 25 ng/mL EGF may be added to the medium.

3.2.4 Synchronous Proliferation of FAO Cells

1. Detach subconfluent cultures of FAO cells by trypsin treatment. Rinse cultures twice with 10 mL PBS and add 2.5 mL trypsin solution. Incubate at 37 °C and 5 % CO_2 for 5 min.
2. Resuspend the cells in medium deprived in fetal calf serum and disrupt cell aggregates by pipetting vigorously.
3. Seed the FAO cells at 2.5×10^4 cells/cm^2 in a medium containing only 0.1 % fetal calf serum.
4. 24 h later, renew the medium with fresh medium without fetal calf serum, but supplemented with 10 mM sodium butyrate. Fetal calf serum starvation and sodium butyrate addition strongly reduce DNA replication as shown by the very low remaining level of [methyl ^{3}H]thymidine incorporation at the end of day 2 (Fig. 5).
5. At day 2, discard medium, wash three times with PBS, and stimulate with NCTC135/HAMF12 medium supplemented with 10 or 20 % fetal calf serum.
6. Follow cell cycle progression over a 28–30 h time course (Fig. 5) (*see* **Note 15**).

3.3 Detection of DNA Replication and Mitosis

3.3.1 Incorporation of BrdU

1. Dilute the 3.3 mg/mL BrdU solution 1/100 in cell culture medium and incubate with the cells for 1–6 h (*see* **Note 16**).
2. Discard the medium, wash the cells with PBS, and fix with the cold 5 % acetic acid, 5 % water, and 90 % ethanol solution for 15 min at 4 °C.
3. Discard the fixation solution and rinse the cells three times with PBS.
4. Permeabilize the cells with 0.3 % Triton X-100 in PBS for 15 min at room temperature.
5. Incubate with 2 N HCl for 30 min at 37 °C.
6. Wash three times 10 min with 0.1 M borate buffer pH 8.5.

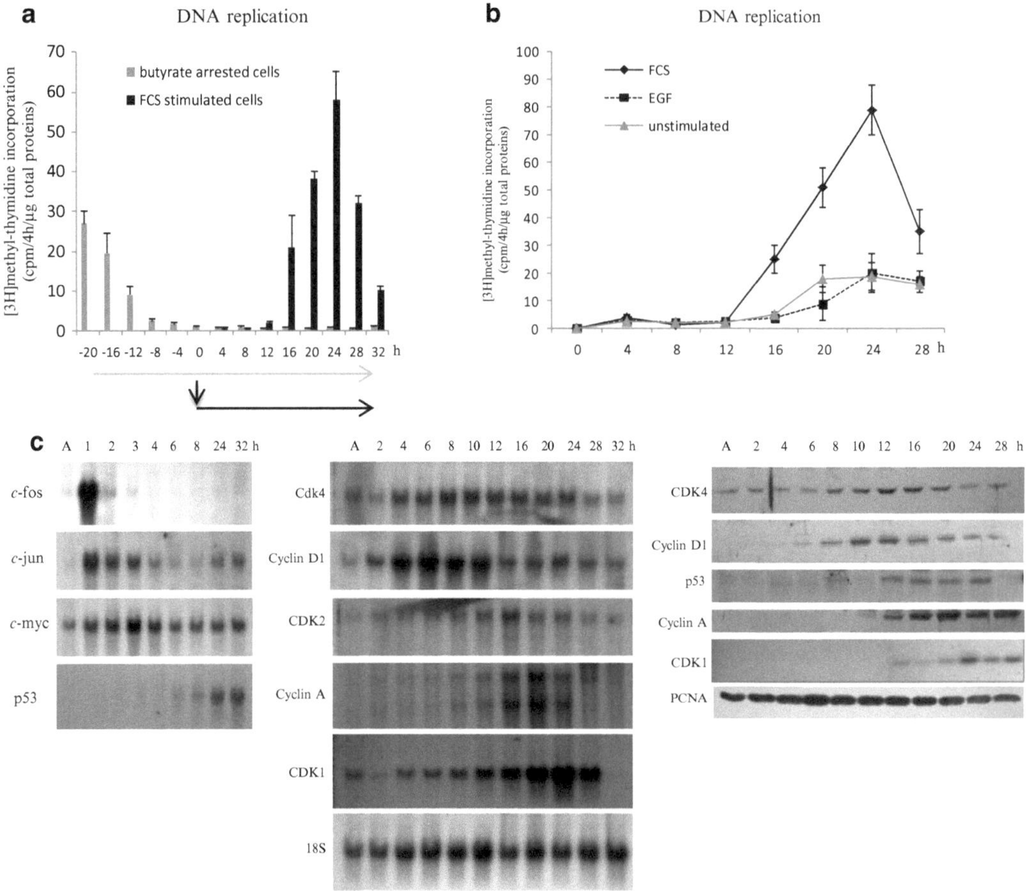

Fig. 5 Protocol of synchronization for FAO hepatoma cells. (**a**) DNA replication measured by incorporation of [methyl ^{3}H]thymidine in genomic DNA of FAO cells during cell cycle arrest induced by fetal calf serum starvation and sodium butyrate treatment and after release by sodium butyrate withdrawal and fetal calf serum readdition. (**b**) Comparison of DNA synthesis measured by incorporation of [methyl ^{3}H]thymidine in FAO cells following mitogenic stimulation by 10 % fetal calf serum versus 25 ng/mL EGF. (**c**) Northern blotting (*left* and *middle panels*) and immunoblotting (*right panel*) analyses of immediate early genes and late markers of the cell cycle in synchronized FAO cells

7. Incubate with the primary anti-BrdU antibody diluted 1/200 in PBS containing 0.05 % Tween 20 and 2 % normal donkey serum for 1 h at 37 °C.
8. Wash three times with PBS containing 0.05 % Tween 20.
9. Incubate with the secondary anti-mouse antibody labeled with Dylight 488 diluted 1/500 in PBS containing 0.05 % Tween 20 and 2 % normal donkey serum for 30 min at 37 °C.
10. Wash five times with PBS.
11. Stain cell nuclei with 1 µg/mL Hoechst 33342 solution in PBS for 20 min at room temperature to counterstain all nuclei.

12. Determine the index of BrdU-positive cell nuclei versus all nuclei by counting either manually or using a microscope or with a dedicated imager system, such as the Array Scan Cellomics device.

3.3.2 Incorporation of [Methyl ^{3}H]thymidine (Figs. 2 and 5)

1. Add 37 kBq/mL of [methyl ^{3}H]thymidine into culture medium of hepatocytes and incubate for 1–6 h (*see* **Note 17**).
2. Discard the medium containing the radioactive thymidine and wash cells twice with the same volume of PBS.
3. Lyze cells in PBS containing 0.2 % SDS. The volume of lysis buffer must be adjusted to the cell culture dish. For 6-well plates, 0.5 mL of lysis buffer should be used. Lysates are collected and frozen during the time course experiment.
4. After thawing, determine the total protein concentration for each sample.
5. Precipitate genomic DNA by adding the same volume of 30 % trichloroacetic acid and incubate overnight at 4 °C.
6. Recover DNA by centrifugation at 1,500 × *g* and room temperature for 10 min.
7. Wash the DNA pellet once with 10 % trichloroacetic acid and twice with 5 % trichloroacetic acid.
8. Dissolve the DNA in 0.2 mL formic acid, transfer to vials containing scintillation agent, and count radioactivity in scintillation counters.
9. Express DNA replication in counts per minute (cpm) of [methyl ^{3}H]thymidine/μg of total protein.

3.3.3 Mitotic Index Determination (Fig. 2) (*See* **Note 18**)

1. Add 0.1 μg/mL KaryoMax Colcemid solution for 12–24 h and fix the cells with 4 % paraformaldehyde in PBS for 30 min at 4 °C.
2. Permeabilize cells with 0.3 % Triton X-100 in PBS for 15 min at room temperature.
3. Stain the cells with 1 μg/mL Hoechst 33342 solution in PBS for 20 min at room temperature to counterstain all nuclei.
4. Count the accumulated mitotic cells during the KaryoMax Colcemid incubation and establish the mitotic index.

3.3.4 DNA Content by Flow Cytometry

1. Detach 5×10^{5} cells by trypsin treatment.
2. Resuspend the cells in 5 mL PBS and disperse cell aggregates vigorously by pipetting up and down.
3. Spin the cells at 4,000 × *g* and room temperature for 1 min and wash the pellet twice with 1 mL of the buffer provided with the Cycle test plus DNA reagent kit.

4. Resuspend the pellet in 125 μL of solution A and incubate 10 min at room temperature.
5. Add 100 μL of solution B, mix well, and incubate 10 min at room temperature. At this step, the cell suspension can be stored for not more than 1 week at 4 °C.
6. Add 100 μL of solution C containing propidium iodide, mix well, and incubate for 10 min *prior* to analysis by flow cytometry.

3.4 Extraction of Total RNA from Normal Liver and Cultured Hepatocytes

1. To prevent RNA degradation, being the major hurdle in this experiment, fresh liver fragments or cultured hepatocytes must be lyzed immediately in lysis buffer provided by the manufacturer (*see* **Note 19**). For frozen samples, thaw the tissue or cells into the lysis buffer and proceed with the extraction.
2. Disperse liver biopsies of less than 200 mg using an ultraturrax device *prior* to proceeding with the extraction that should yield about 300 μg of total RNA. Wash the ultraturrax device with large volumes of sterile water between samples to avoid cross-contamination. Do not sonicate biopsies.
3. Avoid overloading of silica columns by too much homogenized lysate that may affect the binding capacity of the column. Check the amounts of cells or tissue recommended by the manufacturer according to the size of the columns (*see* **Note 20**).
4. Digest traces of genomic DNA using RNase-free DNase.
5. Quantify purified RNAs using the NanoDrop ND-1000 device.

3.5 Northern Blotting (See **Note 21***)*

1. Aliquot volumes corresponding to 10 μg of total RNA and dry using a SpeedVac device.
2. Dissolve RNA pellets in RNA sample buffer. Incubate at 4 °C for 15 min.
3. Heat samples to 65 °C for 10 min before loading onto 1.5 % agarose gel dissolved in 10 mM sodium phosphate pH 7.4 containing 1.1 M formaldehyde.
4. Proceed with the electrophoresis of the RNA samples for 3–4 h at 70 V.
5. Transfer RNA by capillarity using the 20× concentrated saline sodium citrate buffer. Place three layers of 3MM chromatography paper soaked into 20× concentrated standard sodium citrate buffer on a glass or plexiglass plate that lies across a tank filled with 20× concentrated standard sodium citrate buffer. Make sure that the chromatography paper layers drape over the edges of the plate and are in contact with the buffer. Place the gel onto the three layers of blotting paper. On top of the gel, place the nylon membrane, then three layers of dry 3MM

chromatography paper exactly cut at the size of the gel. Add towel paper on top to drain the buffer and to allow the transfer overnight.

6. Crosslink the RNA to the nylon membrane by ultraviolet irradiation at 254 nm for 2 min at 1.5 J/cm^2.
7. Prehybridize the nylon membrane for 2 h at 68 °C in 10–20 mL of prehybridization solution (*see* **Note 22**).
8. Generate cDNA probes by restriction enzyme producing cohesive ends to allow Klenow fragment to perform end-filling of the two strands. Digest for 2 h 1 μg of DNA template with the appropriate restriction enzyme in 25 μL of the enzyme buffer.
9. Label linear double stranded DNA probe using the Klenow fragment of *Escherichia coli* DNA polymerase I to incorporate [α-^{32}P]dNTP at 10 mCi/mL into the cDNA of the mRNA of interest. Add 25 μCi of the [α-^{32}P]dNTP and unlabeled dNTPs to a final concentration of 100 μM and 5 U of Klenow fragment. Incubate for 15 min at room temperature and stop the reaction by denaturing the Klenow fragment at 75 °C for 10 min. Separate the labeled DNA from free dNTPs by spun-column chromatography using Sephadex G-50 at 1,500 × *g* and room temperature for 2 min.
10. Proceed with the hybridization by adding the labeled DNA in the prehybridization solution for 6–12 h.
11. Wash the nylon membrane twice for 15 min with 3× concentrated standard sodium citrate buffer at 68 °C and twice with standard sodium citrate buffer.
12. Place the membrane in a thin plastic bag and expose with X-ray films in a cassette containing intensifying screens.

3.6 Western Blotting

1. Discard culture medium and wash cells with cold PBS. Carefully drain the excess of PBS.
2. Lyze cells with cold protein extraction buffer added to the culture dishes and spread it by the pipette tip over the cell monolayer (*see* **Note 23**). Keep the cell culture dishes on ice. The cell lysis buffer, especially for studying proteins involved in proliferation, cellular signaling through phosphorylation and cell cycling, must be done in a buffer similar as the one described in **Note 6**. The buffer should contain protease and phosphatase inhibitors. The lysis should be performed in the cell culture dish. Do not detach cells using trypsin, which generates proteolytic cleavage per se that is not compatible with immunoblotting. In addition, the disruption of cell–cell interactions induces the expression of many cell cycle genes within 3 min.
3. Use a rubber policeman or a 1 mL syringe plunger to detach cells from the cell culture dish.

4. Collect the lysate and transfer to a microfuge tube *prior* to freezing at −20 °C. Alternatively, when multiple time points are collected, culture dishes can be frozen at −80 °C after washing with PBS and protein extraction is performed only when all samples have been collected.
5. Thaw the lysate and sonicate for 10 s to homogenize the lysate and to break genomic DNA.
6. Proceed with the concentration assessment of total protein in the samples. Protein concentration is determined by spectrophotometry using the Bio-Rad protein assay dye reagent diluted 1/5 in water. The standard curve is prepared using bovine serum albumin with concentrations in the range of 0.5–5 mg/mL. For bovine serum albumin standard dilutions and samples, mix 5 μL to 1 mL of diluted Bio-Rad dye reagent. Read the optical density at 595 nm in 96-well plates.
7. Following determination of protein concentrations, aliquot cell extracts from 5 to 50 μg of total protein (*see* **Note 24**) to be mixed with the denaturing loading buffer.
8. Heat the samples in denaturing loading buffer for 5 min at 95 °C.
9. Set precast Novex bis/tris polyacrylamide gels in the electrophoresis system and dilute the 20× concentrated Novex NuPage MOPS SDS running buffer to final concentration.
10. Load protein samples in the wells of the gel and proceed with the electrophoresis at 200 V for 55 min.
11. Proceed with the loading onto the polyacrylamide gel, electrophoresis, and protein transfer onto nitrocellulose membrane according to the manufacturer's instructions.
12. Transfer the nitrocellulose membrane in saturation buffer such as PBS or TBS containing grade V bovine albumin or 2 % fat free milk for 30 min at room temperature.
13. Add the primary antibody at the recommended concentration. The incubation with the primary antibodies is a crucial step that relies mainly onto the quality of the antibody. Usually, antibodies that have been raised against large peptides and affinity purified with the immunizing antigen give better results. The concentration must be optimized for each antibody. Incubate for 2 h at room temperature or overnight at 4 °C.
14. Wash the membrane at least 3 times with PBS or TBS 0.2 % Tween 20.
15. Incubate the membrane in saturation buffer containing the secondary antibody coupled to horseradish peroxidase at room temperature for 1 h.
16. Proceed with the detection using a chemiluminescent horseradish peroxidase substrate, such as Super Signal West Dura extended duration substrate, and detect the luminescence with the imager.

4 Notes

1. Avoid the use of solvents, fixation reagents (i.e., formaldehyde-based fixators), and cell lysis buffers in cell culture rooms. Perform mycoplasma screening on each batch of hepatoma cells after thawing.
2. Stock solutions of commercial L-glutamine and penicillin–streptomycin solutions are thawed and aliquoted into 5 mL sterile tubes *prior* to freezing. One aliquot of each is thawed and added to each 500 mL bottle of cell culture medium.
3. Bovine serum albumin fraction V must be dissolved in PBS at 10 g/mL under gentle rotation overnight at 4 °C.
4. Porcine or rHu-insulin is solubilized in sterile ultrapure water in which 37 % HCl is added dropwise until complete dissolution to obtain a stock solution at 5 mg/mL. Insulin stock solution can also be aliquoted (i.e., 500 μL/tube) and frozen.
5. Hydrocortisone sodium hemisuccinate, packaged as 100 mg of lyophilized product/vial, is dissolved in 4 mL PBS (i.e., stock solution at 5×10^{-2} M) and aliquots of 500 μL are frozen. For each 500 mL bottle of cell culture medium, 1 aliquot of hydrocortisone sodium hemisuccinate is added to the medium. For plating the hepatocytes, the medium is also supplemented with 10 % fetal calf serum and 5×10^{-7} M hydrocortisone sodium hemisuccinate. After cell attachment, the medium is renewed and fetal calf serum may then be withdrawn from the fresh medium, while the hydrocortisone sodium hemisuccinate concentration is increased up to 5×10^{-7} M. Growth factors can be added according to protocols described below.
6. RLECs are purified from 10-day-old Sprague-Dawley rats [40]. These cells, probably originating from bile ductules, exhibit an unlimited proliferation rate, but are expanded over a limited number of passages due to their spontaneous transformation that makes them unsuitable for the coculture protocol described in this chapter.
7. Fetal calf serum is crucial to properly maintain HepaRG cell proliferation and the capacity to differentiate as well as for RLECs to establish cocultures with primary hepatocytes. We screen new batches of sera when the batch in use becomes low. We obtain small batches from companies specialized in tissue culture reagents and we test these batches over at least 4–5 passages. We monitor cell toxicity, proliferation rate of RLECs and progenitor HepaRG cells, and the ability of HepaRG cells to differentiate in the presence of 2 % DMSO. We also pay particular attention to select sera that limit the expansion of non-fully differentiated small hepatocyte-like cells and fibroblasts. We found great differences between test batches. When an

optimal batch is identified, we purchase a large amount of this serum for long-term storage at −20 °C.

8. Most growth factors and comitogens for normal rodent and human hepatocytes are commercially available from pharmaceutical and biotechnology companies. These soluble factors are usually recombinant proteins that exhibit high activities with respect to promoting hepatocyte proliferation.
9. This buffer prevents the loss of specific protein phosphorylations of key factors of signaling pathways involved in cell cycle regulation. It is suitable for immunoblotting, immunoprecipitation, and protein kinases assays.
10. In all conditions, addition of insulin and glucocorticoids is absolutely required to prevent rapid occurrence of spontaneous apoptosis [42, 43]. Hepatocyte seeding at high cell densities of more than 10^5 hepatocytes/cm^2 and high concentrations of hydrocortisone sodium hemisuccinate (i.e. up to 10^{-5} M) is recommended for maintaining long-term survival and expression of liver-specific functions. In this condition, even in the presence of growth factors, cell proliferation will remain low. Lower plating cell densities at 5×10^4 hepatocytes/cm^2 and low concentrations of hydrocortisone sodium hemisuccinate (i.e., 10^{-7} M) must be used to allow robust cell proliferation after addition of growth factors, such as EGF, TGFα, or HGF. We recommend stimulating cells either during seeding or 4 h after plating when medium is renewed. The presence of growth factors strongly affects culture morphology (Fig. 1). Hepatocytes keep a more cuboidal shape, the cytoplasm appears more granular, and cell limits and nuclei remain well visible, while hepatocytes maintained in nonstimulated conditions used as negative control of proliferation spread, exhibit fragmented nucleoli, and switch to a fibroblast-like shape over time. In addition, between 48 and 96 h, in stimulated conditions, proliferation takes place and cell density increases.
11. Human hepatocytes can also be isolated using a 2-steps procedure by collagenase perfusion adapted to human whole liver or resected biopsies [44]. From our experience, proliferation of human hepatocytes in primary culture is much weaker compared to rodent hepatocytes. However, significant DNA replication can be obtained in human hepatocytes plated at low cell density (i.e., lower than 10^5 hepatocytes/cm^2) upon stimulation with EGF, TGFα, or human serum [45].
12. In coculture, hepatocytes do not replicate DNA in the absence or presence of EGF [33]. The cocktail of growth factors and cytokines does induce robust proliferation. Stimulation with more than one growth factor does not necessarily improve the rate of DNA replication. It is important to note that after 8–10 days of

stimulation, the rate of DNA replication becomes very low (Fig. 3). To reinitiate a new wave of DNA replication, cells must be placed in cell culture medium without cytokines and growth factors for at least 4 days. Then, new stimulation can be performed for a period of 10 days. Switching back and forth between stimulated and nonstimulated conditions allows multiple cell divisions and the expansion of the hepatocyte population.

13. Regarding differentiation of HepaRG cells towards hepatocyte-like cells, cells form a confluent monolayer 14 days after cell splitting. These cells (i.e., $1.2–1.5 \times 10^7$ cells/75 cm^2 flasks) are already committed to hepatocyte or biliary cell lineage. Addition of 2 % DMSO enhances the expression of some hepatocyte-specific functions, especially cytochrome P450 enzymes. Overall, 10 days to 2 weeks of DMSO treatment are required to obtain the highest expression of these markers. The cell number is often slightly reduced from 1.2 to 1.5×10^7 cells before adding DMSO to $1–1.2 \times 10^7$ cells/75 cm^2 flasks after DMSO treatment. During the first week following DMSO addition, extensive cell death and detachment points reflect either too early addition of DMSO or to toxicity induced by the combination of DMSO and serum. During DMSO treatment, the remaining hepatocyte-like cells complete their differentiation and undergo drastic morphological changes to give rise to well-defined colonies of hepatocytes characterized by a dark cytoplasm, a large nucleus with a single nucleolus, and functional neocanaliculi.
14. The procedures for synchronization of HepaRG cells trigger a partially synchronized entry into and progression through the cell cycle that can be followed by analyzing DNA content and expression of cell cycle markers (Fig. 4) at different time points after plating or mitogenic stimulation. HepaRG cells exit quiescence and reenter the cell cycle in less than 1 h, as evidenced by the induction of protein levels of immediate early genes, such as *c*-fos (Fig. 4). Thereafter, genes characterizing the progression throughout the cell cycle are sequentially induced, including CDK4 and cyclin D1 in the mid-late G1 phase, p53 and CDK2 in the late G1 phase, and CDK1 during the S, G2, and M phases. The analysis of the DNA content by flow cytometry indicates that the G1 phase is about 16 h long, while peaks of DNA replication and mitosis take place at 24 and 30 h post-plating, respectively.
15. DNA replication measured by [methyl ^3H]thymidine incorporation takes between 12 and 16 h after mitogenic stimulation by fetal calf serum and is maximal at 24 h. Interestingly, in FAO cells, stimulation by EGF triggers modest DNA replication, which is not significantly higher than that obtained after discarding the cell culture medium containing sodium butyrate

and replacing by cell culture medium without fetal calf serum (Fig. 5). In synchronized FAO cells, the sequential induction of immediate early proto-oncogenes, such as *c*-fos, *c*-jun, and *c*-myc, and the downstream CDKs and cyclins detected by northern blotting and immunoblotting further illustrates the partial synchronous progression in the cell cycle (Fig. 5).

16. The incubation time of BrdU must be at least 1 h to have enough cells entering the S phase. For time course analysis, it is essential to compare identical incubation time periods.
17. We recommend performing the experiments in 24-, 12-, or 6-well plates. Keep in mind that the surface of 1 well in 6-well plates is about 10 cm^2 and that the volume of medium used is usually 2 mL. For 6-well plates, we add 2 μL of 325 GBq/mmole [methyl ^{3}H]thymidine, which corresponds with 74 kBq of thymidine. As for BrdU incorporation, it is important to compare identical incubation time periods.
18. To verify that DNA replication is followed by mitosis and that hepatocytes complete the cell cycle in the presence of growth factors, the mitotic index can be easily determined by counting mitotic figures following the in situ detection of BrdU incorporation and counterstaining with Hoechst 33342. We have optimized this simple protocol to accumulate mitotic cells in the G2/M phase each day and to determine the mitotic index.
19. Liver biopsies and cultured hepatocytes must be lyzed in denaturing conditions with solutions containing high concentrations of chaotropic ions that inactivate RNases present in all cells. These ionic conditions also ensure efficient binding to silica membrane used in the kits to purify nucleic acids.
20. Hepatocytes are relatively large cells that actively transcribe genes to ensure metabolic functions. Thus, these cells contain high amounts of RNA compared to fibroblasts or widely used cell lines, such as HeLa cells. Usually, 10^6 hepatocytes allow the recovery of over 100 μg of total RNA.
21. Most experiments investigating the expression levels of long-coding or noncoding RNA nowadays use quantitative reverse transcriptase polymerase chain reaction analysis. However, many of the initial studies characterizing the expression of cell cycle markers have been performed by northern blotting, which remains an interesting procedure to determine the length of the mRNA of interest and to prove possible alternative splicing generating splice variants with relatively large differences in length.
22. Prehybridization and hybridization of northern blotting can be performed in commercial rotating wheels or tubes that are designed to limit leakage of solutions. However, home-made systems as simple as plastic boxes or heated-sealed plastic bags

work reasonably well at a low cost. Prehybridization and hybridization are then performed in a regular water bath.

23. The volume of lysis buffer depends on the size of the cell culture dish. For 0.5×10^6 hepatocytes or hepatoma cells, we use 200 μL of lysis buffer that should yield a protein concentration for the cell extract in the range of 2–5 μg/μL. This concentration is well adapted for immunoblot analysis, since the amounts of total protein loaded in each well of the polyacrylamide gel should vary between 10 and 50 μg for the detection of most proteins.
24. Manufacturers often provide denaturing and loading buffers that do not contain the classical reducing and denaturing agents, such as beta-mercaptoethanol. In addition, they recommend a denaturation step at 70 °C for 10 min only. While these procedures work well for most proteins, we experienced abnormal apparent molecular weights with some proteins, including CDKs. Thus, for the preparation of electrophoresis in denaturing conditions, we recommend to mix the cell lysate with 4× concentrated loading buffer containing SDS and beta-mercaptoethanol.

References

1. Fausto N (2004) Liver regeneration and repair: hepatocytes, progenitor cells, and stem cells. Hepatology 39:1477–1487
2. Michalopoulos GK, DeFrances M (2005) Liver regeneration. Adv Biochem Eng Biotechnol 93:101–134
3. Avril A, Pichard V, Bralet MP et al (2004) Mature hepatocytes are the source of small hepatocyte-like progenitor cells in the retrorsine model of liver injury. J Hepatol 41:737–743
4. Higgins GM, Anderson RM (1931) Experimental pathology of liver: restoration of the liver of the white rat following partial surgical removal. Arch Pathol 12:186–202
5. Hsu JC, Bravo R, Taub R (1992) Interactions among LRF-1, JunB, *c*-Jun, and *c*-Fos define a regulatory program in the G1 phase of liver regeneration. Mol Cell Biol 12:4654–4665
6. Grisham JW (1962) A morphologic study of deoxyribonucleic acid synthesis and cell proliferation in regenerating rat liver: autoradiography with thymidine-H3. Cancer Res 22: 842–849
7. Fabrikant JI (1968) The kinetics of cellular proliferation in regenerating liver. J Cell Biol 36:551–565
8. Albrecht JH, Hoffman JS, Kren BT et al (1993) Cyclin and cyclin-dependent kinase 1 mRNA expression in models of regenerating liver and human liver diseases. Am J Physiol 265:G857–G864
9. Fausto N, Campbell JS, Riehle KJ (2006) Liver regeneration. Hepatology 43:S45–S53
10. Cressman DE, Greenbaum LE, DeAngelis RA et al (1996) Liver failure and defective hepatocyte regeneration in interleukin-6-deficient mice. Science 274:1379–1383
11. Webber EM, Bruix J, Pierce RH et al (1998) Tumor necrosis factor primes hepatocytes for DNA replication in the rat. Hepatology 28: 1226–1234
12. Seglen PO (1975) Preparation of isolated rat liver cells. Methods Cell Biol 13:29–83
13. Guguen-Guillouzo C, Guillouzo A (1986) Isolation of rat hepatocytes. In: Guillouzo A, Guillouzo C (eds) Isolated and cultured hepatocytes. INSERM, Paris, pp 1–12
14. Vanhaecke T, Rogiers V (2006) Hepatocyte cultures in drug metabolism and toxicological research and testing. Methods Mol Biol 320:209–227
15. Etienne PL, Baffet G, Desvergne B et al (1988) Transient expression of *c*-fos and constant expression of *c*-myc in freshly isolated and cultured normal adult rat hepatocytes. Oncogene Res 3:255–262

16. Loyer P, Cariou S, Glaise D et al (1996) Growth factor dependence of progression through G1 and S phases of adult rat hepatocytes *in vitro*: evidence of a mitogen restriction point in mid-late G1. J Biol Chem 271: 11484–11492
17. McGowan JA, Bucher NL (1983) Pyruvate promotion of DNA synthesis in serum-free primary cultures of adult rat hepatocytes. In Vitro 19:159–166
18. Sawada N (1989) Hepatocytes from old rats retain responsiveness of *c*-myc expression to EGF in primary culture but do not enter S phase. Exp Cell Res 181:584–588
19. Nakamura T, Nishizawa T, Hagiya M et al (1989) Molecular cloning and expression of human hepatocyte growth factor. Nature 342:440–443
20. Garnier D, Loyer P, Ribault C et al (2009) Cyclin-dependent kinase 1 plays a critical role in DNA replication control during rat liver regeneration. Hepatology 50:1946–1956
21. Corlu A, Loyer P (2012) Regulation of the G1/S transition in hepatocytes: involvement of the cyclin-dependent kinase Cdk1 in the DNA replication. Int J Hepatol 2012:689324
22. McGowan JA (1986) Hepatocyte proliferation in culture. In: Guillouzo A, Guillouzo C (eds) Isolated and cultured hepatocytes. INSERM, Paris, pp 13–38
23. Albrecht JH, Hu MY, Cerra FB (1995) Distinct patterns of cyclin D1 regulation in models of liver regeneration and human liver. Biochem Biophys Res Commun 209:648–655
24. Isom HC, Secott T, Georgoff I et al (1985) Maintenance of differentiated rat hepatocytes in primary culture. Proc Natl Acad Sci U S A 82:3252–3256
25. Guguen-Guillouzo C, Clement B, Baffet G et al (1983) Maintenance and reversibility of active albumin secretion by adult rat hepatocytes co-cultured with another liver epithelial cell type. Exp Cell Res 143:47–54
26. Waxman DJ, Morrissey JJ, Naik S et al (1990) Phenobarbital induction of cytochromes P450: high-level long-term responsiveness of primary rat hepatocyte cultures to drug induction, and glucocorticoid dependence of the phenobarbital response. Biochem J 271:113–119
27. Bissell DM, Arenson DM, Maher JJ et al (1987) Support of cultured hepatocytes by a laminin-rich gel: evidence for a functionally significant subendothelial matrix in normal rat liver. J Clin Invest 79:801–812
28. Kojima T, Mitaka T, Paul DL et al (1995) Reappearance and long-term maintenance of connexin32 in proliferated adult rat hepatocytes: use of serum-free L-15 medium supplemented with EGF and DMSO. J Cell Sci 108:1347–1357
29. Corlu A, Ilyin G, Cariou S et al (1997) The coculture: a system for studying the regulation of liver differentiation/proliferation activity and its control. Cell Biol Toxicol 13:235–242
30. Hansen LK, Albrecht JH (1999) Regulation of the hepatocyte cell cycle by type I collagen matrix: role of cyclin D1. J Cell Sci 112: 2971–2981
31. De Smet K, Loyer P, Gilot D et al (2001) Effects of epidermal growth factor on CYP inducibility by xenobiotics, DNA replication, and caspase activations in collagen I gel sandwich cultures of rat hepatocytes. Biochem Pharmacol 61:1293–1303
32. Corlu A, Kneip B, Lhadi C et al (1991) A plasma membrane protein is involved in cell contact-mediated regulation of tissue-specific genes in adult hepatocytes. J Cell Biol 115:505–515
33. Serandour AL, Loyer P, Garnier D et al (2005) TNFalpha-mediated extracellular matrix remodeling is required for multiple division cycles in rat hepatocytes. Hepatology 41:478–486
34. Gripon P, Rumin S, Urban S et al (2002) Infection of a human hepatoma cell line by hepatitis B virus. Proc Natl Acad Sci U S A 99:15655–15660
35. Cerec V, Glaise D, Garnier D et al (2007) Transdifferentiation of hepatocyte-like cells from the human hepatoma HepaRG cell line through bipotent progenitor. Hepatology 45:957–967
36. Aninat C, Piton A, Glaise D et al (2006) Expression of cytochrome P450, conjugating enzymes and nuclear receptors in human hepatoma HepaRG cells. Drug Metab Dispos 34:75–83
37. Kang LI, Mars WM, Michalopoulos GK (2012) Signals and cells involved in regulating liver regeneration. Cells 1:1261–1292
38. Giandomenico AR, Cerniglia GE, Biaglow JE et al (1997) The importance of sodium pyruvate in assessing damage produced by hydrogen peroxide. Free Radic Biol Med 23:426–434
39. Williams GM, Gunn JM (1974) Long-term cell culture of rat liver epithelial cells. Exp Cell Res 89:139–142
40. Williams GM, Weisburger EK, Weisburger JH (1971) Isolation and long-term cell culture of epithelial-like cells from rat liver. Exp Cell Res 69:106–112
41. Deschatrette J, Weiss MC (1974) Characterization of differentiated and dedifferentiated clones from a rat hepatoma. Biochimie 56: 1603–1611

42. Bailly-Maitre B, de Sousa G, Zucchini N et al (2002) Spontaneous apoptosis in primary cultures of human and rat hepatocytes: molecular mechanisms and regulation by dexamethasone. Cell Death Differ 9:945–955
43. Gilot D, Loyer P, Corlu A et al (2002) Liver protection from apoptosis requires both blockage of initiator caspase activities and inhibition of ASK1/JNK pathway *via* glutathione *S*-transferase regulation. J Biol Chem 77: 49220–49229
44. Gerbal-Chaloin S, Duret C, Raulet E et al (2010) Isolation and culture of adult human liver progenitor cells: *in vitro* differentiation to hepatocyte-like cells. Methods Mol Biol 640:247–260
45. Blanc P, Etienne H, Daujat M et al (1992) Mitotic responsiveness of cultured adult human hepatocytes to epidermal growth factor, transforming growth factor alpha, and human serum. Gastroenterology 102:1340–1350
46. Reddy CC, Wells A, Lauffenburger DA (1996) Receptor-mediated effects on ligand availability influence relative mitogenic potencies of epidermal growth factor and transforming growth factor alpha. J Cell Physiol 166:512–522
47. Takatsuki K, Fujiwara K, Hayashi S et al (1990) Acceleration of DNA synthesis in post-hepatectomised regenerating liver of normal rat by insulin and glucagon. Life Sci 29:2609–2615
48. McMahon JB, Richards WL, del Campo AA et al (1986) Differential effects of transforming growth factor-beta on proliferation of normal and malignant rat liver epithelial cells in culture. Cancer Res 46:4665–4671
49. Berasain C, Garcia-Trevijano ER, Castillo J et al (2005) Amphiregulin: an early trigger of liver regeneration in mice. Gastroenterology 128:424–432
50. Ito N, Kawata S, Tamura S et al (1994) Heparin-binding EGF-like growth factor is a potent mitogen for rat hepatocytes. Biochem Biophys Res Commun 198:25–31
51. Cruise JL, Knechtle SJ, Bollinger RR et al (1987) Alpha 1-adrenergic effects and liver regeneration. Hepatology 7:1189–1194
52. Lesurtel M, Graf R, Aleil B et al (2006) Platelet-derived serotonin mediates liver regeneration. Science 312:104–107
53. Sharon Goh YP, Henderson NC, Heredia JE et al (2013) Eosinophils secrete IL-4 to facilitate liver regeneration. Proc Natl Acad Sci U S A 110:9914–9919

Chapter 4

Immortalized Human Hepatic Cell Lines for In Vitro Testing and Research Purposes

Eva Ramboer, Tamara Vanhaecke, Vera Rogiers, and Mathieu Vinken

Abstract

The ubiquitous shortage of primary human hepatocytes has urged the scientific community to search for alternative cell sources, such as immortalized hepatic cell lines. Over the years, several human hepatic cell lines have been produced, whether or not using a combination of viral oncogenes and human telomerase reverse transcriptase protein. Conditional approaches for hepatocyte immortalization have also been established and allow generation of growth-controlled cell lines. A variety of immortalized human hepatocytes have already proven useful as tools for liver-based in vitro testing and fundamental research purposes. The present chapter describes currently applied immortalization strategies and provides an overview of the actually available immortalized human hepatic cell lines and their in vitro applications.

Key words Cell line, Liver, Immortalization, Human

1 Introduction

At present, primary human hepatocytes represent an important tool for research purposes, in particular in the field of in vitro pharmaco-toxicology [1]. However, their use is largely impeded by inadequate supply, high cost, high variability, and limited in vitro proliferation capacity. Alternative cell sources include hepatic cell lines and stem cell-derived hepatocytes [2–4]. Several hepatic cell lines are nowadays available either directly derived from liver tumor tissue or generated from primary hepatocytes in vitro [5, 6]. Although several hepatoma-derived cell lines, such as HepaRG cells, preserve important liver-specific functions, most of them do not exhibit sufficient in vivo-like functionality to be of pharmaco-toxicological relevance [7–12]. Immortalized hepatocytes are usually derived from healthy primary hepatocytes by the use of a defined immortalization strategy. Several fetal and adult hepatic cell lines have already been established, whether or not using a combination of viral oncogenes and the human telomerase reverse

Mathieu Vinken and Vera Rogiers (eds.), *Protocols in In Vitro Hepatocyte Research*, Methods in Molecular Biology, vol. 1250, DOI 10.1007/978-1-4939-2074-7_4, © Springer Science+Business Media New York 2015

transcriptase (hTERT) protein [13–19]. In this chapter, a number of immortalization strategies are discussed and a state-of-the-art overview of the currently available immortalized human hepatic cell lines and their in vitro applications is provided.

2 Hepatocyte Immortalization Strategies

The most common methods for immortalization of primary hepatocytes are (1) overexpression of viral oncogenes, (2) forced expression of hTERT, or (3) a combination of both (Table 1) [14, 20]. A number of additional immortalization genes as well as conditional approaches for hepatocyte immortalization have also been described (Table 2).

2.1 Immortalization Genes

2.1.1 Viral Oncogenes

Several human hepatic cell lines have been generated using viral oncogenes, such as the simian virus 40 large T antigen (SV40 Tag) and the human papillomavirus 16 (HPV16) E6/E7 genes, suggesting that overexpression of these viral oncogenes could be sufficient to surmount the premature in vitro growth arrest of cultured adult hepatocytes [15, 17, 18, 20–29]. Indeed, in contrast to fetal hepatocytes, human adult hepatocytes do not undergo spontaneous cell growth in vitro and possess very limited proliferation capacity, even when cultivated in growth-promoting culture systems [6, 20, 30, 31]. It has been proposed that this in vitro precocious growth arrest could be the result of a telomere-independent senescence mechanism, which possibly involves tumor suppressor proteins and cyclin-dependent kinase inhibitors [32]. Viral oncogenes typically promote cell cycling by inhibiting the p16/retinoblastoma protein (pRB) and p53 pathways [20, 33]. However, while the use of viral oncogenes, such as SV40 Tag, is sufficient to produce immortalized rodent cells, overexpression of these oncogenes probably only extends life span in human cells. Human hepatocytes, like other somatic cells, do not possess telomerase activity and are subject to replicative senescence [20, 32, 34, 35]. Consequently, the immortalization process necessitates telomerase reactivation either through mutations or by involving a second immortalizing gene. In particular, hTERT has been used for this purpose [16, 19, 20, 34, 36–38].

2.1.2 Human telomerase reverse transcriptase

The single use of hTERT as immortalization gene is restricted to a number of human cell types, including fetal and neonatal hepatocytes [6, 13, 14, 34, 36, 39, 40]. These immature cells are in fact still able to proliferate in vitro and do not need cell cycle stimulation for immortalization [6, 13, 14, 34, 40, 41]. However, fetal and neonatal human hepatocytes do not display indefinite growth potential due to telomere-dependent senescence. As a result,

Table 1
Overview of immortalization strategies, hepatic functionality, and some in vitro applications of human adult and fetal hepatic cell lines

Cell line	Immortalization strategy	Hepatic functionality of immortalized cells	In vitro applications	Reference
Adult hepatic cell line				
Fa2N-4	Transfection SV40 Tag	• Possess, in comparison with cryopreserved human hepatocytes: – Significantly lower basal expression level of the nuclear receptor CAR and several drug metabolizing enzymes and transporters, namely, CYP1A2/2D6/2E1/1A1, UGT1A1/1A6/2B15/2B4, sulfotransferase, NTCP, OCT1, OATP1B1/1B3, MRP2, and BSEP. – Markedly higher MDR1 mRNA levels. – Similar basal expression of BCRP, PXR, and AhR. – Apparently higher expression of most transcription factors and coactivators/corepressors that have been associated with PXR and CAR-mediated enzyme induction. • Are incapable of metabolizing compounds due to low basal levels of drug-metabolizing enzymes. • Exhibit, at early passage, inducible CYP1A2/2C9/3A4, UGT1A, and MDR1 mRNA levels as well as CYP1A2/2C9/3A4 activities and could distinguish inducers from noninducers. At higher passages, the cells lose the ability to induce.	Routine screening system for PXR-mediated CYP3A4 induction.	[18, 92, 97]
HepLi5	Retroviral vector SV40 Tag	• Express HBCF-X, GS, GST, ALB, and CYP450 mRNA. • Retain ALB secretion and urea production, though at low levels compared to primary hepatocytes. • Display CYP1A2 activity. • Possess significantly enhanced cellular functions after large-scale culture in roller bottles.		[21]

(continued)

Table 1
(continued)

Cell line	Immortalization strategy	Hepatic functionality of immortalized cells	In vitro applications	Reference
HepLL	Lipid-mediated gene transfer (lipofectamine reagent) SV40 Tag	• Display morphologic characteristics of liver parenchymal cells. • Express HNF4, HBCF-X, GST-Π, and ALB mRNA as well as ALB and CYP2E1 protein but no ASGP mRNA. • Stain positive for human hepatocyte special antigen but negative for AFP. • Secrete ALB and urea at levels not significantly different from primary cultured human hepatocytes. • Synthesize glycogen.	Testing of new drug carriers for anti-HBV drug delivery.	[22, 98]
HepZ	Lipid-mediated gene transfer (lipofectamine reagent) Antisense constructions for Rb and p53 under control of ALB promoter + Cotransfection of E2F transcription factors and cyclin D1	• When grown in bioreactor, cells are able to secrete ALB and A2M and possess inducible CYP450 activity.		[47]
HHE6E7T-1/2	Small hepatocytes Lentiviral and retroviral vectors HPV16 E6/E7 + hTERT	• Display epithelial-like morphology. • Retain characteristics of differentiated hepatocytes, though functions such as ALB secretion as well as mRNA expression levels of ALB, HNF4, and A1AT decrease gradually as the passages progress. CK18 mRNA levels are detected throughout the culture period and no AFP expression is observed. • Are positive for vimentin staining.		[16, 99]

HHL (-5/-7/-16)	Retroviral vector HPV16 E6/E7	• Contain markers of hepatocyte and biliary phenotype (CK7/8/18/19). • Express CYP450 protein at levels comparable to Huh-7 and HepG2 cells. • Produce ALB, though at lower levels than Huh-7 and HepG2 cells. • Stain negative for AFP and do not display elevated nuclear expression of p53 protein. • Possess active gap junctions. • Respond to INF-α stimulation by upregulation of major histocompatibility complex I and II. • Exhibit, in contrast to the Huh-7 and HepG2 cells, increased capacity to bind recombinant hepatitis C virus-like particles.	Used as a cell model to investigate: • Ability of CD8+ T-cells to recognize and kill hepatocytes under cytokine stimulation. • Role of sirtuin 4 in energy homeostasis. • Effect of THCV on insulin signaling in insulin-resistant human hepatocytes. • Gene expression and antigen presentation after adeno-associated viral transduction and effect of proteasome inhibition or capsid mutation.	[15, 89–91, 100–102]
IHH-A5	Lipid-mediated gene transfer (lipofectin reagent) SV40 Tag	• Are morphologically and functionally more similar to hepatoma cell lines than primary hepatocytes in culture. • Secrete different plasma proteins, including ALB, APO-B, and fibrinogen at relatively high rates, within the range observed for early primary human hepatocyte cultures. Addition of IL-6 to the culture medium results in increased fibrinogen secretion and decreased ALB production, demonstrating a proper acute-phase response. • Produce detectable amounts of APO-A1. • Exhibit bile-canalicular structures that, in some cases, accumulated the organic anion glutathione-methylfluorescein. Cell cultures are partly polarized and express the efflux transporters, MDR1 and MRP1, on the membranes of apical vacuoles or on the lateral membranes of adjacent, proliferating cells, respectively. • Do not maintain active Na^+-dependent bile salt uptake. • Display similar lipoprotein metabolism as HepG2 cells.	Used a cell model to investigate: • Molecular mechanisms underlying FXR regulation of ChREBP transcriptional activity in human hepatocytes. • Effect of antipsychotic drugs on SREBP transcription factor pathways. • Role of c-fos expression on hepatocyte cell motility and cell cycle regulation. • Effect of cadmium on nonmalignant human hepatocytes.	[23, 85, 86, 103–105]

(continued)

Table 1 (continued)

Cell line	Immortalization strategy	Hepatic functionality of immortalized cells	In vitro applications	Reference
PH5CH	Lipid-mediated gene transfer (lipofectin reagent) SV40 Tag	• Display epithelial appearance. • Express human CK and ALB protein.	Used a cell model to investigate: • HCV infection, replication, and tropism. • HCC-selective cytotoxicity of HBF-0079. • MicroRNA expression in TGF-β-induced hepatocyte EMT. • Effect of hepatitis B virus proteins on signaling mediated by members of the Toll-like/interleukin 1 superfamily. • Effect of HBV polymerase on IFN beta production in human hepatocytes.	[25, 74, 75, 81, 106–113]

THLE	Retroviral vector SV40 Tag	• Display epithelial morphology. • Secrete ALB and express CK18, TF, A1AT, A2M, GST-Π, and very low levels of GGT at early passages. CK19 expression can only be determined at later passages. Cells are uniformly negative for AFP and factor VIII. The appearance of CK19 and decreased ALB secretion at later passages demonstrate that cells undergo dedifferentiation in culture. • Retain mRNA expression of phase II enzymes such as EH, catalase, GPX, SOD, and GSTs at levels comparable to human liver, with GST-Π and α mRNA as the dominant form in THLE cells or human liver, respectively. • Maintain NADPH CYP reductase expression at a lower steady-state mRNA level than in human liver. • Are able to metabolize three carcinogens, which suggests the presence and activity of CYP1A2/1A1, CYP2E1, and CYP3A4. However, CYP1A2, CYP2E1, CYP3A4, CYP2A3, and CYP2D6 mRNA are not detected. The steady-state mRNA levels of CYP1A1 increase after exposure to Aroclor 1254 or B[α] P. • Overexpression of specific CYP450 gene led to the development of THLE-CYP sublines.	THLE and THLE-CYP cells can be used to study cellular toxicity of compounds. Used a cell model to investigate: • HCC-selective cytotoxicity of HBF-0079. • Hepatoprotective and chemopreventive properties of phytochemicals.	[25, 93, 94, 111, 114–124]
TPH1	Strontium phosphate precipitation HCV core gene	• Exhibit altered cell morphology resembling low-differentiated epithelial cells. • Express no A1AT or AFP mRNA. • Secrete ALB. • Possess G6P activity. • Reactivate telomerase immediately after senescence.	Used a cell model to investigate HCV infection and replication. Induces apoptosis of activated hepatic stellate cells.	[45, 77, 78, 125–129]

(continued)

Table 1
(continued)

Cell line	Immortalization strategy	Hepatic functionality of immortalized cells	In vitro applications	Reference
Fetal and neonatal hepatic cell lines				
FH-TERT	Retroviral vector hTERT	• Express CYP450 mRNA and maintain, in contrast to passaged fetal hepatocytes, liver-enriched differentiation markers, especially C/EBPα and HNF4 as well as elevated levels of HGFR. • Possess glycogen storage and G6P activity, in a pattern similar to primary fetal hepatocytes. • Produce urea and retain level of ALB synthesis equivalent to HepG2 cells.	Used as a stroma to induce human embryonic stem cells differentiation into hematopoietic cells. Used as a cell model to investigate: • Role of RIPK4 as novel tumor suppressor in human hepatocarcinogenesis. • Permissive role of β-catenin signaling in the initial phase of hepatocarcinogenesis.	[34, 83, 84, 130]
Hc3716-hTERT	Retroviral vector hTERT	• Maintain normal mammalian cell morphology. • Exhibit protein expression of ALB, CK8, and CK18, but not AFP. ALB levels are higher than in control, passaged Hc3716 cells. • Possess inducible CYP3A4/7 mRNA levels. • Exhibit wild-type p53 responsiveness.	Used as in vitro model for predicting the side effects of telomere-targeting drugs.	[39]
NeHepLxHT	Retroviral vector hTERT	• Display characteristic morphology of primary fetal liver cells. • Maintain epithelial characteristics as evidenced by immunostaining for epithelial cell markers, the cytokeratins. • Possess gene expression profile similar to human neonatal hepatocytes, with positive expression of A1AT, CKIT, CLDN3, EPCAM, NCAM mRNA and no detection of AFP, ASGPR, or CYP3A4. The very low ALB mRNA levels compared to HepG2 cells and the expression of CK19 in early passages indicate the progenitor nature of the cells.	Used as a cell model to investigate: • Role of FAT10 in promoting malignant cell transformation. • Molecular mechanism responsible for miR-224 overexpression in HCC. • Role of HCV RNA-dependent RNA polymerase in promoting liver inflammation and injury.	[14, 80, 131, 132]

OUMS-29	Lipid-mediated gene transfer (lipofectin reagent) SV40 Tag	• Display epithelial morphology. • Maintain gene expression of ALB, ASGPR, bil-UGT, GS, GST- Π, HBCF-X, AhR, and Arnt. • Secrete ALB, AFP, TF, A1AT, and APO A-1. • Possess inducible CYP1A1/2 mRNA levels and activity. • Overexpression of HNF4α2 led to development of OUMS-29/H-11 cell line with increased liver-specific gene expression, such as A1AT, apolipoproteins, HBCF-X, and HNF1α.	Used as a cell model to investigate: • Role T-cell factor 4 isoforms in promoting hepatic tumorigenicity. • Effect of bile acid species on hepatocyte apoptosis induced by PPARgamma ligands. • Effect of hydrogen peroxide on hepatic pigment epithelium-derived factor levels. • Both morphologic and functional alterations in the mitochondria of troglitazone-treated hepatocytes.	[17, 49, 62, 133–136]

A1AT α1-antitrypsin, *AFP* α-fetoprotein, *AhR* aryl hydrocarbon receptor, *ALB* albumin, *A2M* α2-macroglobulin, *APO* apolipoprotein, *Arnt* AhR nuclear translocator, *ASGP(R)* asialoglycoprotein (receptor), *BCRP* breast cancer resistance protein, *BSEP* bile salt export pump, *CAR* constitutive androstane receptor, *CD* cluster of differentiation, *C/EBP* Ccaat-enhancer-binding protein, *ChREBP* carbohydrate-responsive element-binding protein, *CK* cytokeratin, *CLDN* claudin, *CYP* cytochrome P450, *EH* epoxide hydrolase, *EMT* epithelial–mesenchymal transition, *EPCAM* epithelial cell adhesion molecule, *FXR* farnesoid X receptor, *GGT* glutamyltranspeptidase, *G6P* glucose-6-phosphate, *GPX* glutathione peroxidase, *GS* glutamine synthetase, *GST* glutathione *S*-transferase, *HBCF* human blood coagulation factor, *HBV* hepatitis B virus, *HCC* hepatocellular carcinoma, *HCV* hepatitis C virus, *HGFR* hepatocyte growth factor receptor, *HNF* hepatocyte nuclear factor, *HPV* human papillomavirus, *hTERT* human telomerase reverse transcriptase, *IL* interleukin, *IFN* interferon, *MDR* multidrug resistance protein, *mRNA* messenger ribonucleic acid, *MRP* multidrug resistance-associated protein, *NADPH* nicotinamide adenine dinucleotide phosphate, *NCAM* neural cell adhesion molecule, *NTCP* sodium taurocholate cotransporting polypeptide, *OATP* organic anion transporting polypeptide, *OCT* organic cation transporter, *PPAR* peroxisome proliferator-activated receptor, *PXR* pregnane X receptor, *Rb* retinoblastoma, *RIPK4* receptor-interacting serine-threonine kinase 4, *SOD* superoxide dismutase, *SREBP* sterol regulatory element-binding protein, *SV40 Tag* simian virus 40 large T antigen, *TGF* transforming growth factor, *TF* transferrin, *(bil-) UGT* (bilirubin-)uridinediphosphate-glucuronosyltransferase

Table 2
Overview of conditional immortalization strategies, hepatic functionality, and some in vitro applications of growth-controlled human adult and fetal hepatic cell lines

Cell line	Immortalization strategy	Hepatic functionality of conditional immortalized cells	In vitro applications	Reference
Adult hepatic cell line				
16T-3	Retroviral vector hTERT Tamoxifen-mediated self-excision (Cre-LoxP)	Reverted 16T-3 cells: • Show enhanced mRNA levels of transcriptional factors, C/EBPα and HNF4α as well as increased mRNA expression of hepatocyte-specific genes, including ALB, GST-Π, HBCF-X, bil-UGT, CYP3A4, GS, and ASGPR. • Possess increased ALB production and lidocaine metabolism, though at lower levels than normal human hepatocytes.		[43]
HepLi-4	Retroviral vector SV40 Tag Tamoxifen-mediated self-excision (Cre-LoxP)	Reverted HepLi-4 cells: • Express similar GS and somewhat lower UGT1A1 mRNA levels than adult human liver. ALB and GST-Π mRNA levels are extremely lower or higher, respectively, compared to the human liver. These characteristics indicate that HepLi-4 cells are not fully differentiated after reversion.		[26]
HLTC	Retroviral vector SV40 Tag Temperature-based regulation	• Grow as islands or sheets of cuboidal cells (HLTC-17) or display a more dispersed cuboidal-elongated morphology (HLTC-7/-11/-15/-19). • Secrete fibrinogen at fairly constant rate in all tested cell lines at permissive (33.5 °C) and nonpermissive (39.5 °C) temperature. • Exhibit no ALB, AFP, A1AGP or secretion in any cell line at both temperatures. • Cell lines HLTC-7, -15, and -19 produce A1AT at permissive temperature. However, at nonpermissive temperature the secretion of A1AT is upregulated or become detectable in all the cell lines. • Cell lines HLTC-17 and -11 possess no CYP activity at any temperature even after induction and stain positive for ALB, CK18, CK7, CK19, and vimentin, but negative for CK8, with almost identical patterns at both temperatures.		[28]

IHH10(.3)/12	Lentiviral vector SV40 Tag + hTERT (IHH10) or SV40 Tag + hTERT + Bmi-1 (IHH12) Recombinase- based control (Cre-LoxP)	• Display morphology reminiscent of differentiated hepatocytes. • Express ALB, A1AT, ASGPR, and CYP450 mRNA levels. • Secrete liver-specific proteins, ALB and fibrinogen, at levels similar to Huh-7 cells but lower than primary hepatocytes. The IHH12 cell line do only produce fibrinogen after deimmortalization, suggesting the acquirement of a higher differentiation status in this setting. However, Cre-recombinase treatment of IHH12 cells does not significantly improve the production of ALB. • Possess inducible CYP1A1/2 activity.	A novel in vitro model to investigate the mechanisms and consequences of lipid accumulation in hepatocytes, independently of insulin resistance.	[19, 87]
NKNT-3	Retroviral vector SV40 Tag Recombinase- based control (Cre-loxP)	• Display morphological characteristics of liver parenchymal cells and look more differentiated after reversion. • Express bil-UGT, GS, and GST- Π mRNA levels, which increased substantially after reversion. Contradicting results are published regarding expression of ALB and HBCF-X mRNA levels. One paper demonstrates that ALB and HBCF-X mRNA are newly introduced in the reverted cells whereas several other papers already report expression of these genes and ASGPR mRNA in nonreverted cells. Nevertheless, although reversion does stimulate differentiation, mRNA levels of ALB, A1AT, and TF were maximally 0.1 % of primary human hepatocytes. • Additional experiments reveal that introduction of p21 into human immortalized hepatocytes can increase ALB expression and induce a differentiated morphology. • Cocultivation with immortalized hepatic stellate cells increases urea synthesis and protein expression of CYP3A4/2C9.	Used in combination with HCV like particles as a model system for studying viral binding and entry.	[6, 29, 52, 71, 79]
YOCK-13	Retroviral vector hTERT Tamoxifen-mediated self-excision (Cre-LoxP)	• Display morphological characteristics of normal human hepatocytes. • Express markers of hepatocytic differentiation including ALB, ASGPR, bil-UGT, CYP3A4, GS, GST- Π, and HBCF-X.		[42]

(continued)

Table 2
(continued)

Cell line	Immortalization strategy	Hepatic functionality of conditional immortalized cells	In vitro applications	Reference
Fetal hepatic cell lines				
cBAL111	Lentiviral vector hTERT Transcriptional regulation (Tet-on approach)	• Express relatively high mRNA levels of immature markers, GST-Π and AFP, and very low mRNA levels of mature markers, ALB, A1AT, and TF. Transcript levels of HNF4α increase after prolonged culturing. • Stain positive for GS, ALB, CK18, CK19, vimentin, and the progenitor cell marker CD146 but display CK18 in a pattern characteristic of dedifferentiated human hepatocytes. • Produce urea and ALB, though at lower levels than mature human hepatocytes. • Retain no CYP1A2 and 3A4 activity (no elimination of lidocaine) but are able to eliminate galactose. • cBAL111cells resemble cells with progenitor characteristics rather than fully differentiated hepatocytes. However, there is a trend of increased and decreased expression of mature and immature markers, respectively, with culture time.		[13]
HepCL	Retroviral vector SV40 Tag Temperature-based regulation	• Display morphological characteristics of liver parenchymal cells. • Stain positive for ALB, CK18, and CK19. • Produce amounts of ALB and urea comparable to those of unmodified primary human fetal hepatocytes.		[27]

A1AGP α1- acid glycoprotein, *A1AT* α1-antitrypsin, *AFP* α-fetoprotein, *ALB* albumin, *ASGP(R)* asialoglycoprotein (receptor), *Bmi-1* B lymphoma Mo-MLV insertion region 1 homolog, *C/EBP* Ccaat-enhancer-binding protein, *CD* cluster of differentiation, *CK* cytokeratin, *CYP* cytochrome P450, *GS* glutamine synthetase, *GST* glutathione *S*-transferase, *HBCF* human blood coagulation factor, *HCV* hepatitis C virus, *HNF* hepatocyte nuclear factor, *hTERT* human telomerase reverse transcriptase, *mRNA* messenger ribonucleic acid, *SV40 Tag* simian virus 40 large T antigen, *TF* transferrin, *(bil-)UGT* (bilirubin-)uridinediphosphate-glucuronosyltransferase overexpression of hTERT is needed to immortalize these hepatocytes [13, 14, 34, 40]. Contradicting results have been obtained when merely hTERT is used to immortalize mature hepatocytes [16, 42, 43]. Since telomerase activity most likely does not allow adult hepatocytes to overcome the suggested telomere-independent growth arrest, overexpression of hTERT may not be adequate to stimulate the progression of adult hepatocytes through the cell cycle [5, 13, 20, 44].

2.1.3 Miscellaneous Immortalization Genes

The hepatitis C core protein has been described as a specific immortalization agent for mature human hepatocytes [36, 45, 46]. Another cell line was developed by cotransfecting human adult hepatocytes with p53 and pRB antisense constructs and plasmids that include E2F and cyclin D1 genes [47]. Furthermore, specific combinations of immortalization genes, such as SV40 Tag with hTERT and B lymphoma Moloney Murine Leukemia virus (Mo-MLV) insertion region 1 homolog (Bmi-1), have also proven useful for the immortalization of mature human hepatocytes. Bmi-1 has a similar function as the HPV16E7 oncogene and its expression inactivates the p16/pRB pathway. However, the simultaneous transduction with Bmi-1 and hTERT, like for the combined HPV16E7/hTERT approach, appears to be insufficient to immortalize nonproliferating adult hepatocytes [16, 19].

2.2 Gene Transfer

Appropriate gene transfer is of big importance for hepatocyte immortalization [37]. In this regard, both viral and nonviral transfer methods have been used for the development of immortalized hepatocyte-derived cell lines.

2.2.1 Plasmid Transfection

Since the immortalization process will select cells that stably express the immortalization genes, simple transfection methods can be used [48]. At present, various approaches are available for transfecting plasmids into primary hepatocytes [37, 48]. The strontium phosphate precipitation method has been used to immortalize human hepatocytes [45]. Although this method is cheap and allows robust transfection of primary hepatocytes with low toxicity, it is generally accompanied by limited gene transfer efficiency [37]. Liposomes have also been explored as gene carriers for hepatocyte immortalization [17, 22, 23, 25, 47, 49]. When properly optimized, high gene transfer efficiencies can be obtained with lipid-mediated gene transfer strategies compared to the previously mentioned phosphate-precipitation-based transfection approach [37]. Furthermore, combination with hepatocyte-specific ligands allows more hepatocyte-specific transfections [48].

2.2.2 Viral Transduction

Transduction with viral vectors is a frequently used strategy for gene transfer. Among the available viral vectors, retroviral and lentiviral vectors enable stable integration of the immortalization gene and thus ensure persistent transgene expression in the progeny [48, 50]. These vectors do also not induce harmful immune responses and are able to integrate large genes [51]. Retroviral vectors have often been used to produce human hepatic cell lines [14–16, 21, 24, 26–28, 34, 39, 42, 43, 52]. An important drawback of these vectors, however, is their inability to transduce nondividing cells, which hampers their use for nonproliferating cells, including hepatocytes [51, 53]. Although transduction efficiencies generally remain limited even when growth factors are used, it has been reported that the addition of hepatocyte growth factor to the cell culture medium increases the transduction efficiency in primary human hepatocyte cultures [48, 51, 53–55]. Lentiviral vectors derived from the human immunodeficiency virus can overcome these flaws and effectively transduce both dividing and nondividing cells when applied at a relatively high titer [51, 53, 54, 56]. Several studies are based upon lentiviral gene transfer for immortalization of human adult and fetal hepatocytes [13, 16, 19]. It has been demonstrated that the lentiviral transduction procedure as such does not interfere with the differentiated hepatic phenotype of primary human hepatocytes [57]. Moreover, the addition of growth factors to the cell culture medium can markedly enhance the expression of lentiviral genes in both human adult and fetal hepatocytes when low vector titers are used. This transduction approach therefore offers the opportunity to lower the viral load, which in turn reduces cost and cellular toxicity [56]. Also, the antioxidant vitamin E is known to promote lentiviral transduction rates of human adult hepatocytes [53].

2.2.3 Human Artificial Chromosomes

The successful immortalization of rat hepatocytes and human fibroblasts using human artificial chromosomes vectors expressing SV40 Tag or hTERT, respectively, opens new perspectives for human hepatocyte immortalization [58–60]. Although the transfer efficiency is generally lower compared to viral vectors, human artificial chromosomes have many characteristics of an ideal gene delivery vector. Indeed, they are able to incorporate complete genomic loci and maintain a mitotically stable episomal expression throughout many cell divisions. Moreover, because of this episomal nature, integration-related complications, such as oncogenesis, could be prevented [50].

2.3 Hepatic Functionality of the Immortalized Human Hepatocytes

In general, most human hepatic cell lines possess reduced or only limited liver-specific functionality. Strategies that are typically used to counteract the loss of functionality in primary hepatocyte cultures, including the establishment of coculture systems or the overexpression of liver-enriched transcription factors, were demonstrated beneficial for immortalized human hepatocytes [61, 62]. In this context, overexpression of specific cytochrome P450 (CYP) enzymes lies at the basis of the development of the THLE-CYP sublines [63]. On the other hand, as differentiation and proliferation are mutually exclusive in vitro, overexpression of the cyclin-dependent kinase inhibitor p21 and the use of conditional immortalization strategies were reported to boost to some extent the differentiated phenotype of immortalized hepatocytes [19, 29, 43, 64, 65].

2.4 Conditional Immortalization Strategies

Conditional immortalization supports controlled expansion of cells. At present, three different strategies are applied for human hepatocytes, namely (1) temperature-based regulation, (2) recombinase-based regulation, and (3) transcriptional regulation [36] (Tables 2 and 3).

2.4.1 Temperature-Based Regulation

This approach is based upon the application of a temperature-unstable SV40 Tag mutant. At permissive temperature (i.e., 33 °C), the immortalizing gene is fully active and stimulates hepatocyte proliferation. However, at higher temperatures (i.e., 37–39 °C), the immortalization gene is typically degraded and cell cycle progression is no longer supported [36]. Since no other temperature-sensitive immortalizing genes are yet available, this method is restricted to the use of SV40 Tag [36]. Importantly, the temperature shift related to this strategy could lead to variations in cellular properties, and as such possibly complicate the interpretation of the study outcome [36, 66, 67]. More sophisticated systems based upon recombinase or transcriptional regulation are believed to offer a better solution [68].

2.4.2 Recombinase-Based Control

The site-specific recombinase strategy results in an irreversible reversion of immortalization, due to permanent removal of immortalization genes [36, 69]. The Cre-LoxP site-specific recombination system has often been used to establish reversible immortalization [69, 70]. In this method, the immortalization genes are flanked by two identical DNA sequences, called LoxP sites, and their excision is controlled by Cre recombinase [20, 70]. Hence, proper reversion relies on the effective transfer of this recombinase gene [36]. A new method based on tamoxifen-mediated self-excision has been introduced, which makes secondary virus-mediated transfer of the recombinase gene superfluous [26, 42, 43]. Additionally, the incorporation of a negative selection marker, such as the suicide gene herpes simplex virus thymidine kinase, allows the removal of cells that underwent improper recombination by exposure to ganciclovir [19, 70]. Reversible immortalization of numerous human hepatocyte-derived cell lines, including NKNT-3, IHH, and 16T-3 cells, are based on this recombinase-based control strategy [19, 42, 43, 71].

Table 3
Conditional approaches for human hepatocyte immortalization [36]

Conditional approach	Immortalization construct	Growth curve
Temperature-based regulation	Thermolabile SV40 Tag mutant	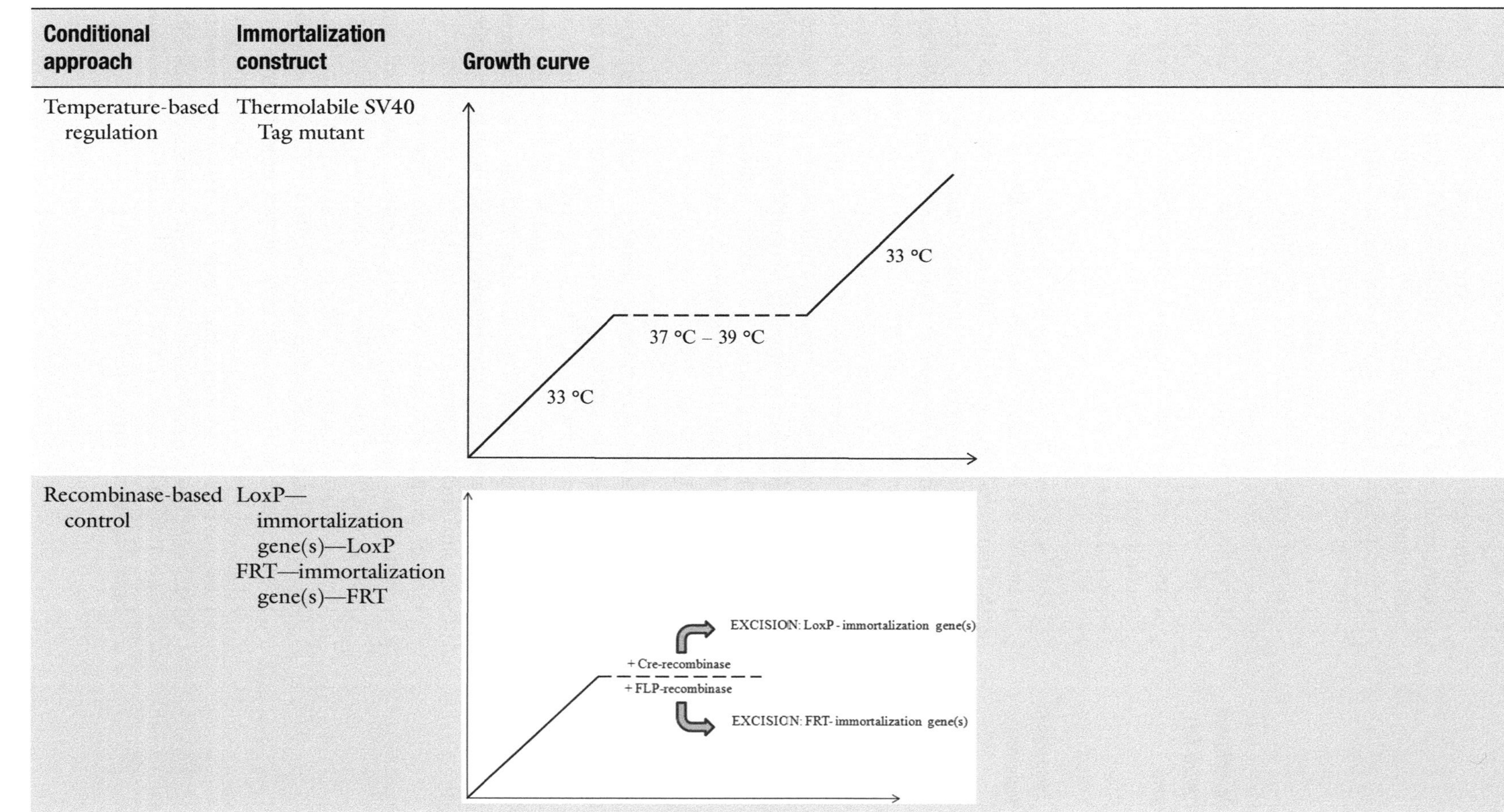
Recombinase-based control	LoxP—immortalization gene(s)—LoxP FRT—immortalization gene(s)—FRT	

Transcriptional regulation	TRE—immortalization gene(s)	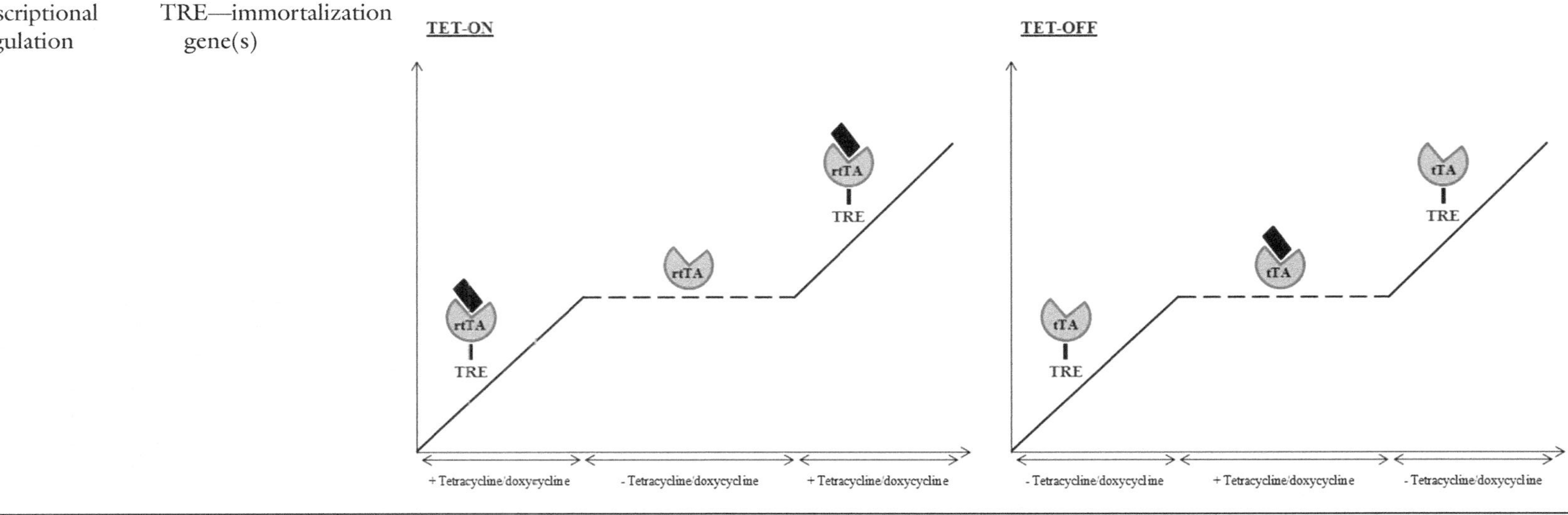

rtTA reverse tetracycline transactivator, *SV40 Tag* simian virus 40 large T antigen, *TRE* tetracycline responsive element, *tTA* tetracycline transactivator

2.4.3 Transcriptional Regulation

In this strategy, the reversibility of immortalization does not rely on recombinase activity, but is achieved through transcriptional control of immortalization gene expression. Consequently, repetitive cycles of hepatocyte proliferation and growth arrest are possible and the risk of chromosomal rearrangement is prevented [36, 68, 72]. The transcriptional control of immortalizing genes can be accomplished by the use of an artificial promoter/transactivator system, such as the well-known tetracycline system [36]. As such, two approaches are now available, namely, the tet-off and the tet-on system, which comprise a tetracycline-regulated promoter and a tetracycline transactivator (tTA) or reverse tetracycline transactivator (rtTA), respectively. In the tet-on system, binding of doxycycline to the transactivator will promote the expression of the regulated gene. This is not the case for the tet-off system, since only unbound tTA is able to bind with the gene promoter [72, 73]. The tet-on approach has been used for the development of the human fetal liver cell line cBAL111 [13].

3 In Vitro Applications of Immortalized Human Hepatocytes

In recent years, both adult and fetal human hepatic cell lines have been explored for research purposes (Tables 1 and 2). Several immortalized human hepatocytes, including PH5CH, TPH1, NKNT-3, and NeHepLxHT cells, have indeed been successfully used as tools in research focused on hepatitis C virus or hepatitis B virus (HBV) [74–81]. A murine model of HBV viremia based upon a human hepatocyte-derived cell line transfected with HBV DNA has been described and allows in vivo HBV research [82]. Human hepatic cell lines have also been applied as cellular models to investigate the cellular mechanisms involved in the processes of hepatocarcinogenesis and steatosis [83–88]. Moreover, the HHL cell line proved useful during the development of adeno-associated viral vectors for liver-directed gene therapy [89–91].

Besides their application in fundamental research, different hepatic cell lines are equally addressed as suitable in vitro tools for screening and safety testing of drug candidates. In this regard, Hc3716-hTERT immortalized hepatocytes constitute an appropriate in vitro model for predicting the side effects of telomere-targeting drugs [39]. Furthermore, Fa2N4 cells may be used as a routine screening system for pregnane X receptor-mediated CYP3A4 induction [92]. Similarly, the hepatic THLE cell line and THLE-CYP sublines have been reported as promising models for investigation of CYP-mediated drug metabolism and liver toxicity [63, 93–95]. However, NKNT-3 cells appeared to be less suitable than the hepatoma cell line HCC1.2 for the development of improved in vitro genotoxicity test systems [96].

4 Conclusion

Over the years, in vitro expansion of human hepatocytes has become more important, as it may serve a plethora of fundamental and applied research and screening purposes. Freshly isolated mature hepatocytes inherently have poor growth potential, a finding that has prompted the search for strategies to immortalize these particular human cells, while preserving their liver-specific functions. The available methods thus far comprise transfection or transduction with prototypical immortalization genes and conditional immortalization by temperature-based regulation, recombinase-based control, and transcriptional regulation. Although hepatocyte immortalization has been explored for decades, cell lines with in vivo-like hepatic functionality are largely lacking. In the upcoming years, more attention should be devoted to the search for culture systems that promote the differentiation status of immortalized human hepatocytes.

Acknowledgements

This work was financially supported by the grants from the University Hospital of Vrije Universiteit Brussel (Willy Gepts Fonds UZ-VUB), the Fund for Scientific Research Flanders (FWO-Vlaanderen), the European Union (FP7/Cosmetics Europe projects HeMiBio and DETECTIVE), and the European Research Council (ERC Starting Grant project CONNECT).

References

1. Hewitt NJ, Lechon MJ, Houston JB et al (2007) Primary hepatocytes: current understanding of the regulation of metabolic enzymes and transporter proteins, and pharmaceutical practice for the use of hepatocytes in metabolism, enzyme induction, transporter, clearance, and hepatotoxicity studies. Drug Metab Rev 39:159–234
2. Sinz M, Kim S (2006) Stem cells, immortalized cells and primary cells in ADMET assays. Drug Discov Today Technol 3:79–85
3. Soldatow VY, Lecluyse EL, Griffith LG et al (2013) *In vitro* models for liver toxicity testing. Toxicol Res 2:23–39
4. Rodrigues RM, De Kock J, Branson S et al (2014) Human skin-derived stem cells as a novel cell source for *in vitro* hepatotoxicity screening of pharmaceuticals. Stem Cells Dev 23:44–55
5. Allen JW, Bhatia SN (2002) Improving the next generation of bioartificial liver devices. Semin Cell Dev Biol 13:447–454
6. Chamuleau RA, Deurholt T, Hoekstra R (2005) Which are the right cells to be used in a bioartificial liver? Metab Brain Dis 20: 327–335
7. Brandon EF, Raap CD, Meijerman I et al (2003) An update on *in vitro* test methods in human hepatic drug biotransformation research: pros and cons. Toxicol Appl Pharmacol 189:233–246
8. Choi S, Sainz B, Corcoran P et al (2009) Characterization of increased drug metabolism activity in dimethyl sulfoxide (DMSO)-treated Huh7 hepatoma cells. Xenobiotica 39:205–217
9. Lim PL, Tan W, Latchoumycandane C et al (2007) Molecular and functional characterization of drug-metabolizing enzymes and transporter expression in the novel spontaneously immortalized human hepatocyte line HC-04. Toxicol In Vitro 21:1390–1401
10. Szabo M, Veres Z, Baranyai Z et al (2013) Comparison of human hepatoma HepaRG

cells with human and rat hepatocytes in uptake transport assays in order to predict a risk of drug induced hepatotoxicity. PLoS One 8:e59432

11. Aninat C, Piton A, Glaise D et al (2006) Expression of cytochromes P450, conjugating enzymes and nuclear receptors in human hepatoma HepaRG cells. Drug Metab Dispos 34:75–83
12. Guillouzo A, Corlu A, Aninat C et al (2007) The human hepatoma HepaRG cells: a highly differentiated model for studies of liver metabolism and toxicity of xenobiotics. Chem Biol Interact 168:66–73
13. Deurholt T, van Til NP, Chhatta AA et al (2009) Novel immortalized human fetal liver cell line, cBAL111, has the potential to differentiate into functional hepatocytes. BMC Biotechnol 9:89
14. Reid Y, Gaddipati JP, Yadav D et al (2009) Establishment of a human neonatal hepatocyte cell line. In Vitro Cell Dev Biol Anim 45:535–542
15. Clayton RF, Rinaldi A, Kandyba EE et al (2005) Liver cell lines for the study of hepatocyte functions and immunological response. Liver Int 25:389–402
16. Tsuruga Y, Kiyono T, Matsushita M et al (2008) Establishment of immortalized human hepatocytes by introduction of HPV16 E6/E7 and hTERT as cell sources for liver cell-based therapy. Cell Transplant 17:1083–1094
17. Fukaya K, Asahi S, Nagamori S et al (2001) Establishment of a human hepatocyte line (OUMS-29) having CYP 1A1 and 1A2 activities from fetal liver tissue by transfection of SV40 LT. In Vitro Cell Dev Biol Anim 37:266–269
18. Hariparsad N, Carr BA, Evers R et al (2008) Comparison of immortalized Fa2N-4 cells and human hepatocytes as *in vitro* models for cytochrome P450 induction. Drug Metab Dispos 36:1046–1055
19. Nguyen TH, Mai G, Villiger P et al (2005) Treatment of acetaminophen-induced acute liver failure in the mouse with conditionally immortalized human hepatocytes. J Hepatol 43:1031–1037
20. Cascio SM (2001) Novel strategies for immortalization of human hepatocytes. Artif Organs 25:529–538
21. Pan X, Li J, Du W et al (2012) Establishment and characterization of immortalized human hepatocyte cell line for applications in bioartificial livers. Biotechnol Lett 34:2183–2190
22. Li J, Li LJ, Cao HC et al (2005) Establishment of highly differentiated immortalized human hepatocyte line with simian virus 40 large tumor antigen for liver based cell therapy. ASAIO J 51:262–268
23. Schippers IJ, Moshage H, Roelofsen H et al (1997) Immortalized human hepatocytes as a tool for the study of hepatocytic (de-)differentiation. Cell Biol Toxicol 13:375–386
24. Pfeifer AM, Cole KE, Smoot DT et al (1993) Simian virus 40 large tumor antigen-immortalized normal human liver epithelial cells express hepatocyte characteristics and metabolize chemical carcinogens. Proc Natl Acad Sci U S A 90:5123–5127
25. Noguchi M, Hirohashi S (1996) Cell lines from non-neoplastic liver and hepatocellular carcinoma tissue from a single patient. In Vitro Cell Dev Biol Anim 32:135–137
26. Zhao L, Li J, Lv G et al (2012) Evaluation of a reversibly immortalized human hepatocyte line in bioartificial liver in pigs. Afr J Biotechnol 11:4116–4126
27. Chen Y, Li J, Liu X et al (2010) Transplantation of immortalized human fetal hepatocytes prevents acute liver failure in 90 % hepatectomized mice. Transplant Proc 42:1907–1914
28. Smalley M, Leiper K, Tootle R et al (2001) Immortalization of human hepatocytes by temperature-sensitive SV40 large-T antigen. In Vitro Cell Dev Biol Anim 37:166–168
29. Kobayashi N, Fujiwara T, Westerman KA et al (2000) Prevention of acute liver failure in rats with reversibly immortalized human hepatocytes. Science 287:1258–1262
30. Ilyin G, Rescan C, Rialland M et al (2000) Growth control and cell cycle progression in cultured hepatocytes. In: Edwards A, Berry M (eds) The hepatocyte review. Kluwer Academic, The Netherlands, pp 263–280
31. Runge DM, Runge D, Dorko K et al (1999) Epidermal growth factor- and hepatocyte growth factor-receptor activity in serum-free cultures of human hepatocytes. J Hepatol 30:265–274
32. Ozturk M, Arslan-Ergul A, Bagislar S et al (2009) Senescence and immortality in hepatocellular carcinoma. Cancer Lett 286:103–113
33. Schafer KA (1998) The cell cycle: a review. Vet Pathol 35:461–478
34. Wege H, Le HT, Chui MS et al (2003) Telomerase reconstitution immortalizes human fetal hepatocytes without disrupting their differentiation potential. Gastroenterology 124:432–444
35. Chiu CP, Harley CB (1997) Replicative senescence and cell immortality: the role of telomeres and telomerase. Proc Soc Exp Biol Med 214:99–106

36. Lipps C, May T, Hauser H et al (2013) Eternity and functionality: rational access to physiologically relevant cell lines. Biol Chem 394:1637–1648
37. McLean J (1999) Immortalization Strategies for Mammalian Cells. In: Jenkins N (ed) Animal cell biotechnology: methods and protocols. Humana, Totowa, pp 61–72
38. Noguchi H, Kobayashi N (2013) Controlled expansion of mammalian cell populations by reversible immortalization. J Biotechnol Biomater 3:158
39. Waki K, Anno K, Ono T et al (2010) Establishment of functional telomerase immortalized human hepatocytes and a hepatic stellate cell line for telomere-targeting anticancer drug development. Cancer Sci 101:1678–1685
40. Wege H, Chui MS, Le HT et al (2003) *In vitro* expansion of human hepatocytes is restricted by telomere-dependent replicative aging. Cell Transplant 12:897–906
41. Curran TR, Bahner RI, Oh W et al (1993) Mitogen-independent DNA synthesis by fetal rat hepatocytes in primary culture. Exp Cell Res 209:53–57
42. Okitsu T, Kobayashi N, Jun HS et al (2004) Transplantation of reversibly immortalized insulin-secreting human hepatocytes controls diabetes in pancreatectomized pigs. Diabetes 53:105–112
43. Totsugawa T, Yong C, Rivas-Carrillo JD et al (2007) Survival of liver failure pigs by transplantation of reversibly immortalized human hepatocytes with tamoxifen-mediated self-recombination. J Hepatol 47:74–82
44. Lee KM, Choi KH, Ouellette MM (2004) Use of exogenous hTERT to immortalize primary human cells. Cytotechnology 45:33–38
45. Ray RB, Meyer K, Ray R (2000) Hepatitis C virus core protein promotes immortalization of primary human hepatocytes. Virology 271: 197–204
46. Basu A, Meyer K, Ray RB et al (2002) Hepatitis C virus core protein is necessary for the maintenance of immortalized human hepatocytes. Virology 298:53–62
47. Werner A, Duvar S, Müthing J et al (1999) Cultivation and characterization of a new immortalized human hepatocyte cell line, HepZ, for use in an artificial liver support system. Ann N Y Acad Sci 875:364–368
48. Wang X, Mani P, Sarkar DP et al (2009) *Ex vivo* gene transfer into hepatocytes. Methods Mol Biol 481:117–140
49. Kobayashi N, Noguchi H, Watanabe T et al (2001) Role of immortalized hepatocyte transplantation in acute liver failure. Transplant Proc 33:645–646
50. Kazuki Y, Oshimura M (2011) Human artificial chromosomes for gene delivery and the development of animal models. Mol Ther 19:1591–1601
51. Zahler MH, Irani A, Malhi H et al (2000) The application of a lentiviral vector for gene transfer in fetal human hepatocytes. J Gene Med 2:186–193
52. Kobayashi N, Noguchi H, Fujiwara T et al (2000) Establishment of a highly differentiated immortalized adult human hepatocyte cell line by retroviral gene transfer. Transplant Proc 32:2368–2369
53. Nguyen TH, Oberholzer J, Birraux J et al (2002) Highly efficient lentiviral vector-mediated transduction of nondividing, fully reimplantable primary hepatocytes. Mol Ther 6:199–209
54. Ohashi K, Park F, Kay MA (2001) Hepatocyte transplantation: clinical and experimental application. J Mol Med 79:617–630
55. Pages JC, Andreoletti M, Bennoun M et al (1995) Efficient retroviral-mediated gene transfer into primary culture of murine and human hepatocytes: expression of the LDL receptor. Hum Gene Ther 6:21–30
56. Selden C, Mellor N, Rees M et al (2007) Growth factors improve gene expression after lentiviral transduction in human adult and fetal hepatocytes. J Gene Med 9:67–76
57. Zamule SM, Strom SC, Omiecinski CJ (2008) Preservation of hepatic phenotype in lentiviral-transduced primary human hepatocytes. Chem Biol Interact 173:179–186
58. Ito M, Ito R, Yoshihara D et al (2008) Immortalized hepatocytes using human artificial chromosome. Cell Transplant 17: 165–171
59. Ito M, Ikeno M, Nagata H et al (2009) Treatment of nonalbumin rats by transplantation of immortalized hepatocytes using artificial human chromosome. Transplant Proc 41:422–424
60. Shitara S, Kakeda M, Nagata K et al (2008) Telomerase-mediated life-span extension of human primary fibroblasts by human artificial chromosome (HAC) vector. Biochem Biophys Res Commun 369:807–811
61. Watanabe T, Shibata N, Westerman KA et al (2003) Establishment of immortalized human hepatic stellate scavenger cells to develop bioartificial livers. Transplantation 75:1873–1880
62. Inoue Y, Miyazaki M, Tsuji T et al (2001) Reactivation of liver-specific gene expression in an immortalized human hepatocyte cell line by introduction of the human HNF4alpha2 gene. Int J Mol Med 8:481–487
63. Soltanpour Y, Hilgendorf C, Ahlström MM et al (2012) Characterization of THLE-

cytochrome P450 (P450) cell lines: gene expression background and relationship to P450-enzyme activity. Drug Metab Dispos 40:2054–2058
64. Kobayashi N, Kunieda T, Sakaguchi M et al (2003) Active expression of p21 facilitates differentiation of immortalized human hepatocytes. Transplant Proc 35:433–434
65. Kunieda T, Kobayashi N, Sakaguchi M et al (2002) Transduction of immortalized human hepatocytes with p21 to enhance differentiated phenotypes. Cell Transplant 11:421–428
66. Allen KJ, Reyes R, Demmler K et al (2000) Conditionally immortalized mouse hepatocytes for use in liver gene therapy. J Gastroenterol Hepatol 15:1325–1332
67. Fox IJ, Chowdhury NR, Gupta S et al (1995) Conditional immortalization of Gunn rat hepatocytes: an *ex vivo* model for evaluating methods for bilirubin-UDP-glucuronosyltransferase gene transfer. Hepatology 21:837–846
68. May T, Hauser H, Wirth D (2004) Transcriptional control of SV40 T-antigen expression allows a complete reversion of immortalization. Nucleic Acids Res 32: 5529–5538
69. Westerman KA, Leboulch P (1996) Reversible immortalization of mammalian cells mediated by retroviral transfer and site-specific recombination. Proc Natl Acad Sci U S A 93:8971–8976
70. Paillard F (1999) Reversible cell immortalization with the Cre-lox system. Hum Gene Ther 10:1597–1598
71. Kobayashi N, Noguchi H, Fujiwara T et al (2000) Establishment of a reversibly immortalized human hepatocyte cell line by using Cre/loxP site-specific recombination. Transplant Proc 32:1121–1122
72. Anastassiadis K, Rostovskaya M, Lubitz S et al (2010) Precise conditional immortalization of mouse cells using tetracycline-regulated SV40 large T-antigen. Genesis 48:220–232
73. Jazwa A, Florczyk U, Jozkowicz A et al (2013) Gene therapy on demand: site-specific regulation of gene therapy. Gene 525:229–238
74. Ikeda M, Kato N, Mizutani T et al (1997) Analysis of the cell tropism of HCV by using *in vitro* HCV-infected human lymphocytes and hepatocytes. J Hepatol 27:445–454
75. Kato N, Ikeda M, Mizutani T et al (1996) Replication of hepatitis C virus in cultured non-neoplastic human hepatocytes. Jpn J Cancer Res 87:787–792
76. Ikeda F, Dansako H, Nishimura G et al (2010) Amino acid substitutions of hepatitis C virus core protein are not associated with intracellular antiviral response to interferon-α *in vitro*. Liver Int 30:1324–1331
77. Raychoudhuri A, Shrivastava S, Steele R et al (2010) Hepatitis C virus infection impairs IRF-7 translocation and alpha interferon synthesis in immortalized human hepatocytes. J Virol 84:10991–10998
78. Raychoudhuri A, Shrivastava S, Steele R et al (2011) ISG56 and IFITM1 proteins inhibit hepatitis C virus replication. J Virol 85: 12881–12889
79. Triyatni M, Saunier B, Maruvada P et al (2002) Interaction of hepatitis C virus-like particles and cells: a model system for studying viral binding and entry. J Virol 76: 9335–9344
80. Yu GY, He G, Li CY et al (2012) Hepatic expression of HCV RNA-dependent RNA polymerase triggers innate immune signaling and cytokine production. Mol Cell 48:313–321
81. Yu S, Chen J, Wu M et al (2010) Hepatitis B virus polymerase inhibits RIG-I- and Toll-like receptor 3-mediated beta interferon induction in human hepatocytes through interference with interferon regulatory factor 3 activation and dampening of the interaction between TBK1/IKKepsilon and DDX3. J Gen Virol 91:2080–2090
82. Brown JJ, Parashar B, Moshage H et al (2000) A long-term hepatitis B viremia model generated by transplanting non-tumorigenic immortalized human hepatocytes in Rag-2-deficient mice. Hepatology 31:173–181
83. Heim D, Cornils K, Schulze K et al (2014) Retroviral insertional mutagenesis in telomerase-immortalized hepatocytes identifies RIPK4 as novel tumor suppressor in human hepatocarcinogenesis. Oncogene. doi:10.1038/onc.2013.551
84. Wege H, Heim D, Lütgehetmann M et al (2011) Forced activation of β-catenin signaling supports the transformation of hTERT-immortalized human fetal hepatocytes. Mol Cancer Res 9:1222–1231
85. Guller MC, Andre J, Legrand A et al (2009) c-Fos accelerates hepatocyte conversion to a fibroblastoid phenotype through ERK-mediated upregulation of paxillin-Serine178 phosphorylation. Mol Carcinog 48:532–544
86. Guller M, Toualbi-Abed K, Legrand A et al (2008) c-fos overexpression increases the proliferation of human hepatocytes by stabilizing nuclear cyclin D1. World J Gastroenterol 14:6339–6346
87. De Gottardi A, Vinciguerra M, Sgroi A et al (2007) Microarray analyses and molecular profiling of steatosis induction in immortalized human hepatocytes. Lab Invest 87: 792–806

88. Martel C, Allouche M, Esposti DD et al (2013) Glycogen synthase kinase 3-mediated voltage-dependent anion channel phosphorylation controls outer mitochondrial membrane permeability during lipid accumulation. Hepatology 57:93–102
89. Pien GC, Basner-Tschakarjan E, Hui DJ et al (2009) Capsid antigen presentation flags human hepatocytes for destruction after transduction by adeno-associated viral vectors. J Clin Invest 119:1688–1695
90. Finn JD, Hui D, Downey HD et al (2010) Proteasome inhibitors decrease AAV2 capsid derived peptide epitope presentation on MHC class I following transduction. Mol Ther 18:135–142
91. Martino AT, Basner-Tschakarjan E, Markusic DM et al (2013) Engineered AAV vector minimizes *in vivo* targeting of transduced hepatocytes by capsid-specific CD8+ T cells. Blood 121:2224–2233
92. LeCluyse E, Sinz M, Hewitt N et al (2010) Cytochrome P450 Induction. In: Lu C, Li A (eds) Enzyme inhibition in drug discovery and development: the good and the bad. Wiley, Hoboken, pp 265–314
93. Rand AA, Rooney JP, Butt CM et al (2014) Cellular toxicity associated with exposure to perfluorinated carboxylates (PFCAs) and their metabolic precursors. Chem Res Toxicol 27:42–50
94. Thompson RA, Isin EM, Li Y et al (2012) *In vitro* approach to assess the potential for risk of idiosyncratic adverse reactions caused by candidate drugs. Chem Res Toxicol 25:1616–1632
95. Dambach DM, Andrews BA, Moulin F (2005) New technologies and screening strategies for hepatotoxicity: use of *in vitro* models. Toxicol Pathol 33:17–26
96. Winter HK, Ehrlich VA, Grusch M et al (2008) Use of four new human-derived liver-cell lines for the detection of genotoxic compounds in the single-cell gel electrophoresis (SCGE) assay. Mutat Res 657:133–139
97. Mills JB, Rose KA, Sadagopan N et al (2004) Induction of drug metabolism enzymes and MDR1 using a novel human hepatocyte cell line. J Pharmacol Exp Ther 309:303–309
98. Wang B, Chen G, Mao Z et al (2012) Preparation and cellular uptake of PLGA particles loaded with lamivudine. Chin Sci Bull 57:3985–3993
99. Tsuruga Y, Kiyono T, Matsushita M et al (2008) Effect of intrasplenic transplantation of immortalized human hepatocytes in the treatment of acetaminophen-induced acute liver failure SCID mice. Transplant Proc 40:617–619
100. Willberg CB, Ward SM, Clayton RF et al (2007) Protection of hepatocytes from cytotoxic T cell mediated killing by interferon-alpha. PLoS One 2:e791
101. Ho L, Titus AS, Banerjee KK et al (2013) SIRT4 regulates ATP homeostasis and mediates a retrograde signaling *via* AMPK. Aging 5:835–849
102. Wargent ET, Zaibi MS, Silvestri C et al (2013) The cannabinoid Δ(9)-tetrahydrocannabivarin (THCV) ameliorates insulin sensitivity in two mouse models of obesity. Nutr Diabetes 3:e68
103. Caron S, Huaman SC, Dehondt H et al (2013) Farnesoid X receptor inhibits the transcriptional activity of carbohydrate response element binding protein in human hepatocytes. Mol Cell Biol 33:2202–2211
104. Lauressergues E, Bert E, Duriez P et al (2012) Does endoplasmic reticulum stress participate in APD-induced hepatic metabolic dysregulation? Neuropharmacology 62:784–796
105. Lasfer M, Vadrot N, Aoudjehane L et al (2008) Cadmium induces mitochondria-dependent apoptosis of normal human hepatocytes. Cell Biol Toxicol 24:55–62
106. Kato N, Ikeda M, Sugiyama K et al (1998) Hepatitis C virus population dynamics in human lymphocytes and hepatocytes infected *in vitro*. J Gen Virol 79:1859–1869
107. Ikeda M, Nozaki A, Sugiyama K et al (2000) Characterization of antiviral activity of lactoferrin against hepatitis C virus infection in human cultured cells. Virus Res 66:51–63
108. Suzuki K, Aoki K, Ohnami S et al (2003) Adenovirus-mediated gene transfer of interferon alpha inhibits hepatitis C virus replication in hepatocytes. Biochem Biophys Res Commun 307:814–819
109. Ikeda M, Sugiyama K, Mizutani T et al (1998) Human hepatocyte clonal cell lines that support persistent replication of hepatitis C virus. Virus Res 56:157–167
110. Tsuchihara K, Tanaka T, Hijikata M et al (1997) Specific interaction of polypyrimidine tract-binding protein with the extreme 3′-terminal structure of the hepatitis C virus genome, the 3′X. J Virol 71:6720–6726
111. Cuconati A, Mills C, Goddard C et al (2013) Suppression of AKT anti-apoptotic signaling by a novel drug candidate results in growth arrest and apoptosis of hepatocellular carcinoma cells. PLoS One 8:e54595
112. Brockhausen J, Tay SS, Grzelak CA et al (2014) miR-181a mediates TGF-β-induced hepatocyte EMT and is dysregulated in cir-

rhosis and hepatocellular cancer. Liver Int. doi:10.1111/liv.12517

113. Wilson R, Warner N, Ryan K et al (2011) The hepatitis B e antigen suppresses IL-1β-mediated NF-κB activation in hepatocytes. J Viral Hepat 18:e499–e507
114. Shah F, Louise-May S, Greene N (2014) Chemotypes sensitivity and predictivity of *in vivo* outcomes for cytotoxic assays in THLE and HepG2 cell lines. Bioorg Med Chem Lett 24:2753–2757
115. Gustafsson F, Foster AJ, Sarda S et al (2014) A correlation between the *in vitro* drug toxicity of drugs to cell lines that express human P450s and their propensity to cause liver injury in humans. Toxicol Sci 137:189–211
116. Rodrigues AV, Rollison HE, Martin S et al (2013) *In vitro* exploration of potential mechanisms of toxicity of the human hepatotoxic drug fenclozic acid. Arch Toxicol 87:1569–1579
117. Foster AJ, Prime LH, Gustafsson F et al (2013) Bioactivation of the cannabinoid receptor antagonist rimonabant to a cytotoxic iminium ion metabolite. Chem Res Toxicol 26:124–135
118. Antolino-Lobo I, Meulenbelt J, Nijmeijer SM et al (2010) Differential roles of phase I and phase II enzymes in 3,4-methylendioxymethamphetamine-induced cytotoxicity. Drug Metab Dispos 38:1105–1112
119. Saha S, New LS, Ho HK et al (2010) Direct toxicity effects of sulfo-conjugated troglitazone on human hepatocytes. Toxicol Lett 195:135–141
120. Saha S, New LS, Ho HK et al (2010) Investigation of the role of the thiazolidinedione ring of troglitazone in inducing hepatotoxicity. Toxicol Lett 192:141–149
121. Ip BC, Hu KQ, Liu C et al (2013) Lycopene metabolite, apo-10′-lycopenoic acid, inhibits diethylnitrosamine-initiated, high fat diet-promoted hepatic inflammation and tumorigenesis in mice. Cancer Prev Res 6:1304–1316
122. Krajka-Kuzniak V, Paluszczak J, Oszmianski J et al (2014) Hawthorn (*Crataegus oxyacantha L.*) bark extract regulates antioxidant response element (ARE)-mediated enzyme expression via Nrf2 pathway activation in normal hepatocyte cell line. Phytother Res 28:593–602
123. Krajka-Kuzniak V, Paluszczak J, Celewicz L et al (2013) Phloretamide, an apple phenolic compound, activates the Nrf2/ARE pathway in human hepatocytes. Food Chem Toxicol 51:202–209
124. Krajka-Kuzniak V, Paluszczak J, Baer-Dubowska W (2013) Xanthohumol induces phase II enzymes *via* Nrf2 in human hepatocytes *in vitro*. Toxicol In Vitro 27:149–156
125. Ait-Goughoulte M, Kanda T, Meyer K et al (2008) Hepatitis C virus genotype 1a growth and induction of autophagy. J Virol 82:2241–2249
126. Kanda T, Basu A, Steele R et al (2006) Generation of infectious hepatitis C virus in immortalized human hepatocytes. J Virol 80:4633–4639
127. Kanda T, Steele R, Ray R et al (2007) Hepatitis C virus infection induces the beta interferon signaling pathway in immortalized human hepatocytes. J Virol 81:12375–12381
128. Basu A, Saito K, Meyer K et al (2006) Stellate cell apoptosis by a soluble mediator from immortalized human hepatocytes. Apoptosis 11:1391–1400
129. Mazumdar B, Meyer K, Ray R (2012) *N*-terminal region of gelsolin induces apoptosis of activated hepatic stellate cells by a caspase-dependent mechanism. PLoS One 7:e44461
130. Qiu C, Hanson E, Olivier E et al (2005) Differentiation of human embryonic stem cells into hematopoietic cells by coculture with human fetal liver cells recapitulates the globin switch that occurs early in development. Exp Hematol 33:1450–1458
131. Gao Y, Theng SS, Zhuo J et al (2014) FAT10, an ubiquitin-like protein, confers malignant properties in non-tumorigenic and tumorigenic cells. Carcinogenesis 35: 923–934
132. Wang Y, Toh HC, Chow P et al (2012) MicroRNA-224 is up-regulated in hepatocellular carcinoma through epigenetic mechanisms. FASEB J 26:3032–3041
133. Tomimaru Y, Xu CQ, Nambotin SB et al (2013) Loss of exon 4 in a human T-cell factor-4 isoform promotes hepatic tumourigenicity. Liver Int 33:1536–1548
134. Nonaka M, Tazuma S, Hyogo H et al (2008) Cytoprotective effect of tauroursodeoxycholate on hepatocyte apoptosis induced by peroxisome proliferator-activated receptor gamma ligand. J Gastroenterol Hepatol 23:e198–e206
135. Nakamura K, Yamagishi S, Yoshida T et al (2007) Hydrogen peroxide stimulates pigment epithelium-derived factor gene and protein expression in the human hepatocyte cell line OUMS-29. J Int Med Res 35:427–432
136. Shishido S, Koga H, Harada M et al (2003) Hydrogen peroxide overproduction in megamitochondria of troglitazone-treated human hepatocytes. Hepatology 37:136–147

Chapter 5

Culture and Functional Characterization of Human Hepatoma HepG2 Cells

María Teresa Donato, Laia Tolosa, and María José Gómez-Lechón

Abstract

Hepatoma cell lines are frequently used as in vitro alternatives to primary human hepatocytes. Cell lines are characterized by their unlimited life span, stable phenotype, high availability, and easy handling. However, their major limitation is the lower expression of some metabolic activities compared with hepatocytes. HepG2 is a human hepatoma that is most commonly used in drug metabolism and hepatotoxicity studies. HepG2 cells are nontumorigenic cells with high proliferation rates and an epithelial-like morphology that perform many differentiated hepatic functions. In this chapter, freezing, thawing, and subculturing procedures for HepG2 cells are described. We further provide protocols for evaluating lipid accumulation, glycogen storage, urea synthesis, and phase I and phase II drug metabolizing activities in HepG2 cells.

Key words Hepatoma cell lines, HepG2 cell culture, Liver functions, Drug metabolism

1 Introduction

In the past decades, huge efforts have been made to establish cell lines that express differentiated hepatic functions. As a result, well-characterized human and rodent liver-derived cell lines have been applied in hepatology research as an alternative to cultured hepatocytes [1, 2]. Among them, a number of human hepatoma cell lines, including HepG2, Hep3B, HuH7, and HepaRG, are commonly used for drug metabolism and hepatotoxicity studies. Although the metabolic functions of hepatoma cells are more limited than those of primary hepatocytes, they offer advantages for in vitro studies, such as high availability, easy handling, nearly unlimited life-span, or stable phenotype that does not depend on donor characteristics. HepG2 is the most widely used human hepatoma cell line in pharmaco-toxicological research. This cell line was derived from liver biopsies of a 15-year-old Caucasian male with a differentiated hepatocellular carcinoma [3, 4]. HepG2 cells are nontumorigenic and highly proliferative cells that have been grown successfully in large-scale culture systems. They show an epithelial morphology when

Mathieu Vinken and Vera Rogiers (eds.), *Protocols in In Vitro Hepatocyte Research*, Methods in Molecular Biology, vol. 1250, DOI 10.1007/978-1-4939-2074-7_5, © Springer Science+Business Media New York 2015

grown on a solid surface and, under certain culture conditions, polarization of HepG2 cells can occur with the formation of bile canaliculi-like structures between adjacent cells [5, 6].

The functional characteristics of HepG2 cells have been extensively examined. They express many differentiated hepatic functions, such as synthesis and secretion of plasma proteins, cholesterol and triglyceride metabolism, lipoprotein metabolism and transport, bile acid synthesis, glycogen synthesis, or insulin signaling [4, 7–10]. However, a major drawback of HepG2 cells is their limited expression of drug metabolizing enzymes and transporters [1, 2, 11, 12]. Important cytochrome P450 (CYP) enzymes involved in phase I drug oxidations in the liver, such as CYP3A4, CYP2C9, CYP2C19, CYP2A6, or CYP2D6, are lacking or barely detectable in HepG2 cells. The abundance of most expressed drug metabolizing CYP genes is markedly lower (i.e., less than 10 %) than in primary hepatocytes [11]. This poor expression of CYP enzymes in HepG2 cells has been related to their altered pattern of so-called liver-enriched transcription factors, a group of transcription-regulatory proteins that control the hepatic expression of various genes, including CYPs [13, 14]. Although the overall expression levels of the phase II enzymes uridine diphosphate glucuronosyltransferase, glutathione *S*-transferase (GST), sulfotransferase, or *N*-acetyltransferase (NAT) are lower in HepG2 cells than in human hepatocytes, the observed differences are even less marked than for CYPs [11]. In particular, the activity or mRNA levels of several GST and NAT enzymes are comparable to those of primary hepatocytes [11, 15–17]. Regarding the expression of transport proteins in HepG2 cells, important qualitative and quantitative differences have been found when compared to liver tissue or primary hepatocytes. In particular, the characteristic hepatic transporters sodium-taurocholate cotransporting polypeptide, bile salt export pump, and organic anion transporting polypeptide C are absent or poorly expressed in HepG2 cells [11, 12].

Different strategies to overcome the functional limitations of HepG2 cells have been proposed, including their transfection with expression vectors encoding genes for specific proteins or transcription factors [2]. As a result, HepG2-derived cells stably or transiently expressing human CYPs and other enzymes have been successfully generated, which provide valuable tools for drug metabolism and hepatotoxicity studies [18–20]. Similarly, genetically engineered HepG2 cells with improved liver-specific functions have been developed for potential applications in bioartificial liver devices [21]. This chapter describes standardized procedures for the optimal maintenance of morphological characteristics and differentiated functions of HepG2 cells. Detailed protocols for routinely culturing, freezing, and thawing HepG2 cells are provided. In addition, a few experimental methods to examine characteristic hepatic functions in this hepatoma cell line are proposed.

2 Materials

All solutions and materials required for HepG2 cell culturing, thawing, and freezing must be sterile. All procedures should be carried out under aseptic conditions (i.e., in a sterile laminar flow cabinet environment).

2.1 HepG2 Culture, Thawing, and Freezing

1. HepG2 cells can be obtained from the European Collection of Cell Cultures (United Kingdom) or from the American Type Culture Collection (United States of America). Frozen vials of HepG2 cells are stored under liquid nitrogen until needed.
2. Culture medium: Dulbecco's Modification of Eagle's Medium (*see* **Note 1**) supplemented with 10 % fetal calf serum (*see* **Note 2**), 4 mM L-glutamine, 50 U penicillin/mL, and 50 μg streptomycin/mL. Store at 4 °C and avoid extended exposure to higher temperatures.
3. Freezing medium: 10 % dimethylsulfoxide (DMSO) in fetal calf serum.
4. 0.25 % trypsin/0.02 % ethylenediaminetetraacetic acid solution.
5. Phosphate-buffered saline (PBS): 137 mM NaCl, 2.7 mM KCl, 10 mM $Na_2HPO_4 \cdot 2H_2O$, 1.8 mM KH_2PO_4 in deionized water. Adjust to pH 7.4 and store for maximum 6 months at 4 °C.
6. 0.4 % trypan blue solution, sterile-filtered, cell culture tested (Sigma-Aldrich, Spain).

2.2 Functional Characterization of HepG2 Cells

1. Free fatty acid (FFA) stock solutions. Prepare 50 mM sodium oleate solution in PBS and 50 mM sodium palmitate solution in methanol. Aliquot and store at −80 °C.
2. FFA mixture. Add 27 μL of 50 mM sodium oleate and 13 μL of 50 mM sodium palmitate stock solutions to 960 μL of culture medium containing 1 % bovine serum albumin (i.e., 2 mM FFA, oleate:palmitate ratio 2:1). For other FFA mixtures (i.e., 0.25–2 mM final FFA concentration), mix adequate volumes of individual FFA stock solutions in culture medium with 1 % bovine serum albumin. FFA mixtures are prepared ex tempore.
3. Nile red solution. Dissolve 50 mg of Nile red (Sigma-Aldrich, Spain) in 1 mL of ethanol. This solution must be stored in dark place at room temperature. Dilute ex tempore to 1 mg/mL in PBS.
4. Insulin solution. Prepare a 6×10^{-5} insulin stock solution by diluting a vial of 100 UI/mL Insulatard (Novo Nordisk, Denmark) in PBS in a ratio 1:10. Store at 4 °C. Dilute this stock solution in culture medium ex tempore to obtain the desired concentration of insulin.

5. 4 % paraformaldehyde solution in PBS (Electron Microscopy Sciences, Spain), 0.5 % periodic acid solution, and Schiff's reagent (Sigma-Aldrich, Spain).
6. NH_4Cl stock solution. Prepare 1 M NH_4Cl in distilled water. Store in aliquots at −20 °C.
7. QuantiChrom™ urea assay kit (BioAssay Systems, Spain). Store at 4 °C.
8. CYP substrate stock solutions. 10 mM 7-ethoxyresorufin, 10 mM 7-methoxyresorufin, and 10 mM 7-benzyloxyresorufin (Sigma-Aldrich, Spain) in DMSO. Store in aliquots at −40 °C.
9. CYP activity incubation media. Incubation medium for each CYP activity is prepared ex tempore by dilution of the corresponding substrate stock solution in culture medium to obtain the final incubation concentration (i.e., 10 μM 7-ethoxyresorufin, 15 μM 7-methoxyresorufin, and 15 μM 7-benzyloxyresorufin). DMSO concentrations in culture medium never exceed 0.5 %.
10. Resorufin stock solution. Dissolve resorufin (Sigma-Aldrich, Spain) in 1 mM methanol. Aliquot in opaque vials and store at −80 °C.
11. Inducer stock solutions. 1 mM 3-methylcholanthrene and 400 mM phenobarbital (Sigma-Aldrich, Spain) solutions prepared in DMSO. Store in aliquots at −20 °C.
12. Deconjugation mixture. β-glucuronidase/arylsulfatase (Roche, Spain) freshly diluted 1:100 in 0.1 M sodium acetate buffer, pH 4.5.
13. Potassium phosphate-buffered solutions for GST activity: 0.1 M potassium phosphate buffer, pH 7.4, for cell homogenization and 0.1 M potassium phosphate buffer, pH 6.5, for activity assay. Store at 4 °C.
14. GST substrate stock solutions. 40 mM chloro-2,4-dinitrobenzene (Sigma-Aldrich, Madrid, Spain) in ethanol and 40 mM reduced glutathione (Sigma-Aldrich, Spain) in distilled water. Chloro-2,4-dinitrobenzene solution can be stored in the dark at 4 °C for a week and glutathione solution is prepared ex tempore.

2.3 Equipments and Laboratory Wares

1. Incubator (37 °C ± 1 °C, 90 % ± 5 % humidity, 5 % ± 1 % CO_2).
2. 96-Well microtiter plates, 3.5 and 6 cm diameter cell culture dishes, and T25 and T75 cell culture flasks.
3. Cryovials and freezing box.
4. Multiplate spectrophotometer
5. Multiplate fluorescence reader.
6. Shaking incubator (37 °C).
7. Ultrasonicator.

3 Methods

3.1 Thawing Procedure for HepG2 Cells

1. Remove the cryogenic vial from the storage liquid nitrogen tank and thaw cells by immersing the ampule into a water bath at 37 °C. Incubate the vial just for a brief time needed to thaw most of its content (i.e., approximately 80 %) (*see* **Note 3**).
2. Using a sterile 2 mL pipette, transfer the cells from the vial into a sterile 15 mL conical tube containing 10 mL of prewarmed culture medium (i.e., 37 °C).
3. Use 2 mL of culture medium to rinse the cryovial. Gently pipette up and down a couple of times and transfer this medium to the 15 mL tube. This ensures that any remaining cells in the vial are recovered.
4. Centrifuge at $350\times g$ and room temperature for 3 min (*see* **Note 4**).
5. Discard medium supernatant without disturbing cell pellet.
6. Resuspend the cells in 4 mL of culture medium by gently pipetting up and down. Complete up to a known volume of culture medium (e.g., 10 mL) (*see* **Note 5**).
7. Transfer the cell suspension to a culture flask and rotate gently to be sure the cells are dispersed evenly over the surface (*see* **Note 6**). Place the flask in an incubator at 37 °C and 5 % CO_2.
8. Following overnight attachment, decant the medium and replace with fresh prewarmed (i.e., 37 °C) medium. This medium change allows to remove nonadherent cells and replenish nutrients.

3.2 Culture and Subculture of HepG2 Cells

1. HepG2 cells are routinely grown in culture grade flasks at 37 °C under a humidified 95 % air and 5 % CO_2 atmosphere. They are adherent cells with high proliferation rates. HepG2 cells are epithelial-like cells that grow as monolayers and form characteristic cell clusters or islands (Fig. 1).
2. Medium is changed every 2 days by aspirating culture supernatant and by adding fresh prewarmed (i.e., 37 °C) culture medium (*see* **Note 7**). Cells reaching 70–80 % confluence must be removed from the flask by trypsinization (*see* **Note 8**).
3. Warm up the culture medium, trypsin–ethylenediaminetetraacetic acid solution and PBS in a water bath at 37 °C.
4. Aspirate the medium from the culture flask and rinse the cell monolayer with PBS (e.g., 5 mL for a T25 flask or 15 mL for a T75 flask). Wash the cells by gentle agitation to remove any remaining serum that might inhibit the action of the trypsin. Discard the washing solution.

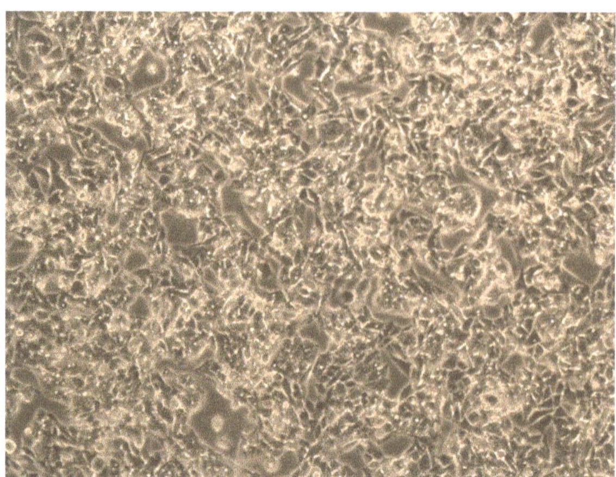

Fig. 1 HepG2 cells in monolayer culture

5. Add the trypsin–ethylenediaminetetraacetic acid solution (e.g., 0.5 mL for a T25 flask or 1 mL for a T75 flask). After 15–30 s, remove the excess trypsin–ethylenediaminetetraacetic acid solution and incubate the cells at 37 °C.
6. After 2–3 min, cells begin to come off from the plastic surface. Lightly tap the side of flask to detach the cells completely.
7. Add prewarmed (i.e., 37 °C) culture medium (e.g., 10 mL) to recover detached cells. Trypsin activity is stopped by adding serum.
8. Centrifuge the cell suspension at 350 × *g* and room temperature for 3 min.
9. Discard supernatant and resuspend the cell pellet in prewarmed (i.e., 37 °C) culture medium. Use a sterile 10 mL pipette to disaggregate cell clumps by gently pipetting up and down.
10. Add appropriate aliquots of cell suspension to new culture flasks as required. For routine subculturing, determination of cell number is not needed. The HepG2 cell line is normally split at a subcultivation ratio of 1:3 to 1:5 (*see* **Note 9**).
11. For cell characterization and functional testing, reseed the cells in the desired culture vessels (i.e., dishes or multiwell formats) at an appropriate cell density (*see* **Note 10**).

3.3 Freezing HepG2 Cells

1. Mycoplasma-free cells with normal morphology and reaching 80 % confluence can be cryopreserved (*see* **Note 11**).
2. Remove the cells from the flasks by trypsinization. Repeat the **steps 3–7** of the culture and subculture protocol.
3. Count the cells using trypan blue exclusion for a viable cell count to determine the number of cryovials that will be needed (i.e., approximately 5×10^6 cells/vial). The viability should be over 90 % to ensure the cells are healthy enough for freezing.

4. Centrifuge the cell suspension at 350×*g* and room temperature for 3 min.
5. Discard the supernatant and resuspend the cell pellet in an appropriate volume of cold-freezing medium by gentle pipetting (*see* **Note 12**).
6. Dispense 1.8 mL of the cell suspension into each freezing tube.
7. Place the cryovials into a freezing box containing isopropanol overnight at −80 °C. This allows the cells to freeze slowly.
8. The day after, transfer the vials into liquid nitrogen freezer for definitive storage.

3.4 Induction of Lipid Accumulation

Nile red is a phenoxazone dye that is strongly fluorescent in a hydrophobic environment. The characteristics of the fluorescent signal are dependent on the degree of hydrophobicity of lipids, which results in a maximum emission shift from red to yellow in the presence of polar and nonpolar lipids, respectively. This probe can be added directly to cultured cells to label the fat accumulated in the cytosol as lipid droplets (i.e., mainly triglycerides). Intracellular accumulation of triglycerides can be analyzed by fluorescence microscopy, flow cytometry, or fluorescence plate readers. Using this procedure, increased fat accumulation in HepG2 cells in response to FFA or drugs known to cause steatosis has been demonstrated and has revealed the potential utility of HepG2 cells as an in vitro liver steatosis model [22–24].

1. Fat overloading in HepG2 cells is induced by incubating cells in 96-well plates at 75 % confluency with culture medium containing a mixture of FFA at different concentrations (i.e., 0.25–2 mM) for a desired period of time (e.g., 12–36 h) (*see* **Note 13**).
2. After treatments, wash cells twice with PBS, add 100 μL of Nile red solution, and incubate at 37 °C for 15 min.
3. Remove the Nile red solution and wash the monolayers thereafter twice with PBS (*see* **Note 14**).
4. Let the wells dry and measure fluorescence intensity at 488 and 550 nm excitation and emission wavelengths, respectively, using a fluorescence plate reader.
5. Fluorescence intensity is directly proportional to intracellular content of neutral lipids (i.e., triglycerides). Fat accumulation is expressed as fold increase in fluorescence intensity compared to controls (i.e., cultured cells not exposed to FFA).
6. Typical results on the ability of HepG2 cells to take FFA up and to accumulate triglycerides in a concentration-dependent way are shown in Fig. 2. The highest lipid content was reached after exposure to 1 or 2 mM FFA for 24 h.

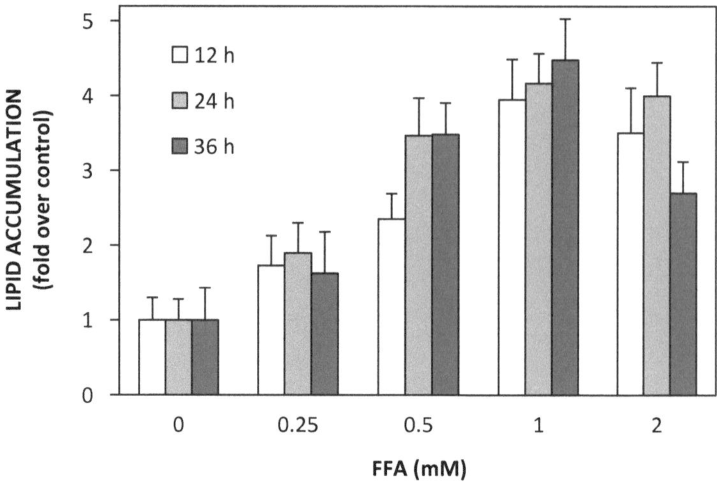

Fig. 2 Accumulation of triglycerides in HepG2 cells. Cells were exposed to increasing concentrations (i.e., 0.25–2 mM) of an FFA mixture (i.e., oleate:palmitate in 2:1 ratio) for 14, 24, or 48 h. Intracellular triglyceride content was determined fluorimetrically using Nile red staining. The results are expressed as a percentage of the control cells. Data are the mean ± standard deviation of 3 independent experiments

3.5 Assessment of Glycogen Storage

The periodic acid-Schiff staining technique is the most widely used procedure to demonstrate glycogen in tissues and cells. This method is based on the oxidation of the carbon to carbon bond by periodic acid with the formation of aldehydes that react to the fuchsin-sulfurous acid present in the Schiff reagent. Cytoplasmic glycogen deposits are revealed as pink-stained areas.

1. Maintain HepG2 cells in standard culture conditions or in medium supplemented with 30 mM glucose and 10^{-7} M insulin for 24 h.
2. After treatments, fix the cells with 4 % paraformaldehyde for 10 min. Wash the cell monolayer with distilled water. All steps are performed at room temperature.
3. Cover the cells with 0.5 % periodic acid solution for 5 min.
4. Rinse the cells with distilled water for 2–3 min.
5. Cover the cells with Schiff solution for 15–30 min in the dark. Check the dark pink color under the microscope.
6. Stop the staining reaction by washing the cells 3 times in tap water for 2 min.
7. Glycogen storage can be identified as periodic acid-Schiff-positive intracellular granules under the light microscope (Fig. 3).

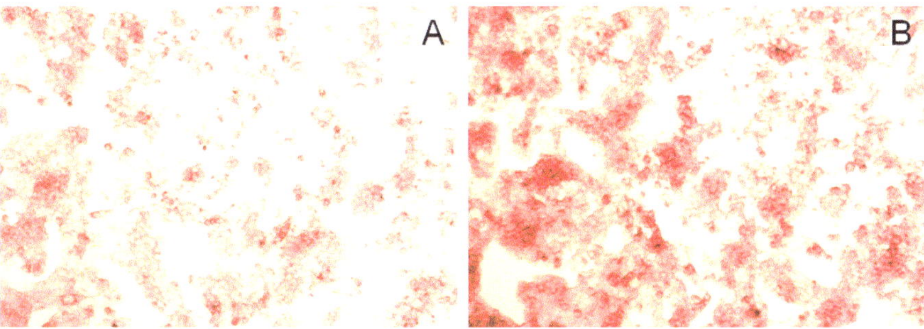

Fig. 3 Periodic acid-Schiff staining to demonstrate glycogen storage in HepG2 cells. (**a**) Cells maintained in the standard culture medium. (**b**) Cells maintained for 24 h in the culture medium supplemented with 30 mM glucose and 10^{-7} M insulin. Periodic acid-Schiff staining reveals intracellular glycogen deposits as pink-stained areas

3.6 Urea Synthesis Assay

The urea cycle is an important detoxification pathway that forms urea from ammonia. As urea is almost exclusively produced by the liver, ureogenic capacity is commonly used as an indicator of the preservation of a differentiated hepatic phenotype of liver-derived cells. Urea production assays consist of the quantification of the urea synthesized and released into the culture medium when cells are incubated with NH_4Cl.

1. Seed 6×10^5 HepG2 cells on 3.5 cm diameter cell dishes in complete culture medium. The medium is refreshed 16–24 h thereafter.
2. After 48 h in culture, remove the medium and add 600 μL of culture medium containing 2 mM NH_4Cl freshly prepared by dilution of 1 M NH_4Cl stock solution (*see* **Note 15**).
3. Incubate the cells in the presence of NH_4Cl in an incubator at 37 °C and 5 % CO_2 for the desired period, bearing in mind that urea formation is linear up to 20 h (Fig. 4). Then, 500 μL of incubation medium is sampled and centrifuged for 2 min at $2{,}000 \times g$ and room temperature. Transfer the supernatants to clean tubes. At this step, samples can be immediately processed or kept frozen (i.e., −20 °C). Samples of incubation medium with NH_4Cl not incubated with cells (i.e., blank samples) are also processed.
4. At this stage, cell monolayers are washed out with PBS and kept frozen (i.e., −20 °C) for further protein quantification (*see* **Note 16**).
5. Place the reagents from the QuantiChrom™ urea assay kit to room temperature and prepare the working reagent solution ex tempore according to the manufacturer's instructions (*see* **Note 17**).

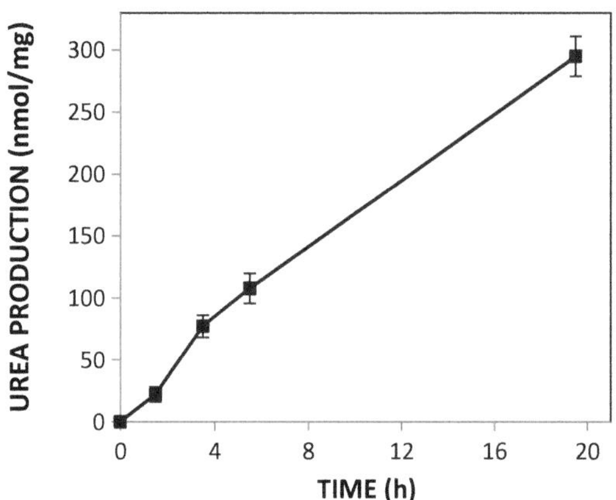

Fig. 4 Urea synthesis by HepG2 cells. Cultured cells were incubated in the presence of 2 mM NH_4Cl. The formed urea was released into the incubation medium and was quantified. For each time point, data correspond to the mean urea production assayed in duplicate

6. Transfer 50 μL clean supernatants to 96-well plates (i.e., each sample in duplicate wells) and add 200 μL working solution. Tap gently to mix and incubate 30 min at room temperature.
7. Read the urea-dependent chromogenic reaction at 430 nm using a microplate spectrophotometer.
8. After subtracting the absorbance of the blank from the absorbance of each sample, the urea concentration in the samples is calculated by interpolation in a standard curve. Rates of urea synthesis by cultured cells are expressed as nmol urea formed/h/mg cell protein.

3.7 Evaluation of CYP Activities

CYP enzymes are mainly responsible for phase I metabolism of xenobiotics in the liver. Although the CYP pattern in most hepatoma cells is characterized by a general shortage of CYP enzymes, HepG2 cells show measurable activity levels of a few CYPs involved in both the metabolic activation and inactivation of many drugs and environmental compounds [1, 17]. CYP activities in cultured cells can be examined by the direct incubation of intact cell monolayers with appropriate probe substrates, which are enzymatically oxidized to render highly fluorescent metabolites. In particular, several *O*-alkyl derivatives of resorufin have been proposed as useful substrates for fluorescence-based CYP reactions, namely, 7-ethoxyresorufin *O*-deethylation (EROD), 7-methoxyresorufin *O*-demethylation (MROD), and 7-benzyloxyresorufin *O*-debenzylation (BROD) [25]. Using this strategy, basal activities of CYP1A (i.e., EROD and MROD) and CYP2B (i.e., BROD) are easily measured in HepG2 cells. Moreover, the response of hepatoma cells to 3-methylcholanthrene and phenobarbital, 2 known CYP inducers, can also be studied.

1. Plate the HepG2 cells on culture dishes or multiwell formats at a density of 5×10^4 cells/cm^2. After 16–24 h, the medium is replaced by fresh culture medium.
2. For induction studies, incubate HepG2 cells with medium containing 2 μM 3-methylcholanthrene, 2 mM phenobarbital, or solvent (i.e., 0.5 % DMSO) for 48 h. Inducer or solvent is renewed 24 h later with medium change.
3. After treatments, remove the culture medium and add 600 μL to 3.5 cm diameter cell dishes of CYP activity incubation medium containing the substrate, namely, 10 μM 7-ethoxyresorufin (i.e., EROD activity), 10 μM 7-methoxyresorufin (i.e., MROD activity), or 15 μM 7-benzyloxyresorufin (i.e., BROD activity). Recommended volumes for multiwell formats are 150 μL for 96-well plates or 300 μL for 24-well plates.
4. Incubate the cells in the presence of the substrates in an incubator at 37 °C and 5 % CO_2) for 60 min. Then, 500 μL of incubation medium is sampled and centrifuged for 2 min at 2,000 × *g* and room temperature. Transfer the supernatants to clean tubes. Samples of incubation medium with substrate not incubated with cells (i.e., blank samples) are also processed. At this step, samples can be immediately processed or kept frozen (i.e., −20 °C).
5. Cell monolayers are washed out with PBS solution and kept frozen (i.e., −20 °C) for further protein quantification.
6. Prior to metabolite (i.e., resorufin) quantification, incubate samples with 0.1 M acetate buffer, pH 4.5, containing β-glucuronidase/arylsulfatase, for 120 min in a shaking incubator at 37 °C to hydrolyze potential conjugates (*see* **Note 18**). The hydrolysis reaction is stopped by adding 500 μL of methanol.
7. Centrifuge for 3 min at 2,000 × *g* and room temperature. Transfer 100 μL of clean supernatants to 96-well plates (i.e., each sample in triplicate wells).
8. Measure the fluorescence intensity at 530 nm excitation and 585 nm emission wavelengths with a fluorescence plate reader.
9. Use a standard curve of resorufin (i.e., 5–200 nM) to interpolate sample fluorescence intensity. Activities are expressed as pmol of metabolite formed/min/mg cell protein (Table 1).
10. In induction studies, activity values in treated cells are referred to those in control (i.e., solvent-treated) cells (Fig. 5).

3.8 Evaluation of GST Activity

In addition to the oxidations catalyzed by CYPs (i.e., phase I metabolism), conjugating reactions by phase II enzymes are key for the detoxification of drugs. Similarly to CYPs, the expression of conjugating enzymes in HepG2 cells is lower than in human liver or primary hepatocytes [1]. Among the phase II reactions,

Table 1
Drug metabolizing activities in HepG2 cells

Enzyme	Reaction	Activity
CYP1A1/2	7-Ethoxyresorufin *O*-deethylation (EROD)	0.85 ± 0.29
CYP1A1/2	7-Methoxyresorufin *O*-demethylation (MROD)	0.35 ± 0.18
CYP2B6	7-Benzyloxyresorufin *O*-debenzylation (BROD)	0.26 ± 0.11
GST	1-Chloro-2,4-dinitrobenzene conjugation	37.3 ± 8.1

EROD, MROD, and BROD are expressed as pmol/mg/min and GST activity is expressed as nmol/mg/min

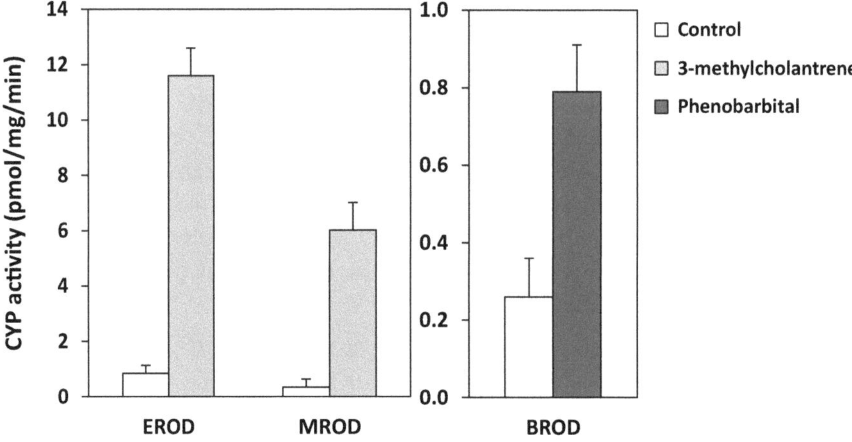

Fig. 5 Effects of model inducers on CYP activities in HepG2 cells. CYP activities were determined in the HepG2 cells exposed for 48 h to 2 μM 3-methylcholantrene (i.e., EROD and MROD) or 2 mM phenobarbital (i.e., BROD). The results are expressed as fold increase over activity in the control cells. Data are mean ± standard deviation of three independent experiments

conjugation of reduced glutathione to electrophilic compounds is among the best preserved in HepG2 cell line. This reaction is catalyzed by a group of GST enzymes that protect cells against the mutagenic, carcinogenic, and toxic effects of reactive metabolites. Several GSTs with distinct specificities to xenobiotic or endogenous substrates are expressed in the human liver. A relatively high expression of some GST enzymes (e.g., GSTA4, GSTM2, or GSTT1) has been identified in HepG2 cells [11]. Typically, 1-chloro-2,4-dinitrobenzene, considered a suitable substrate for a wide range of GSTs, is used to measure total GST activity. The product formed (i.e., 2,4-dinitrophenyl-glutathione) shows a characteristic absorption at 340 nm, which is absent in the unconjugated substrate [26].

1. Seed 6×10^5 HepG2 cells on 3.5 cm diameter cell dishes in complete culture medium. The medium is refreshed 16–24 h later.

2. After 48 h in culture, remove the medium and wash the cell monolayers with PBS. In this step, culture plates can be frozen in liquid nitrogen and stored at −20 °C until processed.
3. Add 500 μL of ice-cold 0.1 M potassium phosphate buffer, pH 7.4, and detach the cells with a cell scraper.
4. Disrupt cells by sonication (i.e., 10–20 s) on ice (*see* **Note 19**).
5. Centrifuge cellular homogenate at 9,000×*g* and 4 °C for 20 min. The supernatant (i.e., S9 fraction) is kept for activity measurement.
6. Transfer 5 μL of the S9 faction to 96-well plates (i.e., duplicate wells). An amount of 5–10 μg protein is recommended for the assay. Alternatively, the activity can be assayed in spectrophotometer cuvettes.
7. Add 185 μL of 0.1 M potassium phosphate buffer, pH 6.5, 5 μL of 40 mM 1-chloro-2,4-dinitrobenzene, and 5 μL of 40 mM glutathione all warmed to room temperature to each well. Mix by gentle shaking for 5 s. A blank sample consisting of reaction mixture without S9 fraction is also prepared.
8. Immediately after preparing the reaction mixtures, read the absorbance at 340 nm using a microplate spectrophotometer every 60 s over a period of 4–6 min (*see* **Note 20**). The activity assay is performed at room temperature (i.e., 20–25 °C).
9. Subtract the (ΔA_{340})/min of the blank from the (ΔA_{340})/min of each sample.
10. Calculate the GST activity using 9.6/mM/cm as the extinction coefficient for 1-chloro-2,4-dinitrobenzene conjugate at 340 nm. The activity is expressed as nmol conjugate formed/min/mg cell protein (Table 1).

4 Notes

1. Eagle's Minimum Essential medium or Roswell Park Memorial Institute 1640 medium can also be used for culturing HepG2 cells.
2. Due to batch-to-batch variability of fetal calf serum, it is recommended to first check a batch of serum for optimal growth properties with HepG2 cells and to then reserve a sufficient amount of fetal calf serum.
3. Thaw cells quickly. As a culture freezes, ice is formed and, upon thawing, ice shards are capable of puncturing cells, which could thus reduce cell viability. By thawing as quickly as possible, these ice shards may be avoided and cell damage is kept at a minimum.

4. Centrifugation is needed to completely remove the cryopreservation medium. This step can be avoided, but it is highly recommended as an excess of cryoprotectant (i.e., DMSO) can induce cytotoxicity to cultured cells.
5. The total volume is required to calculate cell recovery. The recovery rate for frozen cells is usually at least 90 %.
6. Tissue culture vessels (i.e., flasks, cell dishes, or multiwell plates) should be prescreened to ensure that they adequately support the growth of HepG2 cells.
7. Daily examination of cell cultures under a phase-contrast microscope is recommended to identify any signs of contamination or changes in morphology and adhesive properties of cells.
8. The functional characteristics of HepG2 cells can be influenced by the culture environment, growth cycle, and passage number [16, 17]. In order to avoid potential variability and to optimize the results, working conditions (e.g., cell density, time that cells spend in culture within 1 passage or different passages) should be standardized by each laboratory.
9. Cells should be maintained in culture for a limited number of passages (i.e., no more than 15–18 continuous subcultures). A fresh lot of frozen cells from the cell bank should be thawed out periodically. A recently thawed culture may require a certain amount of time before being used in specific assays. The culture is expanded into larger culture vessels (e.g., from 1T25 flask to a larger T75 flask, next from a T75 flask into 4T75 flasks, and so on).
10. A cell suspension sample obtained after trypsinization is used to determine cell viability (e.g., by the trypan blue exclusion test) and cell number (e.g., using a hemocytometer or cell counter). After cell counting, the cell suspension should be conveniently diluted before seeding. Each laboratory needs to determine and adjust the final cell density to their particular requirements (e.g., culture formats or days in culture). For functional characterization, it is important that cells overcome the lag growth phase when they are used to assess metabolic functions that can be altered by the proliferative rate. We recommend 5×10^4 cells/cm^2 for cells to be used after 2–3 days in culture.
11. Cryopreservation is needed for long-term cell storage. The creation of a cell bank of HepG2 cells consisting of stocks of cells stored in freezing tubes in a liquid nitrogen freezer is recommended. This ensures the availability of back-up cells in case of contamination as well as a stock of cells from a low passage number for future studies.

12. Cryopreservation media generally consist of a base medium, a cryopreservative agent (e.g., DMSO or glycerol), and a protein source (e.g., serum). The cryoprotectant and protein protect cells from the stress created by the freeze-thaw process.
13. This FFA combination (i.e., oleate:palmitate in 2:1 ratio) develops dose-dependent intracellular fat accumulation without evident signs of cytotoxicity. Lipid accumulation can also be induced in cells exposed to a single FFA (e.g., oleate or palmitate) or to mixtures containing different combinations of oleate, palmitate, or other FFAs [23, 27]. The potential toxicity of FFA treatment should be examined. As an example, palmitate alone or FFA mixtures with high palmitate content (e.g., 2 mM oleate:palmitate in 1:1 or 1:2 ratios) produce significant decreases in cell viability [23].
14. Lipid accumulation can be visualized by examination of intracellular oil droplet staining using a light microscope. Hematoxylin can be used to stain cell nuclei.
15. Balanced salt solutions supplemented with NH_4Cl can be used for short cell incubations (i.e., up to 2 h). However, culture medium is required for longer incubations. Given the presence of urea, culture medium without serum or with a low serum concentration (e.g., 2 %) is recommended for urea assay incubation.
16. Among the different chromogenic methods available to measure protein concentration in biological samples, the Lowry assay [28], the Bradford method [29], and the bicinchoninic acid assay [30] are the most commonly used for protein quantification in cultured cells.
17. Different enzymatic and chemical methods have been proposed for urea determination in biological samples. Commonly used enzymatic assays are based upon the hydrolysis reaction catalyzed by urease with ammonia formation. Obviously, urease-based assays cannot be used to quantify the urea concentration in culture media containing ammonia. Several chemical methods have been used to determine urea formation in cultured cells, but some have major limitations, such as the lack of suitability for low urea concentrations, interference of the phenol red present in the culture medium with the peak wavelength of the reaction product, or the need for deproteination and heating steps to form the colored product. The proposed colorimetric method uses a modified Jung reagent [31]. The major advantages of this method include its simplicity, high sensitivity, and possibility of measuring the urea concentration in medium samples with high ammonia content.
18. Since intact HepG2 monolayers are used, all the enzymatic systems and cofactors are present, including those involved in

the formation of glucuronide and sulfate conjugates (i.e., phase II reactions). Therefore, in order to avoid underestimation of CYP activities, the potential resorufin conjugates formed during the incubation of HepG2 cells with CYP substrates are enzymatically hydrolyzed before the fluorimetric quantification of resorufin [25].

19. GSTs are cytosolic enzymes and their activity can be measured in crude cell homogenates or in S9 fractions containing both microsomal and cytosolic enzymes. The described procedure has been optimized for HepG2 cells cultured in 3.5 cm diameter cell culture plates. For other plate formats, the protocol should be adjusted. For example, for 96-well cultured HepG2 cells, add 50 μL of ice-cold 0.1 M potassium phosphate buffer, pH 7.4, and disrupt cells using an ultrasonicator. In this case, a crude cell homogenate is preferred rather than an S9 fraction to determine GST activity.
20. GST activity is measured by the increase in absorbance. The linearity of the reaction (i.e., absorbance vs. reaction time) must be determined for each sample. If the (ΔA_{340})/min value for a particular sample is higher than 0.05, GST is too concentrated in the S9 sample and must be diluted or, alternatively, activity can be assayed using a smaller volume of S9.

Acknowledgements

The authors acknowledge the financial support of the Spanish Ministry of Health through grants PI10/0923 and PI13/0986 (Instituto de Salud Carlos III).

References

1. Donato MT, Lahoz A, Castell JV et al (2008) Cell lines: a tool for in vitro drug metabolism studies. Curr Drug Metab 9:1–11
2. Donato MT, Jover R, Gomez-Lechon MJ (2013) Hepatic cell lines for drug hepatotoxicity testing: limitations and strategies to upgrade their metabolic competence by gene engineering. Curr Drug Metab 14:946–968
3. Aden DP, Fogel A, Plotkin S et al (1979) Controlled synthesis of HBsAg in a differentiated human liver carcinoma-derived cell line. Nature 282:615–616
4. Knowles BB, Howe CC, Aden DP (1980) Human hepatocellular carcinoma cell lines secrete the major plasma proteins and hepatitis B surface antigen. Science 209:497–499
5. Sormunen R, Eskelinen S, Lehto VP (1993) Bile canaliculus formation in cultured HEPG2 cells. Lab Invest 68:652–662
6. Fearn RA, Hirst BH (2006) Predicting oral drug absorption and hepatobiliary clearance: human intestinal and hepatic in vitro cell models. Environ Toxicol Pharmacol 21:168–178
7. Javitt NB (1990) Hep G2 cells as a resource for metabolic studies: lipoprotein, cholesterol, and bile acids. FASEB J 4:161–168
8. Dongiovanni P, Valenti L, Ludovica Fracanzani A et al (2008) Iron depletion by deferoxamine up-regulates glucose uptake and insulin signaling in hepatoma cells and in rat liver. Am J Pathol 172:738–747
9. Meier M, Klein HH, Kramer J et al (2007) Calpain inhibition impairs glycogen syntheses in HepG2 hepatoma cells without altering insulin signaling. J Endocrinol 193:45–51
10. Forte TM, McCall MR, Knowles BB et al (1989) Isolation and characterization of lipoproteins produced by human hepatoma-derived

cell lines other than HepG2. J Lipid Res 30:817–829

11. Guo L, Dial S, Shi L et al (2011) Similarities and differences in the expression of drug-metabolizing enzymes between human hepatic cell lines and primary human hepatocytes. Drug Metab Dispos 39:528–538
12. Hilgendorf C, Ahlin G, Seithel A et al (2007) Expression of thirty-six drug transporter genes in human intestine, liver, kidney, and organotypic cell lines. Drug Metab Dispos 35:1333–1340
13. Jover R, Bort R, Gomez-Lechon MJ et al (1998) Re-expression of C/EBP alpha induces CYP2B6, CYP2C9 and CYP2D6 genes in HepG2 cells. FEBS Lett 431:227–230
14. Rodriguez-Antona C, Donato MT, Boobis A et al (2002) Cytochrome P450 expression in human hepatocytes and hepatoma cell lines: molecular mechanisms that determine lower expression in cultured cells. Xenobiotica 32:505–520
15. Wilkening S, Stahl F, Bader A (2003) Comparison of primary human hepatocytes and hepatoma cell line HepG2 with regard to their biotransformation properties. Drug Metab Dispos 31:1035–1042
16. Wilkening S, Bader A (2003) Influence of culture time on the expression of drug-metabolizing enzymes in primary human hepatocytes and hepatoma cell line HepG2. J Biochem Mol Toxicol 17:207–213
17. Lin J, Schyschka L, Mühl-Benninghaus R et al (2012) Comparative analysis of phase I and II enzyme activities in 5 hepatic cell lines identifies Huh-7 and HCC-T cells with the highest potential to study drug metabolism. Arch Toxicol 86:87–95
18. Bai J, Cederbaum AI (2004) Adenovirus mediated overexpression of CYP2E1 increases sensitivity of HepG2 cells to acetaminophen induced cytotoxicity. Mol Cell Biochem 262: 165–176
19. Donato MT, Hallifax D, Picazo L et al (2010) Metabolite formation kinetics and intrinsic clearance of phenacetin, tolbutamide, alprazolam, and midazolam in adenoviral cytochrome P450-transfected HepG2 cells and comparison with hepatocytes and in vivo. Drug Metab Dispos 38:1449–1455
20. Tolosa L, Gomez-Lechon MJ, Perez-Cataldo G et al (2013) HepG2 cells simultaneously expressing five P450 enzymes for the screening of hepatotoxicity: identification of bioactivable drugs and the potential mechanism of toxicity involved. Arch Toxicol 87:1115–1127
21. Tang N, Wang Y, Wang X et al (2012) Stable overexpression of arginase I and ornithine transcarbamylase in HepG2 cells improves its ammonia detoxification. J Cell Biochem 113:518–527
22. McMillian MK, Grant ER, Zhong Z et al (2001) Nile red binding to HepG2 cells: an improved assay for in vitro studies of hepatosteatosis. In Vitro Mol Toxicol 14:177–190
23. Gomez-Lechon MJ, Donato MT, Martínez-Romero A et al (2007) A human hepatocellular in vitro model to investigate steatosis. Chem Biol Interact 165:106–116
24. Donato MT, Martínez-Romero A, Jimenez N et al (2009) Cytometric analysis for drug-induced steatosis in HepG2 cells. Chem Biol Interact 181:417–423
25. Donato MT, Gomez-Lechon MJ, Castell JV (1993) A microassay for measuring cytochrome P450IA1 and P450IIB1 activities in intact human and rat hepatocytes cultured on 96-well plates. Anal Biochem 213:29–33
26. Habig WH, Jakoby WB (1981) Assays for differentiation of glutathione *S*-transferases. Methods Enzymol 77:398–405
27. Malhi H, Barreyro FJ, Isomoto H et al (2007) Free fatty acids sensitise hepatocytes to TRAIL mediated cytotoxicity. Gut 56:1124–1131
28. Lowry OH, Rosebrough NJ, Farr AL et al (1951) Protein measurement with the Folin phenol reagent. J Biol Chem 193:265–275
29. Bradford MM (1976) A rapid and sensitive method for the quantitation of microgram quantities of protein utilizing the principle of protein-dye binding. Anal Biochem 72:248–254
30. Smith PK, Krohn RI, Hermanson GT et al (1985) Measurement of protein using bicinchoninic acid. Anal Biochem 150:76–85
31. Zawada RJ, Kwan P, Olszewski KL et al (2009) Quantitative determination of urea concentrations in cell culture medium. Biochem Cell Biol 87:541–544

Chapter 6

Establishment and Characterization of an In Vitro Model of Fas-Mediated Hepatocyte Cell Death

Mathieu Vinken, Michaël Maes, Sara Crespo Yanguas, Joost Willebrords, Tamara Vanhaecke, and Vera Rogiers

Abstract

Fas-mediated apoptosis underlies a plethora of liver pathologies and toxicities. As a consequence, this process is a major research topic in the field of experimental and clinical hepatology. The present chapter describes the setup of an in vitro model of hepatocellular apoptotic cell death. In essence, this system consists of freshly isolated hepatocytes cultured in a monolayer configuration, which are exposed to a combination of Fas ligand and cycloheximide. This in vitro model has been characterized by using a set of well-acknowledged cell death markers. This experimental system allows the study of the entire course of Fas-mediated hepatocellular cell death, going from early apoptosis to secondary necrosis, and hence can serve a broad range of in vitro pharmaco-toxicological purposes.

Key words Apoptosis, Secondary necrosis, In vitro model, Fas ligand, Primary hepatocyte

1 Introduction

Historically, the liver has gained particular attention of toxicologists, as it represents a key target organ of chemical-induced cell injury. Indeed, hepatocytes drive the majority of the xenobiotic biotransformation machinery in the organism and are therefore frequently involved in toxicity [1–3]. For many years, it was thought that cell death elicited by toxicants primarily occurred through necrosis. It has now become clear that an alternative cell death mode, namely apoptosis, predominates in toxicant-induced cytolethality [3–6]. Apoptosis is a well-orchestrated process, relying on the proteolytic activity of an evolutionarily conserved family of cysteine proteases, called caspases. In fact, two major apoptotic pathways have been described, the extrinsic signaling cascade and the intrinsic pathway [3, 4, 6, 7]. The former is initiated by the binding of a specific subset of ligands, including

Mathieu Vinken and Vera Rogiers (eds.), *Protocols in In Vitro Hepatocyte Research*, Methods in Molecular Biology, vol. 1250, DOI 10.1007/978-1-4939-2074-7_6, © Springer Science+Business Media New York 2015

the Fas ligand, to their corresponding receptors at the cell membrane surface [6, 7]. Hepatocytes highly express the Fas receptor [8] and its ligand binding results in the proteolytic cleavage and autoactivation of procaspase 8. Activated caspase 8 induces caspase 3, which subsequently cleaves a broad spectrum of cellular proteins. In the intrinsic apoptotic pathway, cytochrome C is released from mitochondria, a process that is mediated by members of the B-cell lymphoma-2 protein family. Activation of this pathway occurs upon DNA damage. Cytochrome C triggers caspase 9 activation, which in turn also induces caspase 3 [6, 7].

A number of protocols have been described for studying hepatocellular apoptosis in vivo, including the direct administration of cell death-evoking toxicants to animals [9], the application of genetically modified subjects [10], and the use of partially hepatectomized rodents [11]. The intraperitoneal injection of anti-Fas antibodies in mice, known to activate the Fas receptor, causes fulminant hepatic failure and severe hepatocyte injury, which eventually burgeons into animal death [12]. Such experiments not only raise serious ethical questions, but are also of limited scientific value. Indeed, apoptotic cells are barely detectable in vivo, as they are rapidly engulfed by neighboring phagocytes. During in vitro experimentation, where phagocytosis does not take place, the full course of apoptosis can be monitored, whereby the late apoptotic phase is typically followed by secondary necrosis [4–7]. Cell lines are frequently used experimental tools in in vitro apoptosis research. However, these cells, such as HepG2 cells, Huh7 cells and Hep3B cells, are often derived from tumors and have typically acquired high resistance against apoptosis [13–15]. Primary cells may offer a better alternative, as they display in vivo-like sensitivity to apoptosis, at least during short-term culture [14]. Among the numerous experimental strategies that have been followed to provoke apoptotic cell death in primary hepatocyte cultures, the use of Fas receptor stimuli is a most reasonable approach, as it directly affects the physiological pathway [16]. In vitro models of Fas-mediated hepatocellular cell death have yet been exploited for a wide range of biomedical applications [17–20]. In the majority of such studies, apoptosis is induced without any further investigation of the cell death response as such. However, insight into the underlying mechanisms of cytolethality is of crucial importance, since they may have a direct impact on the outcome of the study. Therefore, the establishment of an in vitro model of Fas-mediated hepatocyte cell death along with its characterization is described in the current chapter.

2 Materials

2.1 Establishment of the In Vitro Model of Fas-Mediated Hepatocellular Cell Death

1. Hepatocyte seeding medium. William's medium E containing 7 ng/mL glucagon, 292 mg/mL L-glutamine, 7.33 IE/mL sodium benzyl penicillin, 50 µg/mL kanamycin monosulfate, 10 µg/mL sodium ampicillin, 50 µg/mL streptomycin sulfate, and 10 % fetal bovine serum (*see* **Note 1**). Prepare in a laminar air flow cabinet and store for maximum 7 days at 4 °C. Prior to use, the hepatocyte seeding medium should be placed for 30 min in a thermostated bath at 37 °C (*see* **Note 2**).
2. Hepatocyte culture medium. Serum-free hepatocyte seeding medium supplemented with 25 µg/mL hydrocortisone sodium hemisuccinate and 0.5 µg/mL insulin (*see* **Note 1**). Prepare in a laminar air flow cabinet and store for maximum 7 days at 4 °C. Prior to use, the hepatocyte culture medium should be placed for 30 min in a thermostated bath at 37 °C (*see* **Note 2**).
3. Hepatocyte cell death medium. Hepatocyte culture medium supplemented with 200 ng/mL Fas ligand (Alexis, Switzerland) (*see* **Note 3**) and 2 µg/mL cycloheximide (Sigma-Aldrich, Belgium) (*see* **Note 4**). Prepare ex tempore in a laminar air flow cabinet. Prior to use, the hepatocyte cell death medium should be placed for 30 min in a thermostated bath at 37 °C.
4. Incubator (37 ± 1 °C, 90 ± 5 % humidity, 5 ± 1 % CO_2).
5. Laminar air flow cabinet.
6. 3.5 cm diameter plastic cell culture dishes.
7. Thermostated bath (37 °C).

2.2 Characterization of the In Vitro Model of Fas-Mediated Hepatocellular Cell Death

1. Phosphate-buffered saline (PBS). 137 mM NaCl, 2.7 mM KCl, 10 mM $Na_2HPO_4{\cdot}2H_2O$, 1.8 mM KH_2PO_4 in deionized water. Adjust to pH 7.4 and store for maximum 6 months at 4 °C. Prior to use, PBS should be placed for 30 min at room temperature.
2. PBS supplemented with divalent cations ($PBSD^+$). PBS containing 1.2 mM $CaCl_2$ and 340 µM $MgCl_2{\cdot}6H_2O$ in deionized water. Adjust to pH 7.4 and store for maximum 6 months at 4 °C. Prior to use, the $PBSD^+$ solution should be placed for 30 min at room temperature.
3. Staining solution. 140 mM NaCl, 5 mM $CaCl_2$, 10 mM *N*-[2-hydroxyethyl]piperazine-*N'*[2-ethaansulfonzuur], 2 % annexin-V-fluos (Roche Diagnostics, Germany), 3 µg/mL Hoechst 33342 (Invitrogen, Belgium) and 1 µg/mL propidium iodide (Invitrogen, Belgium) in deionized water (*see* **Note 5**). Adjust to pH 7.4 and store for maximum 6 months at 4 °C. Prior to use, the staining solution should be placed for 30 min at room temperature.
4. Fluorescence microscope, camera, computer, and software.

3 Methods

3.1 Establishment of a Monolayer Culture of Primary Hepatocytes and Induction of Cell Death

1. Use freshly isolated primary rat hepatocytes [21] (*see* **Note 6**).
2. Evenly plate the hepatocytes on 3.5 cm diameter plastic culture dishes at a density of 0.56×10^5 cells/cm^2 in hepatocyte seeding medium (*see* **Notes 7–9**). Place the cell cultures in an incubator at 37 °C and 5 % CO_2 for 4 h.
3. Remove the hepatocyte seeding medium and replace by identical volumes of hepatocyte culture medium. Place the cell cultures in an incubator at 37 °C and 5 % CO_2 for 20 h.
4. Replace the hepatocyte culture medium. Place the cell cultures in an incubator at 37 °C and 5 % CO_2 for 24 h.
5. Remove the hepatocyte culture medium and replace by identical volumes of hepatocyte cell death medium. Place the cell cultures in an incubator at 37 °C and 5 % CO_2 (*see* **Note 10**).
6. Sample at the start of cell death induction and 2, 4, and 6 h thereafter (*see* **Note 11**).

3.2 In Situ Staining of Cell Death Markers

1. Remove the hepatocyte cell death medium from the 3.5 cm diameter cell culture dishes and wash the cells twice with 2 mL of the PBSD$^+$ solution.
2. Add 2 mL of the staining solution to each culture dish and leave on for 15 min at room temperature in the dark.
3. Remove the staining solution from the culture dishes and wash the cells 4 times with 2 mL of the PBSD$^+$ solution.
4. Analyze the culture dishes by means of fluorescence microscopy at 100× magnification.
5. Take at least 5 images per culture dish by using appropriate filters (Fig. 1) (*see* **Note 12**).
6. Count the number of cells positive for the marker concerned in each image and express relative to the total number of nuclei present (Fig. 1) (*see* **Notes 13–16**).

4 Notes

1. The composition of the cell culture medium is a major determinant of the response of primary hepatocytes to experimentally induced cell death. Several commonly used cell culture additives can counteract the occurrence of apoptosis in primary hepatocyte cultures, including glucocorticosteroids [22, 23], insulin [24], glucagon [18] and serum [25]. The composition of the cell culture medium used in the current protocol has been optimized in such a way that a cell death response,

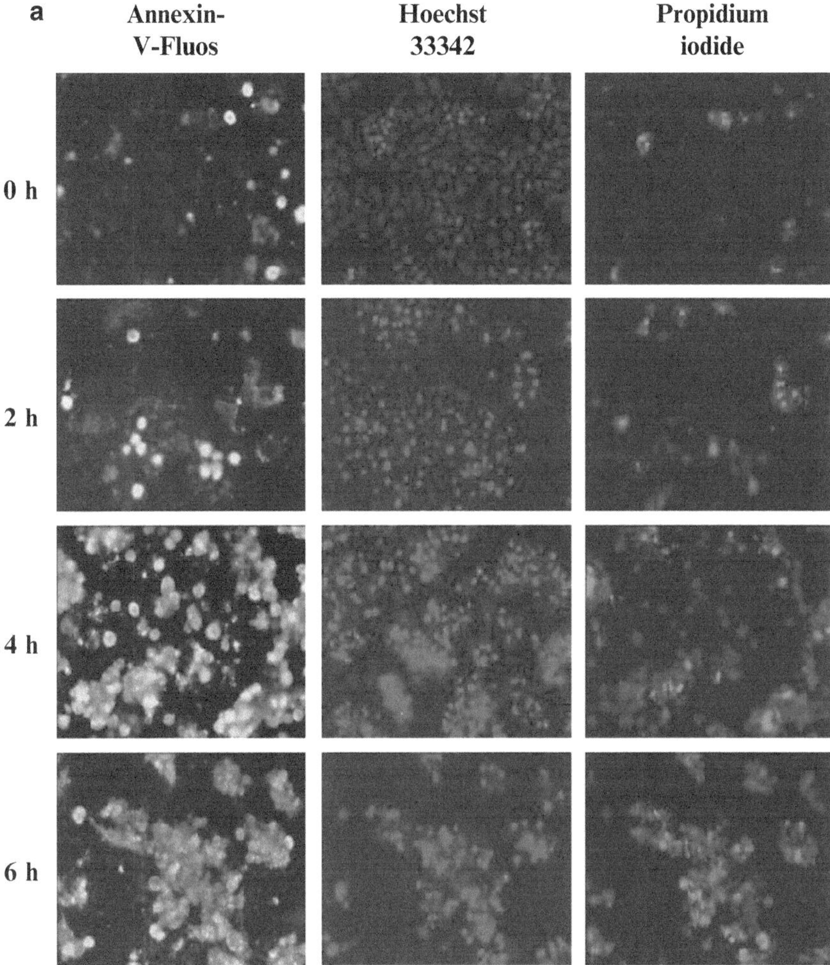

Fig. 1 In situ staining of cell death markers. Freshly isolated rat hepatocytes were cultivated in a monolayer configuration and were exposed to 200 ng/mL Fas ligand and 2 μg/mL cycloheximide, starting at 44 h post-plating. (**a**) Samples were taken at the start of the exposure (0 h), and after 2, 4, and 6 h, and were stained with annexin-V-fluos, Hoechst 33342, and propidium iodide. (**b**) The number of cells, positive for the concerned marker, was counted in each image and expressed relative to the total number of nuclei present. Data are expressed as the mean ± standard deviation of 3 independent experiments. Results were evaluated by 1-way analysis of variance followed by post hoc Bonferroni tests. *filled circle* control, *open triangle* Fas ligand/cycloheximide, $^{**}p<0.01$, $^{***}p<0.001$. Reproduced in modified form with permission from [32]

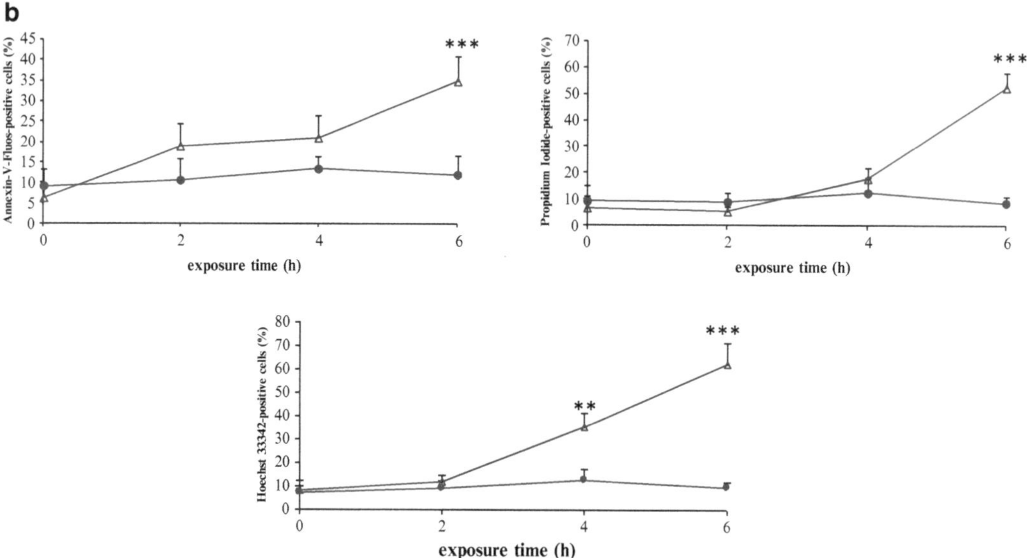

Fig. 1 (continued)

going from early apoptosis to secondary necrosis, is generated and completed in a total time span of 6 h in cultures of primary rat hepatocytes. The composition of the cell culture medium might be subject to optimization when using hepatocytes from other species.

2. The sterility of the cell culture media can be checked by adding 1 mL of the medium to 25 mL sterile thioglycollate medium (Oxoid, Belgium) and placing for 2 days in an incubator.
3. It is strongly recommended to use the soluble Fas ligand and not anti-Fas antibodies to induce cell death in primary hepatocyte cultures. Although both trigger the same pathway, anti-Fas antibodies mediate different effects with different sensitivity in comparison with Fas ligand [26–28].
4. It is strongly recommended to boost the apoptotic effect of Fas ligand by combining with an inhibitor of protein translation, such as cycloheximide. The concentrations of both Fas ligand and cycloheximide have been optimized in such a way that a cell death response, going from early apoptosis to secondary necrosis, is generated and completed in a total time span of 6 h in cultures of primary rat hepatocytes. These concentrations might be subject to optimization when seeding the cells at cell density other than used in the current protocol or when using hepatocytes from other species.
5. The concentrations of annexin-V-fluos, Hoechst 33342 and propidium iodide as well as incubation time with these dyes have been optimized for primary rat hepatocyte cultures cultivated at the indicated cell density. These parameters might be

subject to optimization when seeding the cells at another cell density or when using hepatocytes from other species.

6. It is of utmost importance to assess cell viability following hepatocyte isolation, as this procedure typically causes considerable harm to cells. This is routinely done by means of trypan blue exclusion. Cell viability before plating should be at least 85 %.
7. It is strongly recommended to avoid the use of collagen-coated cell culture dishes, as the presence of an extracellular matrix scaffold counteracts the occurrence of cell death [29].
8. Cell density is a major determinant of the response of cultured hepatocytes to experimentally induced cell death. Hepatocytes cultured at low density (i.e., 0.35×10^5 cells/cm^2) display less apoptotic activity in comparison with counterparts seeded at high density (i.e., 1.4×10^5 cells/cm^2) [30]. The cell density in the current protocol (i.e., 0.56×10^5 cells/cm^2) has been optimized in such a way that a cell death response, going from early apoptosis to secondary necrosis, is generated and completed in a total time span of 6 h in cultures of primary rat hepatocytes. The cell density might be subject to optimization when using hepatocytes from other species.
9. The size of the culture dishes depends on the type of analysis that is intended. For in situ staining of cell death markers, 3.5 cm diameter cell culture dishes are sufficient.
10. The time of cell death induction is a major determinant of the response of cultured hepatocytes to experimentally induced cell death. Considerable spontaneous cell death is observed in the first 24 h following cell plating as a result of insults underwent during the hepatocyte isolation procedure [31]. This might interfere with the experimentally induced cell death process. On the other hand, hepatocytes become resistant to cell death induction as a function of cultivation time, which reflects the progressive dedifferentiation process known to take place in primary hepatocyte cultures [2]. Based on own experience, the best time to induce cell death is between 40 and 48 h postplating.
11. The total duration of sampling in the current protocol (i.e., 6 h) has been optimized in such a way that the entire cell death response, going from early apoptosis to secondary necrosis, can be studied in primary cultures of rat hepatocytes. This timing might be subject to optimization when using hepatocytes from other species.
12. Excitation wavelengths for Hoechst 33342, annexin-V-fluos and propidium iodide are 346, 488, and 535 nm, while emission wavelengths are 497, 518, and 617 nm, respectively.

13. Counting should be done in a double blinded fashion. Alternatively, yet technically more demanding, fluorescence-assisted cell sorting can be used for high-throughput counting.
14. The spots in the cell culture dishes that are used for counting are selected at random. However, since the cells are generally less homogeneously seeded at the borders of the cell culture dish, it is not recommended to consider spots for counting in these areas.
15. At least 6 spots per cell culture dish should be counted. From a statistical point of view, it is strongly recommended to perform the experiments on hepatocytes isolated from the liver of 3 different human donors or animals. Results can be processed and evaluated by analysis of variance followed by post hoc Bonferroni tests.
16. Small groups of apoptotic bodies must be considered as remnants from a single cell.

Acknowledgements

This work was financially supported by the grants of the University Hospital of the Vrije Universiteit Brussel—Belgium (Willy Gepts Fonds UZ-VUB), the University of Sao Paulo—Brazil (USP), the São Paulo Research Foundation—Brazil (FAPESP), the Fund for Scientific Research—Flanders (FWO-Vlaanderen), the European Research Council (project CONNECT), the European Union (FP7) and Cosmetics Europe (projects DETECTIVE and HeMiBio).

References

1. Vanhaecke T, Rogiers V (2006) Hepatocyte cultures in drug metabolism and toxicological research and testing. Methods Mol Biol 320:209–227
2. Vinken M, Papeleu P, Snykers S et al (2006) Involvement of cell junctions in hepatocyte culture functionality. Crit Rev Toxicol 36:299–318
3. Jaeschke H, Gores GJ, Cederbau AI et al (2002) Mechanisms of hepatotoxicity. Toxicol Sci 65:166–176
4. Raffray M, Cohen GM (1997) Apoptosis and necrosis in toxicology: a continuum or distinct modes of cell death? Pharmacol Ther 75:153–177
5. Gomez-Lechon MJ, O'Connor E, Castell JV et al (2002) Sensitive markers used to identify compounds that trigger apoptosis in cultured hepatocytes. Toxicol Sci 65:299–308
6. Gill GH, Dive D (2000) Apoptosis: basic mechanisms and relevance to toxicology. In: Roberts R (ed) Apoptosis in toxicology. Taylor & Francis, London, UK, pp 1–20
7. Yin X-M, Dong Z (2003) Essentials of apoptosis: a guide for basic and clinical research. Humana, Totowa, NJ
8. Feldmann G (1997) Liver apoptosis. J Hepatol 26:1–11
9. Furukawa S, Usuda K, Fujieda Y et al (2000) Apoptosis and cell proliferation in rat hepatocytes induced by barbiturates. J Vet Med Sci 62:23–28
10. Guicciardi ME, Bronk SF, Werneburg NW et al (2005) Bid is upstream of lysosome-mediated caspase 2 activation in tumor necrosis factor alpha-induced hepatocyte apoptosis. Gastroenterology 129:269–284

11. Baier PK, Baumgartner U, Wolff-Vorbeck G et al (2006) Hepatocyte proliferation and apoptosis in rat liver after liver injury. Hepatogastroenterology 53:747–752
12. Ogasawara J, Watanabe-Fukunaga R, Adachi M et al (1993) Lethal effect of the anti-Fas antibody in mice. Nature 364:806–809
13. Valavanis C, Hu Y, Yang Y et al (2001) Model cell lines for the study of apoptosis in vitro. In: Schwartz LM, Ashwell JD (eds) Apoptosis: methods in cell biology. Academic, San Diego, CA, pp 417–436
14. Schulze-Bergkamen H, Untergasser A, Dax A et al (2003) Primary human hepatocytes: a valuable tool for investigation of apoptosis and hepatitis B virus infection. J Hepatol 38:736–744
15. Lei XY, Zhong M, Feng LF et al (2007) siRNA-mediated Bcl-2 and Bcl-xl gene silencing sensitizes human hepatoblastoma cells to chemotherapeutic drugs. Clin Exp Pharmacol Physiol 34:450–456
16. Maeda S (2000) Mechanisms of active cell death in isolated hepatocytes. In: Berry MN, Edwards AM (eds) The hepatocyte review. Kluwer Academic, London, UK, pp 281–300
17. Azzaroli F, Mehal W, Soroka CJ et al (2002) Ursodeoxycholic acid diminishes Fas-ligand-induced apoptosis in mouse hepatocytes. Hepatology 36:49–54
18. Fladmark KE, Gjertsen BT, Doskeland SO et al (1997) Fas/APO-1(CD95)-induced apoptosis of primary hepatocytes is inhibited by cAMP. Biochem Biophys Res Commun 232:20–25
19. Fu T, Blei AT, Takamura N et al (2004) Hypothermia inhibits Fas-mediated apoptosis of primary mouse hepatocytes in culture. Cell Transplant 13:667 676
20. Zhang M, He W, Liu F et al (2004) Inhibition of mouse hepatocyte apoptosis via anti-Fas ribozyme. World J Gastroenterol 10: 2567–2570
21. Papeleu P, Vanhaecke T, Henkens T et al (2006) Isolation of rat hepatocytes. Methods Mol Biol 320:229–237
22. Bailly-Maitre B, de Sousa G, Boulukos K et al (2001) Dexamethasone inhibits spontaneous apoptosis in primary cultures of human and rat hepatocytes *via* Bcl-2 and Bcl-xL induction. Cell Death Differ 8:279–288
23. Bailly-Maitre B, de Sousa G, Zucchini N et al (2002) Spontaneous apoptosis in primary cultures of human and rat hepatocytes: molecular mechanisms and regulation by dexamethasone. Cell Death Differ 9:945–955
24. Bresgen N, Ohlenschlager I, Wacht N et al (2008) Ferritin and FasL (CD95L) mediate density dependent apoptosis in primary rat hepatocytes. J Cell Physiol 217:800–808
25. Ethier C, Raymond VA, Musallam L et al (2003) Antiapoptotic effect of EGF on mouse hepatocytes associated with downregulation of proapoptotic Bid protein. Am J Physiol Gastrointest Liver Physiol 285:G298–G308
26. Fadeel B, Thorpe CJ, Yonehara S et al (1997) Anti-Fas IgG1 antibodies recognizing the same epitope of Fas/APO-1 mediate different biological effects in vitro. Int Immunol 9:201–209
27. Legembre P, Beneteau M, Daburon S et al (2003) Cutting edge: SDS-stable Fas microaggregates: an early event of Fas activation occurring with agonistic anti-Fas antibody but not with Fas ligand. J Immunol 171:5659–5662
28. Thilenius AR, Braun K, Russell JH (1997) Agonist antibody and Fas ligand mediate different sensitivity to death in the signaling pathways of Fas and cytoplasmic mutants. Eur J Immunol 27:1108–1114
29. Vanhaecke T, Henkens T, Kass GE et al (2004) Effect of the histone deacetylase inhibitor trichostatin A on spontaneous apoptosis in various types of adult rat hepatocyte cultures. Biochem Pharmacol 68:753–760
30. Qiao L, Farrell GC (1999) The effects of cell density, attachment substratum and dexamethasone on spontaneous apoptosis of rat hepatocytes in primary culture. In Vitro Cell Dev Biol Anim 35:417–424
31. Vinken M, Decrock E, Doktorova T et al (2011) Characterization of spontaneous cell death in monolayer cultures of primary hepatocytes. Arch Toxicol 85:1589–1596
32. Vinken M, Decrock E, De Vuyst E et al (2009) Biochemical characterization of an in vitro model of hepatocellular apoptotic cell death. Altern Lab Anim 37:209–218

Chapter 7

Serum-Free Directed Differentiation of Human Embryonic Stem Cells to Hepatocytes

Kate Cameron, Baltasar Lucendo-Villarin, Dagmara Szkolnicka, and David C. Hay

Abstract

The increase in human liver disease worldwide is a major concern. At present, the only successful mode of treatment for failing liver function is organ transplantation. While highly successful, donor organs are a limited resource that cannot meet current demands. Therefore, alternative liver support strategies have been explored, including the use of the major and metabolic cell within the liver, the hepatocyte. While current approaches using human hepatocytes are very promising, donor material is still required and therefore suffers from similar limitations to whole organ transplantation. One alternative source of human hepatocytes being actively pursued in the field is pluripotent stem cells. Pluripotent stem cells are a scalable and renewable cell-based resource, which can be efficiently differentiated towards hepatocytes, including pluripotent stem cell lines that have been derived under good manufacturing practice conditions. Therefore, it is believed that this approach provides a promising model system for cell scale-up and differentiation. In the future, pluripotent stem cell-derived hepatocytes could be used in the clinic to support failing liver function if they should be deemed fit for purpose.

Key words Pluripotent stem cell, Hepatocyte, Liver, Human embryonic stem cell, Scale-up

1 Introduction

In the United Kingdom, liver disease is the fifth most common cause of death and is still on the rise [1]. The only established treatment for patients with end-stage liver disease is orthotopic liver transplantation [2]. Given the scarce nature of donor organs, alternative approaches have been devised, including the use of human hepatocytes as a cell-based therapy. Major advantages of this approach are the use of a less invasive procedure and the fact that patients can be given multiple cell infusions if required [3, 4]. There are, however, drawbacks to this approach, which include poor cell engraftment, the requirement for immunosuppression [5], and cell scarcity. In attempts to bypass the issues of tissue availability and

Mathieu Vinken and Vera Rogiers (eds.), *Protocols in In Vitro Hepatocyte Research*, Methods in Molecular Biology, vol. 1250, DOI 10.1007/978-1-4939-2074-7_7, © Springer Science+Business Media New York 2015

immune modulation, other cell-based sources have been explored, including the use of human pluripotent stem cells (hPSCs) [6]. hPSCs represent an unlimited source of human somatic cells for therapy, including hepatocytes. In the past, we have derived hepatocytes from hPSCs using serum-containing medium [7, 8]. Most recently, we have developed a serum-free differentiation procedure to produce functional hepatocytes with high efficiency [9]. Importantly, the serum-free process is scalable and the product can be shipped [10, 11]. This chapter fully describes the method used to differentiate hPSCs into functional hepatocytes.

2 Materials

1. Phosphate-buffered saline (PBS). 137 mM NaCl, 2.7 mM KCl, 10 mM $Na_2HPO_4{\cdot}2H_2O$, 1.8 mM KH_2PO_4 in deionized water. Adjust to pH 7.4 and store for maximum 6 months at 4 °C.
2. mTeSR1 serum-free medium (Stem Cell Technologies, United Kingdom). Add 100 mL of 5× concentrated supplement (Cell Technologies, United Kingdom).
3. Endoderm-priming medium. Add 10 mL of 50× concentrated B27 supplement-containing insulin and 5 mL of penicillin/streptomycin with final concentrations of 100 IU/mL and 100 μg/mL, respectively, to Roswell Park Memorial Institute (RPMI) 1640 medium (Life Technologies, United Kingdom). Filter under vacuum. Add activin A (PeproTech, United Kingdom) to a final concentration of 100 ng/mL. Add recombinant Wnt3a (PeproTech, United Kingdom) to a final concentration of 50 ng/mL (*see* **Notes 1–3**).
4. SR-DMSO hepatocyte differentiation medium. Add 100 mL knock-out serum replacement (KO-SR) (Life Technologies, United Kingdom), 2.5 mL GlutaMAX (Life Technologies, United Kingdom), 5 mL nonessential amino acids (Life Technologies, United Kingdom), 1 mL β-mercaptoethanol (Life Technologies, United Kingdom), 5 mL penicillin/streptomycin, and 5 mL dimethylsulfoxide (DMSO) to 400 mL knock-out Dulbecco's modified Eagle's medium (KO-DMEM) (Life Technologies, United Kingdom). Filter under vacuum (*see* **Notes 1–3**).
5. Hepatocyte maturation medium. Add 5 mL hydrocortisone 21-hemisuccinate, 5 mL GlutaMAX, and 5 mL penicillin/streptomycin to 500 mL HepatoZYME medium (Life Technologies, United Kingdom). Filter under vacuum. Add hepatocyte growth factor (HGF) (PeproTech, United Kingdom) to a final concentration of 10 ng/mL. Add oncostatin M (OSM) (PeproTech, United Kingdom) to a final concentration of 20 ng/mL (*see* **Notes 1–3**).

6. Human basic fibroblast growth factor (bFGF) (PeproTech, United Kingdom).
7. Matrigel (BD Bioscience, United Kingdom).
8. Collagenase (Sigma-Aldrich, United Kingdom).
9. Incubator (37 ± 1 °C, 90 ± 5 % humidity, 5 ± 1 % CO_2).
10. Laminar air flow cabinet.
11. Plastic culture dishes with diameter 3.5 cm (i.e., 6-well plates) and 1.6 cm (i.e., 24-well plates).

3 Methods

3.1 Preparation of Growth Factor Solutions

3.1.1 Human bFGF Solution

1. Prepare 10 % bovine serum albumin (BSA) solution in PBS and filter through a 0.22 μm filter.
2. From the 10 % BSA solution, prepare a 0.2 % BSA solution in PBS.
3. Add 10 mL of the 0.2 % BSA solution to 100 μg human bFGF.
4. Prewet a 0.22 μm filter by filtering 5 mL 10 % BSA solution through the filter. Discard the 10 mL BSA wash.
5. Filter the human bFGF solution through the prewashed filter.
6. Aliquot the human bFGF solution in sterile recipients and store at −20 °C.

3.1.2 Human Activin A Solution

1. Add 1 mL of the 0.2 % BSA solution into a syringe and prewet the filter.
2. Dilute activin A in 0.2 % BSA solution to a concentration of 100 μg/mL.
3. Filter the activin A solution, aliquot in sterile recipients and store at −20 °C.

3.1.3 Mouse Wnt3a Solution

1. Add 200 μL of PBS to a 2 μg of Wnt3a.
2. Aliquot in sterile recipients and store at −20 °C.

3.1.4 Human HGF Solution

1. Dilute HGF in PBS to a stock concentration of 10 μg/mL.
2. Filter the HGF solution, aliquot in sterile recipients, and store at −20 °C.

3.1.5 OSM Solution

1. Dilute OSM in PBS to a stock concentration of 20 μg/mL.
2. Filter the OSM solution, aliquot in sterile recipients, and store at −20 °C.

3.2 Culture of hPSCs

3.2.1 Coating of Plastic Cell Culture Dishes with Matrigel

1. Thaw the 10 mL stock bottle of Matrigel overnight at 4 °C on ice and add 10 mL of KO-DMEM. Mix well using chilled pipettes and store 1 mL aliquots at −20 °C.
2. Thaw an aliquot of Matrigel at 4 °C for at least 2 h or overnight to avoid the formation of a gel.
3. Add 5 mL of cold (i.e., 4 °C) KO-DMEM to the Matrigel and mix well with a pipette.
4. Add to 18 mL with cold KO-DMEM and mix using a pipette.
5. Add 1 mL of Matrigel to each well of a 6-well plate.
6. Incubate the coated plate overnight at 4 °C or room temperature for 1 h before use (*see* **Note 4**).
7. Allow the coated culture container to come to room temperature inside a tissue culture hood before use.
8. Aspirate the Matrigel and add the cell suspension to the well immediately before use.

3.2.2 Resuscitation of hPSC Lines

1. Remove hPSCs from liquid nitrogen storage and thaw quickly in a 37 °C water bath.
2. Carefully transfer the cells to a sterile tube containing 5 mL of warm mTeSR1 serum-free medium.
3. Pellet the cells by centrifugation for 5 min at 300 × *g* and room temperature.
4. Aspirate the supernatant, gently resuspend the cells into warm mTeSR1 serum-free medium, and add to 1 well of a precoated Matrigel plate.
5. Feed cells daily with fresh mTeSR1 serum-free medium. Upon 60–70 % confluence, the cells require passaging.

3.2.3 Passaging Cells with Collagenase

1. Aspirate the media from the well.
2. Wash once with 2 mL PBS.
3. Add 1 mL of 200 U/mL collagenase diluted in KO-DMEM and incubate at 37 °C for 4 min.
4. Aspirate the collagenase.
5. Wash once with 2 mL PBS.
6. Add an appropriate volume of mTeSR1 serum-free medium depending on the split ratio. Use a cell scraper to remove the cells from the surface of the well. Use a 10 mL pipette for up and down pipetting of the cell suspension. It is important that the cells are kept in clumps of cells and that they are not broken up into single cells (*see* **Note 5**).

3.3 Induction of Differentiation

1. An outline of the differentiation protocol is shown in Fig. 1. Culture hPSCs on Matrigel-coated plates in mTeSR1 serum-free medium and maintain in a humidified incubator at 37 °C

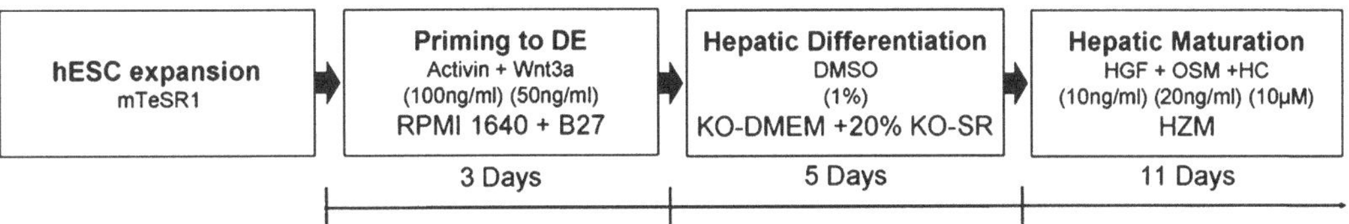

Fig. 1 Schematic representation of differentiation of hESCs to hepatocytes. hESCs are scaled to requirement using mTeSR1 serum-free medium. At the appropriate density, the cells are primed to definitive endoderm (DE) with Roswell Park Memorial Institute (RPMI) 1640 medium containing B27 supplement, 100 ng/mL activin A, and 50 ng/mL Wnt3a for 72 h, changing medium daily. Subsequently, hepatic specification is initiated using knock-out Dulbecco's modified Eagle's medium (KO-DMEM) containing 20 % knock-out serum replacement (KO-SR) and 1 % dimethylsulfoxide (DMSO), changing medium every 48 h. Finally, the hepatocytes are specified in hepatocyte maturation medium, changing medium every 48 h (*HC* hydrocortisone, *HGF* hepatocyte growth factor, *HZM* hepatoZYME medium, *OSM* oncostatin M)

and 5 % CO_2. Until cells reach confluence, passage with collagenase and replate onto a fresh 24-well plate coated with Matrigel (*see* **Notes 6** and 7).

2. Once replated and cells reach 20–30 % confluence, initiate differentiation by replacing culture medium with endoderm-priming medium (*see* **Note 8**). Replace media with fresh media every 24 h for a period of 72 h. After 72 h, replace the media with SR-DMSO hepatocyte differentiation medium for 5 days, replacing every 48 h. After 5 days, replace the media with hepatocyte maturation medium, changing every 48 h until day 20. The cells gradually exhibit morphological changes from a spiky and triangular shape to a characteristic liver morphology displaying a polygonal appearance (*see* **Notes 9–12**).

4 Notes

1. Store all media at 4 °C, aliquot and store at –20 °C if required.
2. The growth factors (i.e., activin A, Wnt3a, OSM, and HGF) should be added fresh each day.
3. RPMI 1640 medium, SR-DMSO medium, and growth factors require filtering before use. HepatoZYME medium does not require filtering.
4. Plates which have been coated with Matrigel can be stored at 4 °C for up to 1 week. They should be clearly labeled with the date they were coated. Discard any plates not used within 1 week.
5. When placing the cells in the incubator, agitate the tissue culture container to ensure as even as possible a distribution of colonies, as the colonies tend to settle in the center of the tissue culture plate affecting cell replating and differentiation.

6. For the preparation of Matrigel, place pipettes in a freezer for at least 10–20 min before use.
7. If cells from multiple wells are required, pool the cells in a centrifuge tube, spin and resuspend in an adjusted volume of mTeSR1 serum-free medium.
8. The differentiation procedure is generally started 24 h after the cells are seeded. To ensure efficient differentiation, the replated hPSC colonies should be small and evenly spaced.
9. If high cell death is observed on days 2 and 3, wash the cells with RPMI 1640 medium without activin A and Wnt3a.
10. On days 4 and 6 of the differentiation protocol, cells may not need to be washed unless cell death is high.
11. Ensure all culture medium has reached room temperature before use.
12. hPSC populations can be characterized by expression of stem cell markers, such as Tra-1-81, Tra 1-60 SSEA-4, and SSEA-1, which can be investigated by a number of techniques, including flow cytometry, immunoblot analysis, and reverse transcriptase polymerase chain reaction analysis. Functional analysis of hPSC-derived hepatocytes is typically done by measuring cytochrome P450 activity, using specific substrates.

Acknowledgements

Kate Cameron was supported by a UKRMP grant award, MR/K026666/1. Baltasar Lucendo-Villarin and Dagmara Szkolnicka were supported by MRC CRM PhD Studentships.

References

1. www.statistics.gov.uk/ (consulted July 2014)
2. Hay DC, Fletcher J, Payne C et al (2008) Highly efficient differentiation of hESCs to functional hepatic endoderm requires activin A and Wnt3a signaling. Proc Natl Acad Sci U S A 105:12301–12306
3. Hughes RD, Mitry RR, Dhawan A (2012) Current status of hepatocyte transplantation. Transplantation 93:342–347
4. Dhawan A, Puppi J, Hughes RD et al (2010) Human hepatocyte transplantation: current experience and future challenges. Nat Rev Gastroenterol Hepatol 7:288–298
5. Forbes SJ, Alison MR (2014) Regenerative medicine: knocking on the door to successful hepatocyte transplantation. Nat Rev Gastroenterol Hepatol 11:277–278
6. Greenhough S, Medine CN, Hay DC (2010) Pluripotent stem cell derived hepatocyte like cells and their potential in toxicity screening. Toxicology 278:250–255
7. Medine CN, Lucendo-Villarin B, Storck C et al (2013) Developing high-fidelity hepatotoxicity models from pluripotent stem cells. Stem Cells Transl Med 2:505–509
8. Sullivan GJ, Hay DC, Park IH et al (2010) Generation of functional human hepatic endoderm from human induced pluripotent stem cells. Hepatology 51:329–335
9. Szkolnicka D, Zhou W, Lucendo-Villarin B et al (2013) Pluripotent stem cell-derived hepatocytes: potential and challenges in pharmacology. Annu Rev Pharmacol Toxicol 53: 147–159

10. Szkolnicka D, Farnworth SL, Lucendo-Villarin B et al (2014) Accurate prediction of drug-induced liver injury using stem cell-derived populations. Stem Cells Transl Med 3:141–148

11. Zhou X, Sun P, Lucendo-Villarin B et al. (2014) Modulating innate immunity improves hepatitis C virus infection and replication in stem cell-derived hepatocytes. Stem Cell Reports 3(1):204–214

Chapter 8

Human Skin-Derived Precursor Cells: Isolation, Expansion, and Hepatic Differentiation

Joery De Kock, Robim M. Rodrigues, Karolien Buyl, Tamara Vanhaecke, and Vera Rogiers

Abstract

Human skin-derived precursor cells are a multipotent stem cell population that resides within the dermis throughout adulthood. Human skin-derived precursor cells can be isolated, purified, and expanded in large quantities from any patient, in health and disease, and differentiated to mesodermal and ectodermal cell types. Recently, it was also found that they can be directed towards hepatic cells with acquired properties of toxicological relevance. As such, they represent a valuable cell source for the further development of human-relevant in vitro models for the identification and quantification of hepatotoxic compounds. In this chapter, a robust basic methodology to isolate, expand, and differentiate human skin-derived precursor cells into hepatic cells in a sequential and time-dependent way is provided.

Key words Adult stem cells, Skin-derived precursor cell, Isolation, Differentiation, Expansion, Hepatocyte, In vitro

1 Introduction

Due to the scarcity of human primary hepatocytes, increasing attention is directed towards human stem cells for their potential use in screening tools for human safety of new chemical entities [1, 2]. Recent research showed that human skin-derived precursor cells (hSKPs) hold great promise for toxicological application [1]. hSKPs are multipotent dermal stem cells that share many properties with embryonic neural crest stem cells [3, 4] and persist in the human dermis throughout adulthood [3, 5]. They can be isolated and expanded in large quantities from small skin biopsies of abdomen [5], breast [5], arm [6], foreskin [7], face [8], and scalp [8, 9] tissue. Furthermore, hSKPs are able to generate peripheral neuronal cells, including Schwann cells [10, 11] and catecholaminergic neurons [12] as well as mesenchymal cell types, such as adipocytes, chondrocytes, osteocytes, and smooth muscle cells [3, 13–15]. Importantly, hSKPs can also be directed towards hepatic cells that exhibit

Mathieu Vinken and Vera Rogiers (eds.), *Protocols in In Vitro Hepatocyte Research*, Methods in Molecular Biology, vol. 1250, DOI 10.1007/978-1-4939-2074-7_8, © Springer Science+Business Media New York 2015

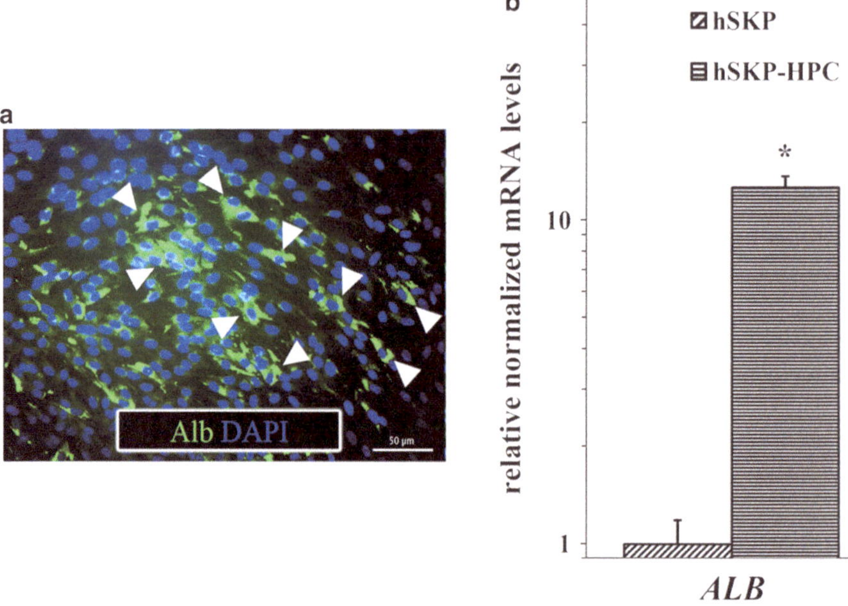

Fig. 1 hSKP-derived hepatic cells express albumin. hSKP-derived hepatic cells (hSKP-HPC) are characterized by their expression of albumin (Alb) both at the (**a**) protein as well as at the (**b**) mRNA level. *Arrowheads* mark strong Alb-positive hepatic cells. *Asterisk* indicates significantly increased expression versus undifferentiated hSKP ($p < 0.05$)

properties of toxicological relevance when sequentially exposed to hepatogenic growth factors and cytokines [1, 16]. In this chapter, we describe how to reproducibly isolate and expand undifferentiated hSKPs. We also provide an efficient technology to differentiate hSKPs towards hepatic cells (Fig. 1). In essence, the protocol includes the sequential exposure of hSKPs to the hepatogenic factors activin A, bone morphogenetic protein (BMP)-4, fibroblast growth factor (FGF)-4, hepatocyte growth factor (HGF), dexamethasone, insulin-transferrin-sodium selenite (ITS), and oncostatin M (OSM), mimicking the liver embryogenesis in vivo [1].

2 Materials

2.1 General Equipment

1. Laminar air flow cabinet biosafety level class II.
2. 0.4 % trypan blue in 0.81 % NaCl and 0.06 % KH_2PO_4 solution.
3. Sterile plasma-treated tissue culture flasks made of high-quality polystyrene.
4. Sterile 96-, 24-, or 6-well plates made of high-quality polystyrene.
5. Thermostated bath at 37 °C.

2.2 Isolation, Dissociation, and Subcultivation of Primary hSKP Cultures

1. Hank's Balanced Salt Solution.
2. hSKP dissociation solution. 0.2 mg/mL Liberase Blendzyme DH (Roche, Belgium) in Hank's Balanced Salt Solution. The solution must be freshly prepared on ice.
3. DNase I solution. 2 mg/mL DNase I (Sigma-Aldrich, Belgium) in Hank's Balanced Salt Solution.
4. Fetal bovine serum, heat inactivated for 30 min at 56 °C (Hyclone, Canada).
5. hSKP wash medium. Prepare three parts Dulbecco's Modified Eagle Medium with GlutaMAX and one part F12 Nutrient Mixture (Life Technologies, Belgium) and add 50 μg/mL streptomycin sulfate, 7.33 IU/mL benzyl penicillin, and 2.5 μg/mL fungizone (Life Technologies, Belgium). The medium is stored at 4 °C and is stable for 4 weeks.
6. 70 μm nylon cell strainer.
7. hSKP growth medium. Prepare three parts Dulbecco's Modified Eagle Medium with GlutaMAX and one part F12 Nutrient Mixture and add 50 μg/mL streptomycin sulfate, 7.33 IU/mL benzyl penicillin, 2.5 μg/mL fungizone, 2 % B27 Supplement (Life Technologies, Belgium), 40 ng/mL FGF-2, and 20 ng/mL epidermal growth factor (EGF). The medium is stored at 4 °C and is stable for 2 weeks.
8. hSKP feeding medium. Prepare three parts Dulbecco's Modified Eagle Medium with GlutaMAX and one part F12 Nutrient Mixture and add 50 μg/mL streptomycin sulfate, 7.33 IU/mL benzyl penicillin, 2.5 μg/mL fungizone, 10 % B27 Supplement, 200 ng/mL FGF-2, and 100 ng/mL EGF. The medium is stored at 4 °C and is stable for 2 weeks.
9. TrypLE™ Express enzyme (Life Technologies, Belgium).

2.3 Collagen Thin Coating

1. 100 μg/mL collagen type I (BD Biosciences, Belgium) in 0.02 N acetic acid. The solution must be freshly prepared on ice.
2. Phosphate-buffered saline (PBS). 137 mM NaCl, 2.7 mM KCl, 10 mM $Na_2HPO_4 \cdot 2H_2O$, 1.8 mM KH_2PO_4 in deionized water. Adjust to pH 7.4, sterilize by passing through a 0.22 μm filter and store for maximum 6 months at 4 °C.

2.4 Hepatic Differentiation of hSKPs

1. hSKP basal medium. Prepare three parts Dulbecco's Modified Eagle Medium with GlutaMAX and one part F12 Nutrient Mixture and add 50 μg/mL streptomycin sulfate, 7.33 IU/mL benzyl penicillin, 2.5 μg/mL fungizone, 0.1 mM l-ascorbic acid, 4 mg/L nicotinamide, 1 mg/mL linoleic acid–bovine serum albumin (Sigma-Aldrich, Belgium), and 27.3 mg/L sodium pyruvate. The medium is stored at 4 °C and is stable for 4 weeks.

2. Hepatic differentiation medium I. hSKP basal medium supplemented with 50 ng/mL activin A. This medium must be prepared ex tempore.
3. Hepatic differentiation medium II. hSKP basal medium supplemented with 25 ng/mL activin A, 5 ng/mL FGF-4, and 10 ng/mL BMP-4. This medium must be prepared ex tempore.
4. Hepatic differentiation medium III. hSKP basal medium supplemented with 10 ng/mL FGF-4 and 20 ng/mL BMP-4. This medium must be prepared ex tempore.
5. Hepatic differentiation medium IV. hSKP basal medium supplemented with 5 ng/mL FGF-4, 10 ng/mL BMP-4, 30 ng/mL HGF, and 0.5 % ITS. This medium must be prepared ex tempore.
6. Hepatic differentiation medium V. hSKP basal medium supplemented with 30 ng/mL HGF, 0.25 % ITS, and 20 μg/L dexamethasone. This medium must be prepared ex tempore.
7. Hepatic differentiation medium VI. hSKP basal medium supplemented with 20 ng/mL HGF and 20 μg/L dexamethasone. This medium must be prepared ex tempore.
8. Hepatic differentiation medium VII. hSKP basal medium supplemented with 20 ng/mL HGF, 20 μg/mL dexamethasone, and 10 ng/mL OSM. This medium must be prepared ex tempore.

3 Methods

3.1 Isolation of hSKPs

Unless stated otherwise, perform all steps on ice (i.e., 4 °C).

1. Collect fresh human skin tissue samples and preserve samples in Hank's Balanced Salt Solution on ice (i.e., 4 °C) until further use (*see* **Note 1**).
2. Cut the human skin tissue samples in small pieces of 5–8 cm^2 (*see* **Note 2**).
3. Bring each sample in a 10 cm tissue culture dish filled with 25 mL of hSKP dissociation solution.
4. Cut each sample in 3–5 mm^2 pieces and float them epidermis side up (*see* **Note 3**).
5. Incubate the samples overnight at 4 °C for approximately 20 h (*see* **Note 4**).
6. Remove the epidermis from the samples and discard it.
7. Cut the remaining dermis into smaller pieces (i.e., 1–2 mm^2) and transfer, including hSKP dissociation solution, to a 50 mL tube.

8. Incubate the samples for 20 min in a thermostated bath at 37 °C (*see* **Note 5**).
9. Add 400 μL of 2 mg/mL DNase I solution to the 50 mL tube and incubate exactly 1 min in a thermostated bath at 37 °C (*see* **Note 6**).
10. Add 10 % fetal bovine serum to the tube to inactivate DNase I and the enzymes present in the hSKP dissociation solution and invert the tube 10–15 times.
11. Incubate on ice (i.e., 4 °C) for 5 min or wait until all large pieces of dermal tissue are sedimented by gravitation (*see* **Note 7**).
12. Gently discard the supernatant, but leave 2 mL and transfer to a 15 mL tube.
13. Start grinding the samples for about 2 min with a sterile 10 mL pipette and, when finished, add 8 mL of hSKP wash medium.
14. Spin for 10 s at 200 × *g* and room temperature to pellet down the large sample pieces.
15. Collect supernatant over a 70 μm cell strainer in a 50 mL tube and keep on ice (i.e., 4 °C) at all times (*see* **Note 8**).
16. Repeat steps 13–15 five times or until all tissue has been processed.
17. Spin the collected supernatant for 6 min at 200 × *g* and room temperature.
18. Remove the remaining medium and add 1 mL of hSKP growth medium.
19. Resuspend the cell pellet with a 1,000 μL pipette tip by triturating 50–100 times (*see* **Note 9**).
20. Use 5 μL of cell suspension and 45 μL of trypan blue solution to count the cells and estimate the cell viability.
21. For a 5–8 cm^2 skin sample, a yield of $5–15 \times 10^6$ cells is typically obtained with a viability between 30 and 80 % (*see* **Note 10**).
22. Use 30 mL of hSKP growth medium in a T75 flask at a density of 50,000 cells/mL.
23. Feed the primary hSKP cultures twice a week by replenishing the medium with 6 mL of fresh hSKP feeding medium until sufficient spheres have formed.
24. Primary hSKP spheres will be visible from day 5–7 onwards (*see* **Note 11**).

3.2 Dissociation of Primary hSKP Spheres

Unless stated otherwise, perform all steps at room temperature.

1. Gently tap the flask to detach any slightly adherent primary hSKP spheres from the surface.
2. Collect the medium and spheres from the culture flask, but discard any strong adherent cell populations.

3. Spin down for 5 min at 325 × g and room temperature.
4. Remove the supernatant and resuspend the spheres in 0.5 mL hSKP dissociation solution.
5. Incubate for exactly 12 min in a thermostated bath at 37 °C.
6. Dissociate spheres by triturating approximately 50 times with a 1,000 μL pipette tip.
7. Add 1 mL of fetal bovine serum and 10 mL of prewarmed (i.e., 37 °C) hSKP wash medium to the dissociated sphere suspension.
8. Spin down for 5 min at 325 × g and room temperature.
9. Remove the supernatant and resuspend the cell pellet in 1 mL of prewarmed (i.e., 37 °C) hSKP wash medium.
10. Use 5 μL cell suspension and 45 μL of trypan blue solution to count the cells and estimate the cell viability.
11. To create hSKP monolayer cultures, typically culture 4×10^3 viable cells/cm^2 in hSKP growth medium supplemented with 5 % fetal bovine serum.
12. The next day, discard the medium and add fresh hSKP growth medium without fetal bovine serum (*see* **Note 12**).
13. Change medium every 2–3 days with hSKP growth medium without fetal bovine serum until 90 % of confluency is reached.

3.3 Subcultivation of hSKPs for Hepatic Differentiation

Unless stated otherwise, perform all steps at room temperature.

1. Discard the medium and add 2.5, 5, or 10 mL of prewarmed (i.e., 37 °C) hSKP wash medium to a T25, T75, or T175 culture flask, respectively.
2. Remove hSKP wash medium and discard.
3. Add 2.5, 5, or 10 mL TrypLE™ Express enzyme to a T25, T75, or T175 culture flask, respectively, and incubate for exactly 5 min at 37 °C.
4. Tap several times to detach cells.
5. Add 10, 20, or 30 mL prewarmed hSKP wash medium to the T25, T75, or T175 culture flask, respectively, and resuspend 15 times with a 10 mL pipette and collect the cell suspension in a 50 mL tube.
6. Centrifuge for 5 min at 325 × g and room temperature.
7. Discard supernatant and resuspend the cell pellet in 1 mL of prewarmed (i.e., 37 °C) hSKP wash medium.
8. Use 5 μL cell suspension and 45 μL of trypan blue solution to count the cells and estimate the cell viability.
9. Typically culture 1.3×10^4 viable cells/cm^2 in hSKP growth medium supplemented with 5 % fetal bovine serum on a collagen thin coated surface.

10. The next day, discard medium and add fresh hSKP growth medium without fetal bovine serum.
11. Hepatic differentiation is started when approximately 90 % of cell confluency is reached.

3.4 Collagen Thin Coating

1. Add 50, 250, or 1.200 μL of 100 μg/mL collagen type I solution per well of a 96-well plate (i.e., 0.3 cm^2), a 24-well plate (i.e., 2.0 cm^2), or a 6-well plate (i.e., 9.6 cm^2), respectively.
2. Swirl the plates to cover the entire area.
3. Incubate the plates for 1 h at room temperature in a laminar air flow cabinet.
4. Discard the collagen type I solution carefully.
5. Wash three times with cold (i.e., 4 °C) PBS to remove any acidic remains.
6. The coated plates can be used immediately or may be air-dried for later use. They can be stored at 4 °C for up to 1 week under sterile conditions.

3.5 Hepatic Differentiation of hSKPs

1. Plate 1.3×10^4 viable cells/cm^2 in hSKP growth medium supplemented with 5 % fetal bovine serum on a collagen thin coated surface.
2. The next day, discard medium and add fresh hSKP growth medium without fetal bovine serum.
3. Refresh media every 2 days until approximately 90 % of cell confluency is reached.
4. Remove hSKP growth medium and initiate hepatic differentiation (i.e., day 0) by exposing the cells for 24 days at 33 °C to hepatogenic factors (*see* **Note 13**). Change media every 2–3 days unless specified otherwise according to the following scheme:
   - Day 0: incubate cells with hepatic differentiation medium I.
   - Days 1 and 2: incubate cells with hepatic differentiation medium II.
   - Days 3–5: incubate cells with hepatic differentiation medium III.
   - Days 6–8: incubate cells with hepatic differentiation medium IV.
   - Days 9–11: incubate cells with hepatic differentiation medium V.
   - Days 12–14: incubate cells with hepatic differentiation medium VI.
   - From day 15 onwards: incubate cells with hepatic differentiation medium VII.

4 Notes

1. Freshly collected human skin tissue samples can be stored at 4 °C in Hank's Balanced Salt Solution for at least 8 h without significant degradation of the stem cell compartment.
2. If necessary, gently scrape off excessive adipose tissue or blood from the skin tissue samples with a forceps to reduce the amount of contaminating cells after isolation.
3. Floating the skin tissue samples epidermis side up allows to efficiently remove the epidermis after overnight incubation. More specifically, gravity, in combination with the dissociation solution, will gently dissociate the epidermis from the dermis. Failing to do so will significantly complicate the removal of the obsolete epidermis and will increase the amount of contaminating epidermis-derived cells in the subsequent hSKP cultures.
4. The overnight incubation time can vary according to the origin of the skin tissue samples. Samples obtained from human foreskin or eyelid will typically be incubated for 20 h, whereas samples obtained from the abdomen will require longer incubation times to efficiently dissociate the epidermis from the dermis with a maximum of 48 h. In addition, the phototype (i.e., I, II, III, or IV) of the skin samples also significantly influences the incubation time. More specifically, samples obtained from individuals that have a darker skin tone (i.e., phototype VI) will require longer incubation times (i.e., more than 25 h) compared to lighter skin tone samples (i.e., phototype I, II, and III) (i.e., less than 20 h).
5. The incubation time depends on the amount of tissue that needs to be dissociated. A typical skin tissue sample of approximately 5 cm^2 will require 20 min, whereas larger tissue samples can require up to 45 min to be sufficiently dissociated for further processing.
6. DNAse I is used to remove contaminating DNA that originates from disrupted cells due to the processing and dissociation of the skin tissue.
7. If necessary, a short centrifugation step (i.e., 10 s) at 200 × *g* may be performed to obtain complete sedimentation.
8. Keeping the cells on ice (i.e., 4 °C) maintains the viability of the isolated stem cells at a higher level.
9. Try to avoid excessive foam formation while resuspending the cells, as this will significantly decrease the viability of the isolated stem cells.
10. The cell viability, based upon trypan blue exclusion, is highly dependent on the presence of contaminating red blood cells and cell debris (*see* **Notes 2** and **3**).

11. Cell aggregates can be found in the primary cultures that are obtained before 5–7 days after isolation. These are not multipotent hSKP spheres and look more like cell clumps.
12. Fetal bovine serum is added to the culture for 24 h in order to make hSKPs adherent to the surface of the culture flask.
13. Hepatic differentiation of hSKPs is performed at 33 °C instead of 37 °C in order to suppress proliferation and to promote differentiation.

Acknowledgements

This work has received funding from Grants of the Fund for Scientific Research in Flanders (FWO-Vlaanderen), the Institute for the Promotion of Innovation through Science and Technology in Flanders (IWT-Vlaanderen), the Research Council (OZR) of the Vrije Universiteit Brussel, Wetenschappelijk Fonds Willy Gepts from the University Hospital of the Vrije Universiteit Brussel (UZ Brussel) and from BRUSTEM-2, an impulse programme of the Institute for the encouragement of Scientific Research and Innovation of Brussels (INNOVIRIS), HEPRO-2, an Interuniversity Attraction Pole programme of the Belgian Science Policy Office (BELSPO), the European Community's Seventh Framework Programme (FP7/2007–2013) under grant agreement No. 20161 (ESNATS), No. 266838 (DETECTIVE), and No. 266777 (HEMIBIO).

References

1. Rodrigues RM, De Kock J, Branson S et al (2014) Human skin-derived stem cells as a novel cell source for *in vitro* hepatotoxicity screening of pharmaceuticals. Stem Cells Dev 23:44–55
2. Szkolnicka D, Farnworth SL, Lucendo-Villarin B et al (2014) Accurate prediction of drug-induced liver injury using stem cell-derived populations. Stem Cells Transl Med 3:141–148
3. Toma JG, McKenzie IA, Bagli D et al (2005) Isolation and characterization of multipotent skin-derived precursors from human skin. Stem Cells 23:727–737
4. Suflita MT, Pfaltzgraff ER, Mundell NA et al (2013) Ground-state transcriptional requirements for skin-derived precursors. Stem Cells Dev 22:1779–1788
5. Gago N, Perez-Lopez V, Sanz-Jaka JP et al (2009) Age-dependent depletion of human skin-derived progenitor cells. Stem Cells 27: 1164–1172
6. Buranasinsup S, Sila-Asna M, Bunyaratvej N et al (2006) *In vitro* osteogenesis from human skin-derived precursor cells. Dev Growth Differ 48:263–269
7. De Kock J, Najar M, Bolleyn J et al (2012) Mesoderm-derived stem cells: the link between the transcriptome and their differentiation potential. Stem Cells Dev 21:3309–3323
8. Hunt DPJ, Jahoda C, Chandran S (2009) Multipotent skin-derived precursors: from biology to clinical translation. Curr Opin Biotechnol 20:522–530
9. Shih DTB, Lee DC, Chen SC et al (2005) Isolation and characterization of neurogenic mesenchymal stem cells in human scalp tissue. Stem Cells 23:1012–1020
10. McKenzie IA, Biernaskie J, Toma JG et al (2006) Skin-derived precursors generate myelinating Schwann cells for the injured and dysmyelinated nervous system. J Neurosci 26:6651–6660

11. Biernaskie J, Sparling JS, Liu J et al (2007) Skin-derived precursors generate myelinating Schwann cells that promote remyelination and functional recovery after contusion spinal cord injury. J Neurosci 27: 9545–9559
12. Fernandes KJL, Kobayashi NR, Gallagher CJ et al (2006) Analysis of the neurogenic potential of multipotent skin-derived precursors. Exp Neurol 201:32–48
13. Lavoie JF, Biernaskie JA, Chen Y et al (2009) Skin-derived precursors differentiate into skeletogenic cell types and contribute to bone repair. Stem Cells Dev 18:893–906
14. De Kock J, Snykers S, Ramboer E et al (2011) Evaluation of the multipotent character of human foreskin-derived precursor cells. Toxicol In Vitro 25:1191–1202
15. Steinbach SK, El-Mounayri O, DaCosta RS et al (2011) Directed differentiation of skin-derived precursors into functional vascular smooth muscle cells. Arterioscler Thromb Vasc Biol 31:2938–2948
16. De Kock J, Vanhaecke T, Biernaskie J et al (2009) Characterization and hepatic differentiation of skin-derived precursors from adult foreskin by sequential exposure to hepatogenic cytokines and growth factors reflecting liver development. Toxicol In Vitro 23:1522–1527

Chapter 9

Generation of Hepatocytes from Pluripotent Stem Cells for Drug Screening and Developmental Modeling

Richard L. Gieseck III, Ludovic Vallier, and Nicholas R.F. Hannan

Abstract

Hepatocytes produced from the differentiation of human pluripotent stem cells can be used to study human development and liver disease, to investigate the toxicological response of novel drug candidates, and as an alternative source of primary cells for transplantation therapies. Here, we describe a method to produce hepatocytes by differentiating human pluripotent stem cells into definitive endoderm, patterning definitive endoderm into anterior definitive endoderm, specifying anterior definitive endoderm into hepatic endoderm, and differentiating hepatic endoderm into immature hepatocytes. These cells are further matured in either two-dimensional or three-dimensional culture conditions to produce cells capable of metabolizing xenobiotics and generating liver-specific proteins, such as albumin and alpha 1 antitrypsin.

Key words Liver, Stem cells, Cytochrome P450, Albumin, Toxicity assay, Endoderm, Three-dimensional scaffold, Hepatoblast, Foregut, Functional hepatocytes

1 Introduction

Human pluripotent stem cells (hPSCs) consist of 2 classes, namely, embryonic stem cells (hESCs) and induced pluripotent stem cells (hiPSCs), both which are characterized by their ability to self-renew and to differentiate into cells of all 3 germ layers [1, 2]. Since their first derivation, these cells have been successfully differentiated into multiple lineages to produce cells that are otherwise rare or difficult to isolate from patients, including hepatocytes [3], pancreatic cells [4], neurons [5], and cardiomyocytes [6], among others. Additionally, hPSCs derived from patients with genetic disorders allow the unique opportunity to study the pathogenesis and mechanisms of human disease in vitro [7, 8].

Mathieu Vinken and Vera Rogiers (eds.), *Protocols in In Vitro Hepatocyte Research*, Methods in Molecular Biology, vol. 1250, DOI 10.1007/978-1-4939-2074-7_9, © Springer Science+Business Media New York 2015

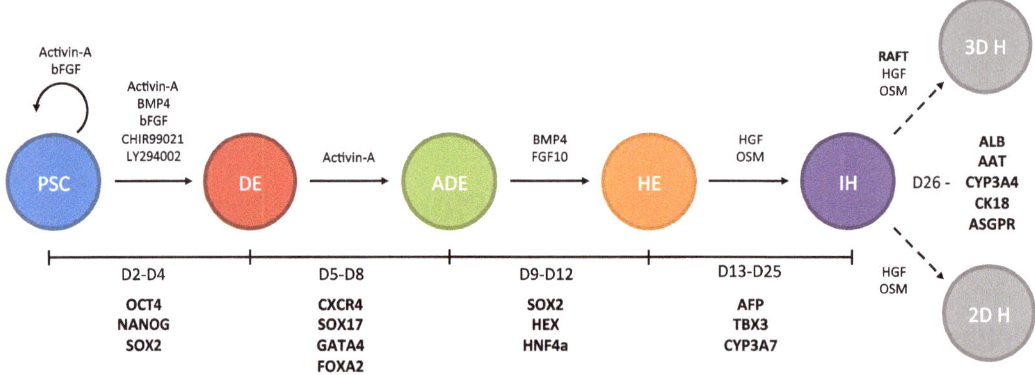

Fig. 1 Overview of the differentiation protocol. Human pluripotent stem cells (PSCs) are differentiated in a stepwise fashion to obtain hepatocytes. The pluripotent stem cells are first differentiated into definitive endoderm (DE) expressing CXC chemokine receptor 4 (CXCR4), SRY-related high-mobility-group box 17, GATA binding protein 4 (GATA4), and forkhead box A2 (FOXA2). This population is patterned into anterior definitive endoderm (ADE), a population marked by the upregulation of SRY-related high-mobility-group box 2 (SOX2), hematopoietically expressed homeobox protein (HEX), and hepatocyte nuclear factor 4 alpha (HNF4a). The anterior definitive endoderm is patterned into hepatic endoderm (HE) in order to produce immature hepatocytes (IH) with a fetal identity. These cells can be further matured in either two-dimensional (2D) or three-dimensional (3D) conditions to produce hepatocytes (H), which express several markers of maturation, including albumin (ALB), alpha 1 antitrypsin (AAT), cytochrome P450 3A4 (CYP3A4), cytokeratin 18 (CK18), and asialoglycoprotein receptor (ASGPR). The three-dimensional hepatocyte population also expresses markers of hepatocyte polarization, including multidrug resistance-associated protein 2 and dipeptidyl peptidase 4. (*AFP* alpha fetoprotein, *bFGF* basic fibroblast growth factor, *BMP4* bone morphogenetic factor 4, *CYP3A7* cytochrome P450 3A7, *FGF10* fibroblast growth factor 10, *HEX* hematopoietically expressed homeobox protein, *HGF* hepatocyte growth factor, *NANOG* homeobox transcription factor, *OCT4* octamer-binding transcription factor 4, *OSM* oncostatin M, *RAFT* real architecture for 3D tissue, *SOX17* SRY-related high-mobility-group box 17, *TBX3* T-box 3)

In this chapter, we describe a multistep protocol for producing hPSC-derived hepatocytes by mimicking the developmental changes occurring during natural development of the liver [9] (Fig. 1). hPSCs are first differentiated into definitive endoderm, the common progenitor of the thyroid, lung, pancreas, gut, and liver. The resulting progenitor cells are further differentiated by patterning into a population known as the anterior definitive endoderm, blocking this population from becoming cells of the gut. The anterior definitive endoderm population is then specified into hepatic endoderm that contains hepatoblasts, which are competent to become cholangiocytes (i.e., the epithelium of the bile ducts within the liver) and hepatocytes. The early-stage hepatocytes produced from the hepatic endoderm represent a cell type phenotypically similar to hepatocytes found within human fetuses, expressing the fetal proteins alpha fetoprotein and cytochrome P450 (CYP) 3A7 (Fig. 2). These fetal-like cells can be matured in 1 of 2 ways

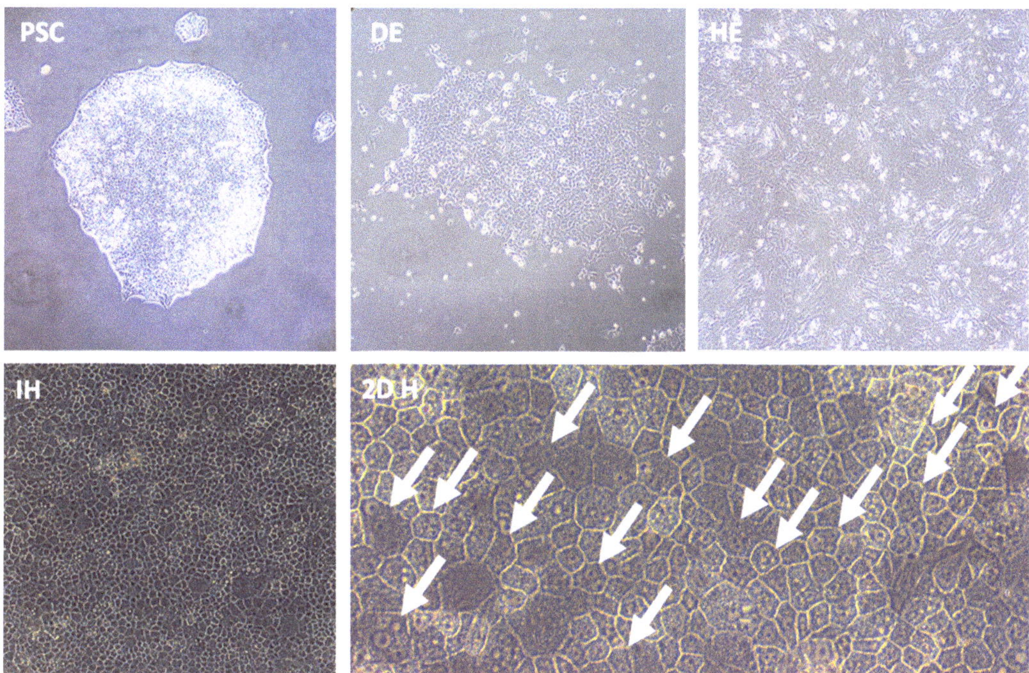

Fig. 2 Morphological progression to immature hepatocytes. Human pluripotent stem cells (PSCs) grow as compact colonies of cells with very high nuclear to cytoplasmic ratios. As PSCs differentiate into definitive endoderm (DE), cells enlarge and migrate outward from the center of the colony. Cells continue to proliferate, reach confluence, and develop a fibroblastic morphology during the anterior definitive endoderm and hepatic endoderm (HE) stages. Immature (IH) and mature hepatocytes (H) exhibit a polygonal morphology with clear cell–cell boundaries. Mature hepatocytes have prominent nucleoli and sometimes contain multiple nuclei, marked by arrows. (2D, two-dimensional)

depending on the intended end use of the cells. For cells being used in short-term experiments that do not rely upon the polarization of hepatocytes for active xenobiotic transport, in situ maturation on two-dimensional plastic cell culture plates, on which the initial differentiation occurred, is most efficient. These cells downregulate expression of the fetal markers alpha fetoprotein and CYP3A7, and simultaneously express several markers associated with mature hepatocytes, such as albumin, alpha 1 antitrypsin, and low levels of CYP3A4. However, these cells lack expression of several drug transporters and polarization markers, including multidrug resistance-associated protein 2, and also have a limited lifetime. Dissociated cells matured in three-dimensional collagen matrices spontaneously polarize and can be maintained in culture for much longer culture periods compared to cells matured in a two-dimensional configuration [10] (Fig. 3). Overall, this method provides a robust and efficient approach for producing mature hepatocytes for modeling embryonic development, disease studies, and toxicological investigations.

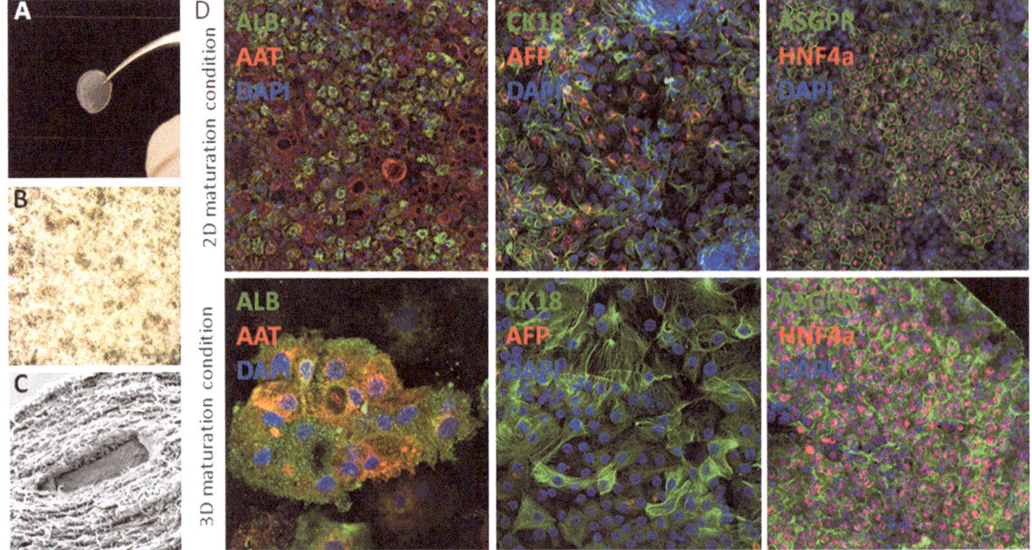

Fig. 3 Progression to maturing hepatocytes. (**a**) Three-dimensional (3D) hepatocyte culture removed from a 96-well plate. (**b**) Micrograph of the clump nature of hepatocytes in three-dimensional culture. (**c**) Electron micrograph demonstrating the density and porosity of the three-dimensional collagen matrix surrounding a hepatocyte. (**d**) Immunofluorescent micrographs demonstrating the typical expression and localization of several hepatic markers in both two-dimensional (2D) and three-dimensional maturation conditions. (*AAT* alpha 1 antitrypsin, *AFP* alpha fetoprotein, *ALB* albumin, *ASGPR* asialoglycoprotein receptor, *CK18* cytokeratin 18, *DAPI* diamidinophenylindole, *HNF4a* hepatocyte nuclear factor 4 alpha)

2 Materials

2.1 General Equipment

1. Inverted microscope.
2. Horizontal laminar flow hood or biosafety cabinet suitable for cell culture.
3. Incubator with control of humidity, temperature (i.e., 37 °C), CO_2 (i.e., 5 %), and O_2 (i.e., 5 % and ambient).
4. 10 cm diameter, 6-well, 12-well, 24-well, and 96-well cell culture plates.
5. 15 and 50 mL tubes.
6. Gelatin/mouse embryonic fibroblast-coated cell culture plates.
7. Real architecture for 3D tissue starter kit with 96-well plates (TAP Biosystems, United Kingdom).

2.2 Cell Culture

1. Calcium-free and magnesium-free Dulbecco's phosphate-buffered saline (PBS) (Gibco, United Kingdom).
2. Activin A (R&D Systems, United Kingdom).
3. Basic fibroblast growth factor (bFGF) (R&D Systems, United Kingdom).
4. Bone morphogenetic factor 4 (BMP4) (R&D Systems, United Kingdom).

5. Oncostatin M (OSM) (R&D Systems, United Kingdom).
6. Hepatocyte growth factor (HGF) (Peprotech, United Kingdom).
7. Fibroblast growth factor 10 (FGF10) (Autogen Bioclear, United Kingdom).
8. 4 % paraformaldehyde (Alfa Aesar, United Kingdom).
9. Donkey serum (Serotec, United Kingdom).
10. Cell dissociation buffer enzyme-free Hanks-based (*see* **Note 1**).
11. Serum-containing medium for coating cell culture plates. Mix 450 mL Dulbecco's Modified Eagle's Medium/F12, 50 mL fetal bovine serum (Biosera, United Kingdom), 5 mL L-GLUTAMINE, 5 mL penicillin/streptomycin solution (Gibco, United Kingdom), and 3.5 μL beta mercaptoethanol (Sigma-Aldrich, United Kingdom). Store at 4 °C for up to 1 month (*see* **Note 2**).
12. Chemically defined medium with bovine serum albumin for maintenance of hESCs. Mix 250 mL Dulbecco's Modified Eagle's Medium/F12, 250 mL Iscove's Modified Dulbecco's Medium, 5 mL concentrated lipids (Gibco, United Kingdom), 20 μL 97 % thioglycerol (Sigma-Aldrich, United Kingdom), 350 μL insulin (Roche, United Kingdom), 250 μL transferrin (Roche, United Kingdom), and 5 mL penicillin/streptomycin solution. Add 2.5 g bovine serum albumin fraction V (Sigma-Aldrich, United Kingdom). Store at 4 °C for up to 1 month (*see* **Note 3**).
13. Chemically defined medium with polyvinylalcohol for maintenance of hiPSCs and differentiation to endoderm. Mix 250 mL Dulbecco's Modified Eagle's Medium/F12, 250 mL Iscove's Modified Dulbecco's Medium, 5 mL concentrated lipids, 20 μL 97 % thioglycerol, 350 μL insulin, 250 μL transferrin, and 5 mL penicillin/streptomycin solution. Add 0.5 g polyvinylalcohol (Sigma-Aldrich, United Kingdom). Store at 4 °C for up to 1 month (*see* **Note 4**).
14. Roswell Park Memorial Institute-B27 differentiation medium for differentiation of anterior definitive endoderm and hepatoblast specification. Mix 490 mL Roswell Park Memorial Institute Medium 1640, 10 mL B27 supplement containing insulin (Gibco, United Kingdom), 5 mL Minimal Essential Medium nonessential amino acids (Gibco, United Kingdom), and 5 mL penicillin/streptomycin solution. Store at 4 °C for up to 3 weeks.
15. Gelatin solution for coating cell culture plates. Dissolve 0.5 g porcine gelatin (Sigma-Aldrich, United Kingdom) in 500 mL embryo transfer water. Heat to 56 °C until the gelatin dissolves. Cool to room temperature (*see* **Note 5**).
16. Collagenase for dissociation of hESC and hiPSC colonies. Mix 400 mL Dulbecco's Modified Eagle's Medium/F12, 100 mL Knock-out Serum Replacer Medium (Life Technologies, United Kingdom), 5 mL L-glutamine, 3.5 μL beta mercaptoethanol. Add 500 mg collagenase type IV (Invitrogen, United Kingdom). Mix by inversion. Store at 4 °C and use within 4 weeks (*see* **Note 6**).

17. HepatoZYME serum-free medium with supplements. Mix 500 mL HepatoZYME serum-free medium (Gibco, United Kingdom), 5.3 mL L-glutamine, 10.6 mL Minimal Essential Medium nonessential amino acids, 10.6 mL concentrated lipids, 750 μL insulin, 1.5 mL transferrin, and 5.3 mL penicillin/streptomycin solution. Store at 4 °C in the dark and use within 4 weeks.
18. Dispase for dissociation of hiPSCs. Add 500 mg dispase (Gibco, United Kingdom) to 500 mL Dulbecco's Modified Eagle's Medium. Store at 4 °C for up to 1 month (*see* **Note 7**).
19. Dispase/collagense for passaging hiPSCs or hESCs. Mix dispase and collagenase at a ratio of 1:1. Warm to 37 °C prior to use.
20. 10 mM LY294002 (Promega, United Kingdom) solution in dimethylsulfoxide.
21. 10 mM CHIR99021 (StemGent, United Kingdom) in dimethylsulfoxide (*see* **Note 8**).

2.3 Cell Characterization

1. Alpha fetoprotein primary antibody (Dako, United Kingdom).
2. Albumin primary antibody (R&D Systems, United Kingdom).
3. Alpha 1 antitrypsin primary antibody (Dako, United Kingdom).
4. CXC chemokine receptor 4 primary antibody (R&D Systems, United Kingdom).
5. SRY-related high-mobility-group box 17 primary antibody (R&D Systems, United Kingdom).
6. Hepatocyte nuclear factor 4 alpha primary antibody (Santa Cruz, United States of America).
7. Cytokeratin 18 primary antibody (AbCam, United States of America).
8. Dipeptidyl peptidase 4 primary antibody (AbCam, United States of America).
9. Multidrug resistance protein 2 primary antibody (AbCam, United States of America).
10. Alexa Fluor 488 secondary antibody (Life Technologies, United Kingdom).
11. Alexa Fluor 647 secondary antibody (Life Technologies, United Kingdom).
12. Periodic acid Schiff kit (Sigma-Aldrich, United Kingdom).
13. Indocyanine green (Sigma-Aldrich, United Kingdom).
14. Low density lipoprotein kit (Cayman Chemicals, United Kingdom).
15. Albumin blue (Active Motif, United Kingdom).
16. Alpha 1 antitrypsin enzyme-linked immunosorbent assay kit (AbCam, United Kingdom).
17. P450 Glo Assay (Promega, United Kingdom).
18. Primers for quantitative reverse transcriptase polymerase chain reaction analysis (Table 1).

Table 1
Primer sequences for differentiation validation

Gene	Full name	Accession	Forward	Reverse
AAT	Homo sapiens serpin peptidase inhibitor (alpha 1 antiproteinase, antitrypsin), member 1, mRNA	NM_000295.4	CCATTGCTGAAGACCTTAGTGATG	AGACCCTTTGAAGTCAAGCGACC
ABCB4	Homo sapiens ATP-binding cassette, subfamily B (MDR/TAP), member 4 (ABCB4), transcript variant A, mRNA	NM_000443.3	CGGGGACAGTGCTTCTCGAT	AACCCGGCTGTTGTCTCCAT
ABCC1	Homo sapiens ATP-binding cassette, subfamily C (CFTR/MRP), member 1 (ABCC1), mRNA	NM_004996.3	ATCGCTCACCCCTGTTCTCG	ACAGGCACGACTTGTTCCGA
ABCC2	Homo sapiens ATP-binding cassette, subfamily C (CFTR/MRP), member 2 (ABCC2), mRNA	NM_000392.3	AGAGTCTTCGTTCCAGACGCA	TCTGTGAGTACAAGGGCCAGC
ABCC4	Homo sapiens ATP-binding cassette, subfamily C (CFTR/MRP), member 4 (ABCC4), transcript variant 1, mRNA	NM_005845.3	TTGCTGCCGCTGACGTTTTT	GGGGAACCAAGGGTATTGCG
AFP	Homo sapiens alpha fetoprotein (AFP), mRNA	NM_001134.2	AGAACCTGTCACAAGCTGTG	GACAGCAAGCTGAGGATGTC
AHR	Homo sapiens aryl hydrocarbon receptor (AHR), mRNA	NM_001621.4	TGTGCCGAGTCCCATATCCG	GCAGGCGTGCATTAGACTGG
ALB	Homo sapiens albumin (ALB), mRNA	NM_000477.5	CCTTTGGCACAATGAAGTGGGTAACC	CAGCAGTCAGCCATTTCACCATAG
CYP1A1	Homo sapiens cytochrome P450, family 1, subfamily A, polypeptide 1 (CYP1A1), mRNA	NM_000499.3	AAACAGGGCCACATAGATGC	AGGGTCCTGGTTTGGCTAGT
CYP1A2	Homo sapiens cytochrome P450, family 1, subfamily A, polypeptide 2 (CYP1A2), mRNA	NM_000761.3	AGGTCAACCATGACCCAGAG	ATGGCCAGGAAGAGGAAGAT
CYP1B1	Homo sapiens cytochrome P450, family 1, subfamily B, polypeptide 1 (CYP1B1), mRNA	NM_000104.3	CACCAAGGCTGAGACAGTGA	GATGACGACTGGGCCTACAT
CYP2C19	Homo sapiens cytochrome P450, family 2, subfamily C, polypeptide 19 (CYP2C19), mRNA	NM_000769.1	AACTCCCTCCTGGCCCCACT	TGCAGCACCACCATGCGTTC

(continued)

Table 1
(continued)

Gene	Full name	Accession	Forward	Reverse
CYP2C9	Homo sapiens cytochrome P450, family 2, subfamily C, polypeptide 9 (CYP2C9), mRNA	NM_000771.3	GCCGGCATGGAGCTGTTTTTAT	GCCAGGCCATCTGCTCTTCTT
CYP3A4	Homo sapiens cytochrome P450, family 3, subfamily A, polypeptide 4 (CYP3A4), transcript variant 1, mRNA	NM_017460.5	TGTGCCTGAGAACACCAGAG	GTGGTGGAAATAGTCCCGTG
CYP3A5	Homo sapiens cytochrome P450, family 3, subfamily A, polypeptide 5 (CYP3A5), mRNA	NM_000777.2	CCCCTTTGTGGAGAGCACTA	TCGTTGAGGCGACTTTTCTT
CYP3A7	Homo sapiens cytochrome P450, family 3, subfamily A, polypeptide 7 (CYP3A7), mRNA	NM_000765	ATTCCAAGCTATGTTCTTCATCAT	AATCTACTTCCCCAGCACTGA
FXR	Homo sapiens nuclear receptor subfamily 1, group H, member 4 (NR1H4), transcript variant 1, mRNA	NM_001206979.1	GCAAAGAGATGGGAATGTTGGCT	TGTCGAGGTCACTTGTCGCA
GAPDH	Homo sapiens glyceraldehyde-3-phosphate dehydrogenase (GAPDH), transcript variant 1, mRNA	NM_002046.4	GAGTCAACGGATTTGGTCGT	TTGATTTTGGAGGGATCTCG
GSTA4	Homo sapiens glutathione S-transferase alpha 4 (GSTA4), mRNA	NM_001512.3	GATGGGTTTTAGCTGCCGCC	GCAGATCCAGTGTCCCCTCC
GSTT1	Homo sapiens glutathione S-transferase theta 1 (GSTT1), mRNA	NM_000853.2	GTCGGTCGGTCCCCACTATG	GGCACCTTCTTGAGGGGGTT
HDAC	Homo sapiens histone deacetylase 2 (HDAC2), transcript variant 1, mRNA	NM_001527.3	AGTCAAGGAGGCGGCAAAA	TGCGGATTCTATGAGGCTTCA
HNF4	Homo sapiens hepatocyte nuclear factor 4, alpha (HNF4A), transcript variant 1, mRNA	NM_178849.2	CATGGCCAAGATTGACAACCT	TTCCCATATGTTCCTGCATCAG
HNMT1	Homo sapiens N-myristoyltransferase 1 (NMT1), mRNA	NM_021079.3	ACAATATCCGCCAGGAGCCC	GGATGTTTGCTGGGATGGCG
LXRα	Homo sapiens nuclear receptor subfamily 1, group H, member 3 (NR1H3), transcript variant 1, mRNA	NM_005693.3	GTCCCTCCCCTCAGCCTTTC	CTCCACCGCAGAGTCAGGAG

Gene	Full name	Accession	Forward	Reverse
MAOA	Homo sapiens monoamine oxidase A (MAOA), nuclear gene encoding mitochondrial protein, transcript variant 1, mRNA	NM_000240.3	CGGAGGTGGCATTTCAGGACT	AAGGCGCCCCGAAATGGATA
MAOB	Homo sapiens monoamine oxidase B (MAOB), nuclear gene encoding mitochondrial protein, mRNA	NM_000898.4	ACAAATGCGACGTGGTCGTG	TCCCGGGCTTCCAGAACAAC
MGST1	Homo sapiens microsomal glutathione S-transferase 1 (MGST1), transcript variant 5, mRNA	NM_001260511.1	TCGGCCTCACCACCAAAATTGA	TCAAGGTCATTCAGGTGGGCTC
MGST2	Homo sapiens microsomal glutathione S-transferase 2 (MGST2), transcript variant 1, mRNA	NM_002413.4	ATTGTGGATGGCTGGGTGGT	ACCTTCATGGTGGGAACAGCTT
NAT1	Homo sapiens N-acetyltransferase 1 (arylamine N-acetyltransferase) (NAT1), transcript variant 1, mRNA	NM_001160170.1	GGCCCAACAGGCTTTCTACC	GTGCCCTTGCCTTGAGTCTG
NNMT	Homo sapiens nicotinamide N-methyltransferase (NNMT), mRNA	NM_006169.2	GTCTACAATCCCTGGCGGCT	AAGGGAGCTTAGGCGAACGG
OCT1	Homo sapiens solute carrier family 22 (organic cation transporter), member 1 (SLC22A1), transcript variant 1, mRNA	NM_003057.2	GGTGTGTTCCTCCCTGTGTGA	AAAGCGACCCCCTTGGTCTC
PBDG	Homo sapiens hydroxymethylbilane synthase (HMBS), transcript variant 1, mRNA	NM_000190.3	GGAGCCATGTCTGGTAACGG	CCACGCGAATCACTCTCATCT
RPLP0	Homo sapiens ribosomal protein, large, P0 (RPLP0), transcript variant 1, mRNA	NM_001002.3	GGCGTCCTCGTGGAAGTGAC	GCCTTGCGCATCATGGTGTT
RXRα	Homo sapiens retinoid X receptor, alpha (RXRA), mRNA	NM_002957.4	GTCGCAGACATGGACACCAA	GGGGAGCTGATGACCGAGAAA
SLCO2B1	Homo sapiens solute carrier organic anion transporter family, member 2B1 (SLCO2B1), transcript variant 2, mRNA	NM_001145211.2	TCAGGGAGAGCCAGAGGTGA	GCTTTGCCTCCAGGTGTGTT
SULT1A1	Homo sapiens sulfotransferase family, cytosolic, 1A, phenol-preferring, member 1 (SULT1A1), transcript variant 1, mRNA	NM_001055.3	CTCAGAGAACAACCCTGCAT	GCTCCATGTTCCTGAGCTCTT

(continued)

Table 1
(continued)

Gene	Full name	Accession	Forward	Reverse
SULT1A2	Homo sapiens sulfotransferase family, cytosolic, 1A, phenol-preferring, member 2 (SULT1A2), transcript variant 1, mRNA	NM_001054.3	TTCCCCACACAACACCCACA	TTCACGTACTCCAGTGGCGG
TPMT	Homo sapiens thiopurine S-methyltransferase (TPMT), mRNA	NM_000367.2	AACCAGCTGTAAGCGAGGCA	TGCCGTTCACCCACTTGTCT
UGT1A1	Homo sapiens UDP glucuronosyltransferase 1 family, polypeptide A1 (UGT1A1), mRNA	NM_000463.2	TGATCCCAGTGGATGGCAGC	CAACGAGGCGTCAGGTGCTA
UGT1A6	Homo sapiens UDP glucuronosyltransferase 1 family, polypeptide A6 (UGT1A6), transcript variant 1, mRNA	NM_001072.3	GGAGCCCTGTGATTTGGAGAGT	GACCCCGGTCACTGAGAACC
UGT1A9	Homo sapiens UDP glucuronosyltransferase 1 family, polypeptide A9 (UGT1A9), mRNA	NM_021027.2	AGGAACATTTATTATGCCACCGTTT	GTTGGGCATCACGGGTTTGG
UGT2B4	Homo sapiens UDP glucuronosyltransferase 2 family, polypeptide B4 (UGT2B4), mRNA	NM_021139.2	AAAGGAGCAGCAACTGGAAAACA	TGGGCTGTTGGGATCGAAAGAA
UGT2B7	Homo sapiens UDP glucuronosyltransferase 2 family, polypeptide B7 (UGT2B7), mRNA	NM_001074.2	TCCCAACAACTCATCCGCTC	CAGAACTTTCTAGTTATGTCACCAA

3 Methods

3.1 Coating Cell Culture Plates with Gelatin and Serum

1. Remove the desired number of 10 cm diameter, 6-well, 12-well, 24-well, or 96-well cell culture plates from their packing and place inside a negative pressure laminar flow cell culture safety cabinet.
2. Coat the desired numbers of cell culture plates for maintenance and differentiation using gelatin solution at room temperature (i.e., approximately 18–25 °C) for 30 min.
3. Aspirate the gelatin and replace with serum-containing medium.
4. Place in an incubator at 37 °C and 5 % CO_2 for at least 24 h (*see* **Note 9**).

3.2 Maintenance of hPSCs

1. Maintain hPSCs in chemically defined medium containing 10 ng/mL activin A and 12 ng/mL bFGF in a humidified incubator at 37 °C, 5 % CO_2, and ambient O_2.
2. Replace the cell culture medium daily. All growth factors should be added directly to the medium prior to use (*see* **Note 10**).

3.3 Passing hPSCs (Day 0)

1. Passage the hPSCs when reaching 70–80 % confluency (*see* **Note 11**).
2. Wash the cells with PBS.
3. Aspirate the PBS and replace with 5 mL dispase/collagenase solution.
4. Place the cells in an incubator at 37 °C and 5 % CO_2 for 60 min or until colonies of cells detach from the cell culture plate and float in the dispase/collagenase solution (*see* **Note 12**).
5. Add 5 mL chemically defined medium to the plate containing dispase/collagenase and the floating colonies.
6. Transfer the floating colonies to a 15 mL tube along with all medium (i.e., 10 mL) from the cell culture plate.
7. Centrifuge at 213 × g and room temperature for 2 min.
8. Aspirate the supernatant and wash the cells with 10 mL chemically defined medium.
9. Centrifuge at 213 × g and room temperature for 2 min.
10. Aspirate the supernatant and resuspend the cells in the desired volume of chemically defined medium (*see* **Note 13**).

11. Remove serum-coated plates from the incubator at 37 °C, aspirate the medium, and wash with 10 mL PBS.
12. Aspirate PBS and add 10 mL chemically defined medium containing 10 ng/mL activin A and 12 ng/mL bFGF.
13. Resuspend clumps of hPSCs and add 1 mL cell suspension/plate (*see* **Note 14**).
14. Ensure even distribution of the cells by gently rocking the cell culture plate side to side and back and forth.
15. Place the cell culture plate in an incubator at 37 °C and 5 % CO_2.
16. Replace the cell culture medium after 24 h (*see* **Note 15**).

3.4 Differentiation of hPSCs into Definitive Endoderm (Day 2–4)

1. 48 h after plating the hPSCs, visually inspect the cell culture plates for correct cell density.
2. Prepare chemically defined medium with polyvinylalcohol containing 100 ng/mL activin A, 20 ng/mL bFGF, 10 ng/mL BMP4, and 10 μM LY294002 (*see* **Note 16**).
3. Aspirate the maintenance medium and replace with chemically defined medium with polyvinylalcohol containing 100 ng/mL activin A, 20 ng/mL bFGF, 10 ng/mL BMP4, and 10 μM LY294002.
4. The following day, prepare fresh chemically defined medium with polyvinylalcohol containing 100 ng/mL activin A, 20 ng/mL bFGF, 10 ng/mL BMP4, and 10 μM LY294002.
5. Aspirate the medium and replace with fresh chemically defined medium with polyvinylalcohol containing 100 ng/mL activin A, 20 ng/mL bFGF, 10 ng/mL BMP4, and 10 μM LY294002.
6. On the third day, prepare fresh chemically defined medium with polyvinylalcohol containing 100 ng/mL activin A, 20 ng/mL bFGF, 10 ng/mL BMP4, and 10 μM LY294002.
7. Aspirate the medium and replace with fresh chemically defined medium with polyvinylalcohol containing 100 ng/mL activin A, 20 ng/mL bFGF, 10 ng/mL BMP4, and 10 μM LY294002 (*see* **Note 17**).

3.5 Patterning of Definitive Endoderm into Anterior Definitive Endoderm (Day 5–8)

1. Prepare an appropriate volume of Roswell Park Memorial Institute-B27 differentiation medium containing 50 ng/mL activin A.
2. Aspirate the chemically defined medium with polyvinylalcohol containing 100 ng/mL activin A, 20 ng/mL bFGF, 10 ng/mL BMP4, and 10 μM LY294002, and replace with Roswell Park Memorial Institute-B27 differentiation medium containing 50 ng/mL activin A.
3. Repeat this process daily for 3 days (*see* **Note 18**).

3.6 Specification of Anterior Definitive Endoderm to Hepatic Endoderm (Day 9–12)

1. Prepare Roswell Park Memorial Institute-B27 differentiation medium containing 10 ng/mL BMP4 and 50 ng/mL FGF10.
2. Aspirate the Roswell Park Memorial Institute-B27 differentiation medium containing 50 ng/mL activin A and replace by Roswell Park Memorial Institute-B27 differentiation medium containing 10 ng/mL BMP4 and 50 ng/mL FGF10.
3. Repeat until day 12 (*see* **Note 19**).

3.7 Differentiation of Hepatic Endoderm into Immature Hepatocytes (Day 13–25)

1. Prepare HepatoZYME serum-free medium with supplements containing 50 ng/mL HGF and 10 ng/mL OSM.
2. Aspirate the Roswell Park Memorial Institute-B27 differentiation medium containing 10 ng/mL BMP4 and 50 ng/mL FGF10, and wash the cells with PBS.
3. Aspirate the PBS and replace with H HepatoZYME serum-free medium with supplements containing 50 ng/mL HGF and 10 ng/mL OSM.
4. Repeat every other day until day 25 (*see* **Note 20**).
5. At this point, continue to either 3.8. or 3.9 depending upon the desired use of the cells (*see* **Note 21**).

3.8 Maturation of Immature Hepatocytes in Two-Dimensional Culture (Day 26 to the Desired Endpoint)

1. Prepare HepatoZYME serum-free medium with supplements containing 50 ng/mL HGF and 10 ng/mL OSM.
2. Aspirate the medium and replace by HepatoZYME serum-free medium with supplements containing 50 ng/mL HGF and 10 ng/mL OSM.
3. Repeat every other day until the desired endpoint (*see* **Note 22**).

3.9 Maturation of Immature Hepatocytes in Three-Dimensional Culture (Day 26 to the Desired Endpoint)

1. Prepare HepatoZYME serum-free medium with supplements containing 50 ng/mL HGF and 10 ng/mL OSM.
2. Aspirate the medium and wash the cells with PBS.
3. Aspirate the PBS and wash the cells with the cell dissociation buffer.
4. Aspirate the cell dissociation buffer and replace with fresh cell dissociation buffer.
5. Place the cell culture plate in an incubator at 37 °C and 5 % CO_2 for 15 min.
6. Break the epithelial sheet apart by using the tip of a 5 mL pipette (*see* **Note 23**).
7. Add HepatoZYME serum-free medium with supplements to the cells to create a 1:1 mixture of HepatoZYME serum-free medium with supplements and cell dissociation buffer.

8. Pipette cells into 15 mL tubes and centrifuge at 213×*g* and room temperature for 5 min.
9. Resuspend the cells in the appropriate volume of HepatoZYME serum-free medium with supplements to the desired cell density (*see* **Notes 24** and **25**).
10. Place the cells on ice until needed.
11. Determine the total volume of three-dimensional matrix needed depending on the number of wells desired. Generally, this is 240 μL/well plus 10 % extra to ensure sufficient volume (*see* **Note 25**).
12. Calculate 10.0 % of the total matrix volume and pipette the calculated amount of 10× concentrated Minimal Essential Medium into a sterile beaker and keep on ice. 10× concentrated Minimal Essential Medium is provided with the Real architecture for 3D tissue starter kit.
13. Calculate 80.0 % of the total matrix volume and pipette the calculated amount of 2 mg/mL rat tail collagen type I solution into the beaker with the 10× concentrated Minimal Essential Medium. Collagen solution is provided with the Real architecture for 3D tissue starter kit.
14. Gently swirl until the mixture has a homogeneous yellow color and keep on ice.
15. Calculate 5.8 % of the total matrix volume and pipette the calculated amount of neutralizing solution into the 10× concentrated Minimal Essential Medium-collagen mixture. Neutralizing solution is provided with the Real architecture for 3D tissue starter kit.
16. Gently swirl until the mixture has a homogeneous orange color and keep on ice.
17. Calculate 4.2 % of the total matrix volume and pipette the cell suspension into the neutralized 10× concentrated Minimal Essential Medium-collagen mixture.
18. Gently swirl until the mixture is homogenized and keep on ice.
19. Pipette 240 μL of the mixed solution into each well of a 96-well plate on ice.
20. Gently tap the sides of the cell culture plate to ensure homogenization of the cells within the wells.
21. Place the cells on the cell culture plate heater at 37 °C for 15 min. The plate heater is provided with the Real architecture for 3D tissue starter kit.
22. Carefully apply the provided cell absorbers to the cell culture plate for 15 min, keeping the cell culture plate on the plate heater (*see* **Note 26**).

23. Remove the cell absorbers.
24. Quickly replace the medium with fresh HepatoZYME serum-free medium with supplements containing 50 ng/mL HGF and 10 ng/mL OSM every other day until the desired endpoint (*see* **Notes 27–29**).

4 Notes

1. All cell culture reagents listed are stored, reconstituted, and used as specified in the product sheets supplied by the manufacturers unless specifically stated. Kits used for functional validations should be used as specified by the manufacturers at or after day 25 of differentiation. All suppliers mentioned provide the actual products used to develop and optimize the protocol described in this chapter. We recommend that initially the same products and suppliers are used before changing to alternative suppliers.
2. Sterilize the serum-containing medium using a 0.22 μm filtration device. Serum coating can be replaced by 15 g/mL human fibronectin (Millipore, United Kingdom) to remove all animal products from the system.
3. Sterilize the chemically defined medium with bovine serum albumin using a 0.22 μm filtration device. The medium should be warmed to 37 °C prior to use. Bovine serum albumin batches must be screened for their ability to support the undifferentiated growth on hESCs as well as the attachment of cells and proliferation following transfection. Not all batches of bovine serum albumin are suitable and must therefore be prescreened. Suitable suppliers include Sigma-Aldrich (United Kingdom), PAA Laboratories (United Kingdom), and Europa Bioproducts (United Kingdom). Chemically defined medium with bovine serum albumin should be compared to standard feeder cultures with particular emphasis on the maintenance of colony morphology and undifferentiated growth characteristics as well as continued expression of POU domain class 5 transcription factor 1, homeobox transcription factor NANOG, and SRY-related high-mobility-group box 2 with early differentiation markers, such as caudal type homeobox 2, Brachyury, mix1 homeobox-like 1, and SRY-related high-mobility-group box 17 at levels comparable to the feeder cultures.
4. Polyvinylalcohol needs to be dissolved in 50 mL of Iscove's Modified Dulbecco's Medium by adding 0.5 g polyvinylalcohol to a 50 mL tube with Iscove's Modified Dulbecco's Medium and placing on a tube roller overnight at 4 °C. Sterilize the chemically defined medium with polyvinylalcohol using a 0.22 μm filtration device. The medium should be warmed to 37 °C prior to use.

5. Sterilize the gelatin solution using a 0.22 μm filtration device. The gelatin solution may be stored for 4–8 weeks at room temperature.
6. Sterilize collagenase using a 0.22 μm filtration device and aliquot into 50 mL tubes. Collagenase should be warmed to 37 °C prior to use.
7. Sterilize dispase using a 0.22 μm filtration device and aliquot into 50 mL tubes. Dispase should be warmed to 37 °C prior to use.
8. Check that LY294002 and CHIR99021 are well dissolved, aliquot, and then store at −20 °C.
9. Serum-coated cell culture plates must be prepared at least 24 h prior to use and can be stored at 37 °C for 7–14 days. Do not let serum-coated cell culture plates dry out.
10. Differentiation efficacy can be increased by splitting hPSCs grown on feeders or in chemically defined medium directly to culture conditions inductive for mesoderm differentiation. Fluorescence-assisted cell sorting analysis can be performed to confirm that 80–90 % of the cells express the pluripotency markers octamer-binding transcription factor 4 and tumor rejection antigen 1–60. Growth factors cannot be kept at 4 °C for longer than 1 week. HepatoZYME serum-free medium with supplements cannot be kept more than 4 weeks at 4 °C. All growth factors should be added directly to the fresh medium each day. Making bulk batches of medium containing growth factors is not advisable unless all medium will be used within 2–3 days.
11. Cells should be at least 70 % confluent to ensure that there are sufficient cells for maintenance as well as for differentiation. Typical volumes are 10 mL for 10 cm diameter cell culture plates, 1 mL/well for 6-well plates, 0.5 mL/well for 12-well plate, 300 μL/well for 24-well plates, and 150 μL/well for 96-well plates. The volumes given in the methodological section of this chapter have been optimized for 10 cm diameter cell culture plates. Adjust volumes accordingly when using other cell culture plate formats.
12. The dissociation time can vary from cell line to cell line. Cells should be checked every 20 min to ensure that cells are not overdigested.
13. hPSCs cannot be split according to conventional single cell methods and must be passaged as clumps. A ratio of growth surface area is used as a guide for splitting. For example, when splitting cells at a ratio 1:10 using 10 cm diameter cell culture plates, cells from a single 10 cm diameter cell culture plate could be divided evenly among 10 new 10 cm diameter cell culture plates. When resuspending cells, a volume of 10 mL

would be used, thus 1 mL of cells could be added to each of the new 10 cm diameter cell culture plates.

14. Large clumps of undifferentiated cells should be broken down into clumps of approximately 20–50 cells. This can be achieved using a 10 mL pipette and pipetting the whole volume of cells and medium up and down 6–12 times until the desired clump size is achieved.

15. Successful splitting can be determined by either quantitative reverse transcriptase polymerase chain reaction analysis or immunocytochemistry for pluripotency markers POU domain class 5 transcription factor 1, the homeobox transcription factor NANOG and SRY-related high-mobility-group box 2, and the absence of differentiation markers, such as caudal type homeobox 2, mix1 homeobox-like 1, SRY-related high-mobility-group box 17, CXC chemokine receptor 4, and paired box 6. More than 90 % of the cells should be positive for POU domain class 5 transcription factor 1 and the homeobox transcription factor NANOG at this stage. The cells surface marker and tumor rejection antigen 1–60 can also be used as a marker of pluripotency. However, its expression is not directly linked to the pluripotent state.

16. When differentiating hiPSCs, 10 μM CHIR99021 must be added on the first day only. After 24 h, undifferentiated colonies should have undergone significant morphological changes, including the migration of the outermost cells away from the undifferentiated colony. The colony of cells itself should have also begun to separate into individual cells. Quantitative reverse transcriptase polymerase chain reaction analysis and immunocytochemistry can be performed after 24 h of culture to demonstrate downregulation of POU domain class 5 transcription factor 1, the homeobox transcription factor NANOG and SRY-related high-mobility-group box 2, and upregulation of mix1 homeobox-like 1, Brachyury, and eomesodermin. Almost 90 % of the cells should be positive for Brachyury at this stage.

17. When differentiating hiPSCs on day 3, use Roswell Park Memorial Institute-B27 differentiation medium supplemented with 100 ng/mL activin A and 100 ng/mL bFGF. After 3 full days of differentiation into definitive endoderm, cells should be forming a monolayer. Quantitative reverse transcriptase polymerase chain reaction analysis and immunocytochemistry can be performed to show complete downregulation of POU domain class 5 transcription factor 1, the homeobox transcription factor NANOG and SRY-related high-mobility-group box 2, loss of Bachyury, mix1 homeobox-like 1 and eomesodermin, and upregulation of CXC chemokine receptor 4, SRY-related high-mobility-group box 17, GATA binding protein 4, forkhead box A2, and hepatocyte nuclear factor 4 alpha.

Cells should be at least 70–80 % positive for SRY-related high-mobility-group box 2 at this stage of differentiation.

18. As cells develop an anterior endoderm identity, they will continue to upregulate CXC chemokine receptor 4, SRY-related high-mobility-group box 17, GATA binding protein 4, hepatocyte nuclear factor 4 alpha, and hematopoietically expressed homeobox protein and SRY-related high-mobility-group box 2. Importantly, cells should be negative for the posteriorized endoderm marker caudal type homeobox 2 as well as more committed foregut lineage markers, such as alpha fetoprotein, pancreatic, and duodenal homeobox 1 and NK2 homeobox 1. This identity can be confirmed by both quantitative reverse transcriptase polymerase chain reaction analysis and immunocytochemistry. More than 90 % of the cells should be positive for CXC chemokine receptor 4 and SRY-related high-mobility-group box 17 by fluorescence-assisted cell sorting at this stage.

19. Following specification to hepatic endoderm, cells will upregulate alpha fetoprotein and T-box 3 shown by both quantitative reverse transcriptase polymerase chain reaction analysis and immunocytochemistry. More than 90 % of the cells should be alpha fetoprotein-positive at this stage by fluorescence-assisted cell sorting.

20. As the cells progress from hepatic endoderm through the bipotential hepatoblast stage and subsequently commit to the hepatocyte lineage, they should homogenously adopt a polygonal morphology by day 20. By day 25, downregulation of alpha fetoprotein along with upregulation of albumin and alpha 1 antitrypsin should be apparent by quantitative reverse transcriptase polymerase chain reaction analysis, enzyme-linked immunosorbent assay, and immunocytochemistry. Cells should exhibit active CYP-based xenobiotic metabolism as evaluated using the P450 Glo assays. Additionally, the cells should stain positive for glycogen, as evaluated by periodic acid Schiff staining, exhibit low density lipoprotein uptake, and actively clear indocyanine green. Cells also should upregulate cytokeratin 18, hepatocyte nuclear factor 4 alpha and CYP3A4, and downregulate CYP3A7 as evaluated by quantitative reverse transcriptase polymerase chain reaction analysis and immunocytochemistry. Cells may proliferate during this period and produce outgrowths from the plastic cell culture plates. This is normal and does not affect subsequent differentiation and maturation.

21. The choice of two-dimensional versus three-dimensional maturation should be based upon the intended application of the cells, as each culture method has benefits and drawbacks. Hepatocytes cultured in a two-dimensional configuration are more suited for multiplex optimization studies due to ease of handling, lower cost, and higher peak CYP activity. Hepatocytes

cultured in a three-dimensional configuration polarize and express drug transporters that are absent in a two-dimensional configuration, making this culture method more suitable for complex toxicology studies that rely on active transport of metabolites. Additionally, three-dimensional cultures have dramatically increased longevity, generally over twice that of cells cultured in a two-dimensional configuration.

22. Cells cultured in a two-dimensional configuration may begin to detach from the plastic cell culture plate and initiate apoptosis from day 40 onwards. It is rare for two-dimensional cultures to last beyond day 55.
23. Cell dissociation buffer causes the hepatocyte epithelial sheet to detach from the plastic cell culture plate, but does not separate the cells from each other. In order to produce epithelial clumps small enough to transfer to the three-dimensional culture, one should manually perturb the cells by scraping the tip of a pipette across the cells to shear the sheet into smaller clumps. The cells are fairly hardy. However, one should use the minimum force and time necessary to produce clumps that are around 100–200 μm in diameter.
24. Determining the cell density can be difficult and may yield inconsistent results when counting clumps of cells. To overcome this, a small aliquot of clumps can be fully dissociated to single cells using trypsinization before being counted.
25. The protocol for producing the three-dimensional collagen matrices is included with the real architecture for 3D tissue starter kit and can also be found online by courtesy of TAP Biosystems (United Kingdom) at the https://www.raft3dcell-culture.com/applications.php.
26. The top of the absorber plate can be removed and unused absorbers saved if not all wells are being used. To use the absorbers in future experiments, keep sterile and simply place the absorbers on top of the wells without the aligning plate.
27. When changing the cell culture medium in three-dimensional cultures, be careful not to aspirate the construct out of the well. Constructs may fully detach if the aspirating pipette comes in direct contact. To avoid this, tilt the plate and slowly aspirate down the side of the well. A slight amount of cell culture medium may be left in the well to avoid aspirating the culture. Additionally, using a 20 μL pipette tip placed on the end of the aspirating pipette eliminates the risk of the construct being pulled into the vacuum flask should it separate from the bottom. In the rare occurrence that the construct still pulls off of the well, quickly remove the pipette tip and use a micropipette to apply pressure to dislodge the construct back into the well. Gently use the tip to flatten the construct back into the bottom of the well.

28. Three-dimensional cultures have been consistently maintained for over 75 days and maintain expression of CYP isoenzymes over this period. Longer culture periods have not been attempted and may or may not remain viable. Cells should express the polarization markers multidrug resistance-associated protein 2 and dipeptidyl peptidase 4 along canalicular boundaries as evaluated by immunocytochemistry.
29. Quantitative reverse transcriptase polymerase chain reaction analysis of three-dimensional cultures requires the enzymatic degradation of the matrix before mRNA isolation. To degrade the matrix, transfer the matrix to a 15 mL tube using sterile forceps and incubate the matrix with collagenase type IV at 37 °C checking every 20 min until the matrix is dissolved. Centrifuge at $590 \times g$ and room temperature for 10 min, carefully aspirate the collagenase and wash with PBS before proceeding with subsequent mRNA isolation. Cells should express CYP isoenzymes and uridine 5′-diphosphoglucuronosyltransferases along with the polarization marker multidrug resistance-associated protein 2. A comprehensive list of markers is presented with the primer sequences in Table 1.

References

1. Takahashi K, Tanabe K, Ohnuki M et al (2007) Induction of pluripotent stem cells from adult human fibroblasts by defined factors. Cell 131:861–872
2. Thomson JA, Itskovitz-Eldor J, Shapiro SS et al (1998) Embryonic stem cell lines derived from human blastocysts. Science 282:1145–1147
3. Touboul T, Hannan NR, Corbineau S et al (2010) Generation of functional hepatocytes from human embryonic stem cells under chemically defined conditions that recapitulate liver development. Hepatology 51:1754–1765
4. Lumelsky N, Blondel O, Laeng P et al (2001) Differentiation of embryonic stem cells to insulin-secreting structures similar to pancreatic islets. Science 292:1389–1394
5. Lee SH, Lumelsky N, Studer L et al (2000) Efficient generation of midbrain and hindbrain neurons from mouse embryonic stem cells. Nat Biotechnol 18:675–679
6. Rohwedel J, Maltsev V, Bober E et al (1994) Muscle cell differentiation of embryonic stem cells reflects myogenesis *in vivo*: developmentally regulated expression of myogenic determination genes and functional expression of ionic currents. Dev Biol 164:87–101
7. Yusa K, Rashid ST, Strick-Marchand H et al (2011) Targeted gene correction of alpha1-antitrypsin deficiency in induced pluripotent stem cells. Nature 478:391–394
8. Carvajal-Vergara X, Sevilla A, D’Souza SL et al (2010) Patient-specific induced pluripotent stem-cell-derived models of LEOPARD syndrome. Nature 465:808–812
9. http://www.stembook.org/ (consulted July 2014).
10. Gieseck RL III, Hannan NR, Bort R et al (2014) Maturation of induced pluripotent stem cell derived hepatocytes by 3D-culture. PloS One 9:e86372

Chapter 10

Differentiation-Promoting Medium Additives for Hepatocyte Cultivation and Cryopreservation

Varvara Gouliarmou, Olavi Pelkonen, and Sandra Coecke

Abstract

Isolated primary hepatocytes are considered as the reference system for in vitro hepatic methods. Following the isolation of primary hepatocytes from liver tissue, an unfavorable process named dedifferentiation is initiated leading to the attenuation of the hepatocellular phenotype both at the morphological and functional level. Freshly isolated hepatocytes can be used immediately or can be cryopreserved for future purposes. Currently, a number of antidedifferentiation strategies exist to extend the life span of isolated hepatocytes. The addition of differentiation-promoting compounds to the hepatocyte culture medium is the oldest and simplest antidedifferentiation approach applied. In the present chapter, the most commonly used medium additives for cultivation and cryopreservation of primary hepatocytes are reviewed.

Key words Hepatocyte, Differentiation, Medium additives, Cryopreservation, Plated hepatocytes, CYP induction

1 Introduction

Recent pieces of European legislation promote the use of in vitro test methods to increase human safety and to reduce or replace animal studies needed for risk assessment. Following the exposure to xenobiotics, the liver is the main organ for their biotransformation and detoxification. Consequently, it is of paramount importance and relevance for safety assessment to establish reliable and robust in vitro hepatic systems. In vitro hepatic systems enable studies of hepatotoxicity [1–5], hepatic metabolic clearance [6–9], and prediction of liver metabolic induction or inhibition due to xenobiotic exposure [10–13].

The machinery of any in vitro hepatic system is its biological test system. For this reason, special consideration should be paid to the latter in order to ensure good quality of the produced in vitro data based on good scientific and good quality practices [14]. Various biological test systems of human or animal origin are currently employed, ranging from simple subcellular liver fractions

Mathieu Vinken and Vera Rogiers (eds.), *Protocols in In Vitro Hepatocyte Research*, Methods in Molecular Biology, vol. 1250, DOI 10.1007/978-1-4939-2074-7_10, © Springer Science+Business Media New York 2015

to isolated perfused livers. Primary hepatocyte and liver cell lines are generally preferred, since they offer a good compromise between in vivo relevance and complexity. Cell lines usually originate from liver cancer cells and have the advantage of being easy to culture and to provide high cell supply due to their high proliferation capacity [15–17]. Their main disadvantage is that they usually lack basic hepatocyte characteristics [17]. For instance, the human hepatoma HepG2 cell line undergoes substantial loss of cytochrome P450 (CYPs) enzymes and drug transporters [18]. The HepaRG cell line, which is currently the most promising cellular system, shows an intermediate/slow metabolizer CYP2C9 and CYP2D6 isoform status [15]. Also, during recent years, attempts have focused on producing metabolically active hepatocytes from human embryonic stem cells and hepatic stem cells, with several reports describing the development and optimization of various differentiation protocols [19–23]. However, the applicability of human stem cell-derived hepatocyte-like cells as a reliable tool to substitute primary hepatocytes, especially for studying metabolism, is under question, since previous studies have shown that the key CYP3A4 isoform is present at 1,000-fold lower amounts in differentiated stem cells than in human primary hepatocytes [24]. Recent progress has also been made in generating hepatocyte-like cells from human induced pluripotent stem cells [25–27]. A direct comparison of differentiated stem cells and induced pluripotent stem cells with human primary hepatocytes has shown that liver-specific functions, including glycogen synthesis, urea production, albumin secretion, and CYP activity, of the stem cell-derived cells were present, but much lower compared to human hepatocytes [27]. At present, it is not clear to what degree stem cell-derived hepatocytes need to resemble primary hepatocytes to ensure a reliable metabolically competent system. Furthermore, critical issues must be resolved before large numbers of metabolic competent hepatocytes can be generated from such cell lines and used for metabolism testing [28].

Primary hepatocytes are generally accepted as the system of choice for hepatic in vitro methods. Isolated hepatocytes offer the full complement of xenobiotic metabolizing enzymes, drug transporters, cofactors at physiological relevant concentrations, and membrane integrity [29, 30]. Their main disadvantage is their limited availability, especially of human hepatocytes, which hampers the planning and performance of studies. However, the improvement of cryopreservation techniques during the last years has facilitated a wider use of isolated hepatocytes in in vitro studies. Hepatocyte isolation from human livers is performed with the established 2-step collagenase technique [31]. This technique was originally developed for the isolation of rat hepatocytes, but has been optimized for the isolation of hepatocytes from different animal species, including human [32, 33]. The first step of this procedure comprises liver perfusion with a warm (i.e., 37 °C), calcium-free, ethylene glycol tetraacetic acid-containing and isotonic buffer to remove blood and to break cell–cell contacts, followed by perfusion with a warm, isotonic, collagenase-containing solution to dissociate the liver parenchyma into single cells. In general, a higher amount of collagenase is required for the isolation of hepatocytes from human livers than from rat livers [34]. As collagenase is a mixture of proteases, its composition can affect its effectiveness in the dissociation of the hepatocytes as well as its cytotoxicity. After digestion, the cells are harvested by low-speed centrifugation. Currently, isolating 3.3–10 % viable cells from the total number of hepatocytes present in perfused liver is considered a satisfying yield for human hepatocytes. This clearly shows that the isolation technique of human hepatocytes is far from being optimized [34].

Among the pitfalls of primary hepatocytes is that they readily dedifferentiate already during the isolation process, resulting in a progressive loss of their hepatocellular phenotype at the morphological and functional level. The mechanisms underlying dedifferentiation have been previously described [35, 36]. Initially, the isolation disrupts the vast array of cellular contacts and considerable ischemia–perfusion injury takes place. These negative events initiate an inflammatory reaction, mediated by nuclear factor kappa beta, and a proliferative response, driven by mitogen-activated protein kinase, both negatively affecting liver-specific gene expression patterns. This leads to the downregulation of CYP enzymes and liver-enriched transcription factors, such as CCAAT/enhancer binding protein alpha, which eventually results in the deterioration of the differentiated hepatocellular status. The latter then triggers the initiation of spontaneous cell death by apoptosis, thereby limiting the overall cultivation time of primary hepatocytes to a maximum of 3 days [37, 38].

Isolated hepatocytes can be used in suspension for experiments requiring a relative short-time duration due to the progression of the dedifferentiation process. For instance, for metabolism studies, usually a duration of 4 h is recommended, since CYP-associated activities drastically drop beyond this timeframe [39]. Another alternative is to cryopreserve hepatocytes under appropriate conditions to maintain their differentiation status and to enable their future use [39, 40]. Finally, another common practice is to culture fresh or thawed hepatocytes. Cell culture formats commonly used for in vitro xenobiotic metabolism and CYP induction studies comprise conventional and sandwich cultures. The conventional culture format refers to plated or attached hepatocytes maintained on a rigid substratum (e.g., collagen) [41, 42], whereas in the second culture format, cells are sandwiched between two layers of extracellular matrix components, the upper one being either collagen or a composite of extracellular matrix proteins, such as Matrigel [43].

Both cultivation and cryopreservation of hepatocytes aim at maintaining the hepatocellular phenotype of the isolated cells. During the last decades, various strategies have been used to maintain the differentiation status of the isolated hepatocytes [35, 36]. These strategies are briefly presented in the current chapter with the main focus on the strategy that employs differentiation-promoting compounds in the culture medium, being the simpler and one of the first antidedifferentiation approaches used. Additionally, the most commonly used cryopreservation medium additives will be reviewed, since these chemicals can affect the postthaw differentiation status of hepatocytes, their viability, and their plating efficiency.

2 Markers of the Hepatocyte Differentiation Status

The hallmark hepatocyte expression and functional markers that characterize their differentiation status have been previously described [44]. Here, these differentiation markers are reviewed briefly, since the performance and success of the applied antidedifferentiation strategies are judged by their ability to maintain these markers.

2.1 Cell Morphology

Lactate dehydrogenase leakage and trypan blue exclusion from cells are two commonly used methods to assess the status of the hepatocyte plasma membrane surface associated with morphological disruption. Generally, differentiated hepatocytes maintain their polygonal shape, while dedifferentiated hepatocytes in culture exhibit a more flattened appearance, weakly defined borders, and fibroblast-like spinous processes [44]. Previous studies have shown that omission of dexamethasone from the cell culture medium leads to perturbation of cuboidal networks, with cells exhibiting condensed cytoplasm, abnormal rounding structures, and formation of fibroblast-like protrusions [45]. In optimal cultivation conditions, such as the presence of Matrigel, many of the morphological features of hepatocytes, like the cuboidal and three-dimensional structure, are well maintained [46].

2.2 Plasma Proteins

The synthesis of plasma proteins, such as albumin, transferrin, transthyretin, and alpha 1 antitrypsin, is among the most commonly used markers of hepatocyte differentiation [44]. In particular, albumin secretion has been used as a differentiation marker in numerous studies [47–50]. Additional markers are the synthesis of alpha fetoprotein and glutathione *S*-transferase P1, which both are repressed in differentiated hepatocytes, but that become increased in hepatocytes undergoing dedifferentiation [51, 52].

2.3 Expression of Connexins and Cytokeratins

Intercellular connections control a wide variety of liver-specific functions, including xenobiotic metabolism, albumin secretion, and urea production, and rely on the expression of connexins, the molecular building stones of gap junctions [53]. Cultures of differentiated hepatocytes display high levels of connexin32. However, upon dedifferentiation, connexin32 amounts gradually decline, while connexin43, expressed mainly in fetal liver in vivo, becomes progressively detectable [54]. Thus, connexin32 is a feature of fully differentiated hepatocytes. Additionally, the expression of cytokeratins 18 and 19 is a widely accepted hallmark of differentiated hepatocytes [44].

2.4 CYP Activity and Liver-Enriched Transcription Factors

Xenobiotic metabolism is a critical liver-specific function. Hence, the activity of phase I and phase II metabolizing enzymes is a crucial hallmark of differentiated hepatocytes [55–57]. The metabolic competence of cells is usually assessed by measuring mRNA levels and enzymatic activities of various enzymes. The most commonly tested enzymes are the CYPs. Special cocktails containing probe substrates for drug metabolizing enzymes have been used for this purpose [58, 59]. The expression of many liver-specific genes in adult liver is regulated by liver-enriched transcription factors, including hepatocyte nuclear factor 1 (HNF1), HNF3, HNF4, HNF6, and CCAAT/enhancer binding protein in adult liver [44]. Previous studies have shown that HNF1 alpha is a direct transactivator of CYP expression [60]. Likewise, HNF3 and CCAAT/enhancer binding protein play important roles in the regulation of CYPs [61, 62]. This illustrates that liver-enriched transcription factors are at the basis of the differentiated hepatocellular phenotype.

2.5 CYP Induction

Induction of xenobiotic metabolizing enzymes, such as CYPs, has vast toxicological significance [63]. Some induction pathways, such as those governed by the aryl hydrocarbon receptor, are rather robust and are maintained in both established cell lines and hepatocytes that are maintained suboptimally in culture [44]. A conspicuous exception is the induction by phenobarbital. In fact, the cellular capacity to respond to phenobarbital is considered as a uniquely sensitive indicator of hepatocyte differentiation [45]. However, it has been demonstrated that the regulation of CYP metabolic activity in cultured hepatocytes is dependent on the composition of the culture medium [64]. For instance, Yoshida and colleagues studied the effect of insulin as a medium additive on phenobarbital-mediated induction of CYP2B2 and CYP2B1 expression in cultured adult rat hepatocytes. It was found that, although the induction by phenobarbital remained absent in monolayers of hepatocytes cultured on type I collagen, it became manifested when insulin was omitted from the cell culture medium [65].

2.6 Perspectives

It is clear that the discussed markers of the hepatocyte differentiation status reflect the plethora of physiological functions that liver performs. However, there are currently no established criteria to assess the effectiveness of existing or new antidedifferentiation strategies. Additionally, a key question is if all of the discussed markers should be characterized or only those that are relevant for the applied in vitro hepatic method. Also, it is unclear whether these markers are sufficient to provide a full picture of the hepatocellular phenotype. For instance, an additional marker could be the presence of drug transporters. Indeed, besides xenobiotic metabolizing enzymes, the level of efflux and influx transporters expressed in hepatocytes is a decisive factor for hepatobiliary disposition of most drugs [66] and for the clearance of slow turn-over compounds [67].

Generally speaking, the establishment of standardized hepatocyte status markers or criteria can facilitate the objective assessment of the performance of antidedifferentiation strategies and of other hepatocyte-related methods, such as cryopreservation methods. Most important, these markers can be applied as test system characterization criteria, thus a kind of quality control that enables the assessment and comparison of existing and novel liver-based test systems. In turn, this could assist in the harmonization of in vitro hepatic methods.

3 Overview of Strategies to Counteract Hepatocyte Dedifferentiation In Vitro

Several antidedifferentiation strategies have been developed to retain liver-specific functions in isolated hepatocytes for 2 weeks or longer [35]. These strategies fall into two categories, namely classical and novel approaches.

3.1 Classical Approaches

Classical antidedifferentiation strategies typically reduce the consequences and thus attenuate the dedifferentiation process as such [35, 36]. These methods intend to mimic the natural hepatocyte microenvironment in an artificial in vitro setting [37, 68]. The first approach, which is further discussed in this chapter, is the enrichment of standard cell culture media with both physiological (e.g., nicotinamide) [69] and nonphysiological additives, such as dimethylsulfoxide (DMSO) [70], which improve hepatocellular functionality and morphology. Another classical antidedifferentiation strategy is targeted towards the reestablishment of heterogeneous cell–cell interactions by cocultivating hepatocytes with another cell type, which can be of either hepatic (e.g., sinusoidal endothelial cells) [71] or nonhepatic (e.g., fibroblasts) [9] origin. Furthermore, cell–extracellular matrix contacts can be reestablished by using sandwich cultures that employ natural (e.g., collagen) [72] or synthetic (e.g., polyethersulfone) [73] scaffolds as rigid substrata.

3.2 Novel Approaches

Innovative antidedifferentiation strategies attack the causes of dedifferentiation by interfering with its driving molecular mechanisms. These include genetic approaches (i.e., transfection with liver-specific genes and liver-specific transcription factors) and epigenetic approaches (i.e., interfering with posttranslational histone modifications, DNA methylation, and noncoding RNA-related mechanisms) [35, 36].

4 Differentiation-Promoting Culture Medium Additives for Primary Hepatocytes

As indicated, the addition of medium additives to the culture medium of primary hepatocytes is the oldest and simplest approach used to counteract the dedifferentiation process. Research on hepatocyte cell biology over the years has uncovered the biological role of several extracellular growth and differentiation factors [74–76]. In order to reach optimal conditions during cellular and metabolic experimentation and processing, it is particularly important to target the intracellular concentration levels of these factors. In addition, other molecules, like carbohydrates, lipids, vitamins, and amino acids, exert supplementary effects on hepatocyte differentiation and viability. The intracellular concentrations of these factors are directly and indirectly affected by their extracellular medium concentrations, namely, by their transport into the cytoplasm as well as by their intracellular biotransformation. Consequently, the composition of the culture medium is a main experimental factor for controlling intracellular processes and functional activities of hepatocytes in culture [77]. In Table 1, the most commonly used differentiation-promoting medium additives used for rat hepatocytes are described together with medium additives aimed at protecting cells and increasing their attachment to cell culture plates. However, it should be kept in mind that cell culture medium additives applied to improve hepatocyte viability and metabolic functions, specifically the expression of CYP enzymes and their induction potential, must be used with caution [44]. Indeed, their ability to change the activities of individual CYP enzymes may mask actions of test compounds, resulting in misinterpretation of toxicological endpoints under investigation. For instance, dexamethasone, a popular differentiation-promoting medium additive for primary hepatocytes, alters the expression of a number of drug metabolizing enzymes, especially when used in micromolar concentrations (i.e., more than 1 μM) [35]. Dexamethasone induces the synthesis of specific CYPs, glutathione *S*-transferases, uridine diphosphate glucuronosyl transferases [81–83], and sulfotransferases in rat, but not in human hepatocytes [84]. Most importantly, dexamethasone equally potentiates the induction of CYP1A by beta-naphthoflavone and CYP2B enzymes by phenobarbital, although these effects are dependent on the type of basal cell

Table 1
Commonly used differentiation-promoting medium additives and their effects on liver-specific functions of cultured rat hepatocytes [65, 78–80]

Medium additives	Positive effects	Negative effects
Dexamethasone (25–100 nM)	• Increases cell attachment and cell survival • Improves morphology • Increases preservation of liver-specific functions, including albumin synthesis • Preserves liver-specific functions by stabilizing mRNA levels and inhibiting proteinase synthesis • Increases extracellular matrix secretion • Promotes ordered arrangement of the cytoskeleton and gap junction expression and function • Supports CYP activity and maintains responsiveness to CYP inducers • Reduces the decrease in protein synthesis during the first 24 h of culture • Enhances formation of elaborate networks of bile canaliculi	• Suppresses the expression of P-glycoprotein • Changes in GST isoenzyme pattern expression, nitric oxide synthase, and collagen type I • Permissive and synergistic effects on the actions of other hormones • Induces certain drug-metabolizing enzymes, including uridine diphosphate glucuronosyl transferases, CYPs, and GSTs
Insulin (0.5–10 μg/L)	• Increases cell survival and cell attachment • Enhances amino acid transport, protein synthesis, glycogenesis, and lipogenesis • Inhibits protein degradation and potentiates inhibition of RNA degradation by amino acids • Exhibits growth factor-like properties by binding to growth factor receptors in addition to its specific receptors and stimulates G-protein activity and DNA synthesis • Prolongs cell survival	• Suppresses the induction of CYP2B1 and CYP2BA gene expression by phenobarbital
Glucagon (0.0007–35 μg/mL)	• Acts on various hepatic enzyme activities, such as pyruvate kinase, lactate dehydrogenase and glucokinase, as well as enzymes involved in the synthesis of urea and gluconeogenesis	• Enhances protein degradation • Decreases protein synthesis • Increases the expression of certain CYPs in murine hepatocytes
Growth hormone	• Regulates several CYPs in mature rats	• Suppresses CYP4A-mediated lauric acid hydroxylation • Inhibits induction of CYP2B • Exerts only a minor effect on the induction of CYP3A by pregnenolone 16α-carbonitrile or dexamethasone • Alters GST isoenzyme pattern expressions

Epidermal growth factor and hepatocyte growth factor (5–25 ng/mL)	• Induce pronounced stimulation of DNA synthesis accompanied by specific induction of proto-oncogene mRNA levels • Activate phospholipase A2 • Epidermal growth factor induces CYP1A1 and CYP2B1 • Stimulate entrance into the G1 and S phases of the cell cycle • Stimulate growth and tissue regeneration • Stimulate progression from the G0 state to the G1 and S phases • Induce DNA synthesis	• Epidermal growth factor suppresses the constitutive expression of CYP2C11
Nicotinamide	• Increases survival time for hepatocytes to over 1 month • Causes higher albumin expression • Increases total CYP content, aryl hydrocarbon hydroxylase and 7-ethoxycoumarin-*O*-deethylation activities, and tryptophan 2,3-dioxygenase mRNA levels • Enhances induction of CYP1A2 and GST • Inhibits overexpression of the P-glycoprotein • Inhibits the gradual loss of constitutive levels of CYP1A2	
Selenium (6.25 ng/mL)	• Necessary for normal cell survival and growth • Stimulates the constitutive expression and induction of CYP-dependent enzymes in the presence of phenobarbital	
Alpha-tocopherol	• Effectively maintains total CYP content	• Cannot restore glutathione synthetase expression
Dimethylsulfoxide (maximum 2 %)	• Prolongs hepatocyte survival, morphology, CYP content, and GST levels • Sustains albumin synthesis • Acts as a scavenger of oxygen radicals • Stabilizes mRNA levels and restricts protein degradation by direct binding to CYPs • Penetrates membrane structures and induces specific CYPs	• Inhibits CYP enzymes in higher concentrations
Phenobarbital	• Decreases the degradation rate of total CYP content • Delays the loss of liver-specific functions • Stabilizes the hepatocyte membranes and reduces the mitochondrial formation of active oxygen species resulting in the prolonged survival of hepatocytes in primary culture	• May cause apoptotic cell death

(continued)

Table 1
(continued)

Medium additives	Positive effects	Negative effects
Proinflammatory cytokines	• Interleukin 6 induces the expression of fibronectin, alpha 1 acid glycoprotein, fibrinogen, and procollagen without affecting the expression of albumin • Stimulate growth and tissue regeneration • Stimulate progression from the G0 state to the G1 and S phases • Induce DNA synthesis	• Interleukin 1 alpha/beta, interleukin 6, tumor necrosis factor alpha, and interferon gamma generally inhibit the induction of CYPs in hepatocyte cultures treated with inducers • Interleukin 1 beta and interleukin 6 inhibit phenobarbital induction of CYP2B • Interleukin 1 beta suppresses induction of CYP1A
Amino acids	• Decrease protein degradation and stimulate protein synthesis • Reduce autophagic protein degradation • High levels of leucine and phenylalanine increase the stability of liver-specific mRNAs • Leucine and phenylalanine together with galactose and pyruvate increase levels of CYPs and triglycerides	
Serum (2–10 %)	• Increases cell attachment and cell survival • Improves morphology	• Promotes dedifferentiation • Reduction of CYP expression, activity, and induction
Pyridines	• Maintain total CYP content by permitting the synthesis and reducing the degradation of CYPs	
Pyruvate (1 nM)	• Prolongs survival and increases albumin secretion, glucokinase, and adenosine triphosphate levels • Sustains levels of CYPs, triglyceride, and urea synthesis • Maintains GST expression • Reduces oxidative stress	• Increases fetal-like GSTs
Alpha 1 acid glycoprotein (0.1–0.2 mg/L)	• Reduces cell death	

CYP cytochrome P450, *GST* glutathione *S*-transferase

culture medium [45, 85]. Another example, as discussed, is the suppression of phenobarbital-mediated induction of CYP2B1 and CYP2B2 by insulin [65].

5 Cryopreservation Medium Additives for Primary Hepatocytes

Various parameters determine successful cryopreservation of isolated hepatocytes, including liver source, composition of the cryopreservation medium (e.g., cryoprotectants), cooling and thawing rates [29], and the degree of membrane damage during hepatocyte isolation [34]. As such, 2 types of cryoprotectants are routinely used, namely those that permeate the cell membrane, like DMSO [86] and glycerol [87], and those that do not, including oligosaccharides (e.g., trehalose) [88], sugars (e.g., glucose, sucrose, or fructose) [89], and polymers (e.g., dextran) [90]. Lactobionate (i.e., 100 mM), raffinose (i.e., 30 mM), and dexamethasone are the basic additives of the university of Wisconsin solution [91, 92], which is considered the golden standard cryopreservation medium for isolated hepatocytes. Lactobionate is a large molecular weight anion that is impermeable to most membranes and which is supposed to suppress hypothermia-induced cell swelling. Raffinose offers additional osmotic support and dexamethasone is added to stabilize cell membranes. DMSO is a most effective and commonly used cryoprotectant. It is usually added at 10 or 12 % and acts by delaying and reducing ice formation during the freezing process [93]. It has been reported that DMSO stabilizes membranes during freezing and thawing through electrostatic interactions with phospholipid membranes [94, 95] and that it gives the best plating efficiency [90, 96–98]. It has been shown that CYPs of human hepatocytes cryopreserved in a medium containing DMSO are inducible by xenobiotics [99] and that the combination of DMSO with 2.3 % polyvinylpyrrolidone, another cryoprotective agent, significantly increases the viability and maintains the capacity to respond to CYP inducers [100]. Also, the combination of DMSO with oligosaccharides with higher molecular weights results in higher postthaw cell viabilities [101]. Moreover, plating efficiency and survival rates of rat hepatocytes in plastic dishes are greater after freezing in the presence of di-, tri-, and tetrasaccharides. However, this positive effect is observed only for rat, but not for human hepatocytes [102]. When trehalose is combined with DMSO for the cryopreservation of human hepatocytes, a significant increase in total protein levels and secretion of albumin is observed after thawing [88]. A recent study showed preserved liver-specific functionality after thawing by adding 1 % DMSO to the preincubation medium prior to the cryopreservation of human hepatocytes [100].

Fetal calf serum and human albumin are also typical ingredients of hepatocyte cryopreservation solutions with concentrations ranging from 10 to 90 %. The concentration of these ingredients generally has little effect on cell viability or drug metabolizing enzyme activity [101]. However, a study that focused on the cryopreservation of rat hepatocytes showed optimal viability and CYP activity when 90 % fetal calf serum and 10 % DMSO were added to the cryopreservation medium [103].

Another interesting cryoprotectant includes wheat protein extract. It seems that wheat protein extracts enable long-term storage and recoveries of viable cells, and that they maintain the hepatocellular phenotype through an osmotic modulation effect [104]. Differentiation markers, such as albumin secretion and urea formation, are well maintained during 4 days postplating after thawing of hepatocytes kept in wheat protein extract-containing cryopreservation media. In addition, the induction of CYP1A1 and CYP2B in cryopreserved rat hepatocytes is similar to that of freshly isolated counterparts when wheat protein extract is included in the cryopreservation medium [105].

6 Conclusions

This chapter has reviewed the most commonly used differentiation-promoting medium additives for cultured and cryopreserved hepatocytes, thereby showing the importance to optimize the chemical composition of culture and cryopreservation media. The addition of soluble additives is the simplest antidedifferentiation method that, in combination with other approaches (e.g., hepatocyte sandwich cultures), can maintain in vivo-like liver-specific functions of hepatocytes for extended periods of time. However, it should be kept in mind that everything is context dependent and typically the simplest solutions are the most effective ones. For instance, Di and group have developed a novel relay method using cryopreserved human hepatocytes to measure intrinsic clearance of low-clearance compounds [106]. The relay method involves transferring the supernatant from hepatocyte incubations to freshly thawed hepatocytes at the end of the 4 h incubation period to prolong the exposure time of a studied compound to active enzymes in hepatocytes. In this way, an accumulative incubation time of 20 h or longer in hepatocytes is achieved [106]. Nevertheless, even in this simple alternative, the chemical composition of the medium is crucial for optimal performance of the test system. Another key point is that in most studies performed thus far, improvement of culture media for mammalian cells has been conducted by using empirical adjustments and was aided by statistical design methodologies only in a few cases [77, 107]. This possibly means that the full potential of the combinatorial effects of medium additives has not been explored yet and that there might be a window for improvement towards the optimal condition. Finally, in most studies, rat hepatocytes have been used, which makes the extrapolation to human hepatocytes challenging due to interspecies differences. Future research should be focused on human liver-based in vitro methods to be fully applicable for human risk assessment purposes.

References

1. Sahu SC (2003) Hepatocyte culture as an in vitro model for evaluating the hepatotoxicity of food-borne toxicants and microbial pathogens: a review. Toxicol Mech Methods 13:111–119
2. Shen C, Zhang G, Meng Q (2007) An in vitro model for long-term hepatotoxicity testing utilizing rat hepatocytes entrapped in micro-hollow fiber reactor. Bioch Eng J 34:267–272
3. Shen C, Zhang G, Qiu H et al (2006) Acetaminophen-induced hepatotoxicity of gel entrapped rat hepatocytes in hollow fibers. Chem Biol Interact 162:53–61
4. Swift B, Pfeifer ND, Brouwer KL (2010) Sandwich-cultured hepatocytes: an in vitro model to evaluate hepatobiliary transporter-based drug interactions and hepatotoxicity. Drug Metab Rev 42:446–471
5. Yeon JH, Na D, Park JK (2010) Hepatotoxicity assay system using suspended human hepatocytes trapped in microholes of a microfluidic device. Electrophoresis 31:3167–3174
6. Shibata Y, Kuze J, Chiba M (2014) Utility of cryopreserved hepatocytes suspended in serum to predict hepatic clearance in dogs and monkeys. Drug Metab Pharmacokinet 29:168–176
7. Richardson VM, Ferguson SS, Sey YM et al (2014) In vitro metabolism of thyroxine by rat and human hepatocytes. Xenobiotica 44:391–403
8. Coe KJ, Koudriakova T (2014) Metabolic stability assessed by liver microsomes and hepatocytes. In: Caldwell GW, Yan Z (eds) Methods in pharmacology and toxicology. Springer Science and Business Media, New York, pp 87–99
9. Chan TS, Yu H, Moore A et al (2013) Meeting the challenge of predicting hepatic clearance of compounds slowly metabolized by cytochrome P450 using a novel hepatocyte model HepatoPac. Drug Metab Dispos 41:2024–2032
10. Singer M, Sensenhauser C, Dallas S (2014) In vitro CYP induction using human hepatocytes. In: Caldwell GW, Yan Z (eds) Methods in pharmacology and toxicology. Springer Science Business Media, New York, pp 237–253
11. Kaneko A, Kato M, Endo C et al (2010) Prediction of clinical CYP3A4 induction using cryopreserved human hepatocytes. Xenobiotica 40:791–799
12. Fahmi OA, Boldt S, Kish M et al (2008) Prediction of drug-drug interactions from in vitro induction data: application of the relative induction score approach using cryopreserved human hepatocytes. Drug Metab Dispos 36:1971–1974
13. Meneses-Lorente G, Pattison C, Guyomard C et al (2007) Utility of long-term cultured human hepatocytes as an in vitro model for cytochrome P450 induction. Drug Metab Dispos 35:215–220
14. Bal-Price A, Coecke S (2011) Guidance on good cell culture practice (GCCP). In: Ascher M, Sunol C, Bal-Price A (eds) Cell culture techniques. Springer Science Business Media LLC, New York, pp 1–25
15. Andersson TB, Kanebratt KP, Kenna JG (2012) The HepaRG cell line: a unique in vitro tool for understanding drug metabolism and toxicology in human. Expert Opin Drug Metab Toxicol 8:909–920
16. Lubberstedt M, Muller-Vieira U, Mayer M et al (2011) HepaRG human hepatic cell line utility as a surrogate for primary human hepatocytes in drug metabolism assessment in vitro. J Pharmacol Toxicol Methods 63:59–68
17. Donato MT, Lahoz A, Castell JV et al (2008) Cell lines: a tool for in vitro drug metabolism studies. Curr Drug Metab 9:1–11
18. Guo L, Dial S, Shi L et al (2011) Similarities and differences in the expression of drug-metabolizing enzymes between human hepatic cell lines and primary human hepatocytes. Drug Metab Dispos 39:528–538
19. Farzaneh Z, Pakzad M, Vosough M et al (2014) Differentiation of human embryonic stem cells to hepatocyte-like cells on a new developed xeno-free extracellular matrix. Histochem Cell Biol 30:1–10
20. Hay DC, Zhao D, Ross A et al (2007) Direct differentiation of human embryonic stem cells to hepatocyte-like cells exhibiting functional activities. Cloning Stem Cells 9:51–62

21. Subramanian K, Owens DJ, Raju R et al (2014) Spheroid culture for enhanced differentiation of human embryonic stem cells to hepatocyte-like cells. Stem Cells Dev 23:124–131
22. Liu ZF, Xing ZG, Ding YN et al (2008) Research progress in directional differentiation from bone marrow stem cells into hepatocyte-like cells. World Chi J Digestol 16:658–662
23. Waclawczyk S, Buchheiser A, Flogel U et al (2010) In vitro differentiation of unrestricted somatic stem cells into functional hepatic-like cells displaying a hepatocyte-like glucose metabolism. J Cell Physiol 225:545–554
24. Wobus AM, Loser P (2011) Present state and future perspectives of using pluripotent stem cells in toxicology research. Arch Toxicol 85:79–117
25. Chen YF, Tseng CY, Wang HW et al (2012) Rapid generation of mature hepatocyte-like cells from human induced pluripotent stem cells by an efficient three-step protocol. Hepatology 55:1193–1203
26. Si-Tayeb K, Noto FK, Nagaoka M et al (2010) Highly efficient generation of human hepatocyte-like cells from induced pluripotent stem cells. Hepatology 51:297–305
27. Song Z, Cai J, Liu Y et al (2009) Efficient generation of hepatocyte-like cells from human induced pluripotent stem cells. Cell Res 19:1233–1242
28. Pelkonen O, Turpeinen M, Hakkola J et al (2013) Preservation, induction or incorporation of metabolism into the *in vitro* cellular system: views to current opportunities and limitations. Toxicol In Vitro 27:1578–1583
29. Gomez-Lechon MJ, Donato MT, Castell JV et al (2003) Human hepatocytes as a tool for studying toxicity and drug metabolism. Curr Drug Metab 4:292–312
30. Gomez-Lechon MJ, Donato MT, Castell JV et al (2004) Human hepatocytes in primary culture: the choice to investigate drug metabolism in man. Curr Drug Metab 5:443–462
31. Berry MN, Friend DS (1969) High-yield preparation of isolated rat liver parenchymal cells: a biochemical and fine structural study. J Cell Biol 43:506–520
32. Lee SM, Schelcher C, Demmel M et al (2013) Isolation of human hepatocytes by a two-step collagenase perfusion procedure. J Vis Exp 79:1–10
33. Hang HL, Zhang L, Shi XL et al (2011) Isolation, culture and cryopreservation of adult human hepatocytes. World Chi J Digestol 19:2016–2021
34. Li AP (2007) Human hepatocytes: isolation, cryopreservation and applications in drug development. Chem Biol Interact 168:16–29
35. Fraczek J, Bolleyn J, Vanhaecke T et al (2013) Primary hepatocyte cultures for pharmaco-toxicological studies: at the busy crossroad of various anti-dedifferentiation strategies. Arch Toxicol 87:577–610
36. Vinken M, Vanhaecke T, Rogiers V (2012) Primary hepatocyte cultures as in vitro tools for toxicity testing: *quo vadis*? Toxicol In Vitro 26:541–544
37. Elaut G, Henkens T, Papeleu P et al (2006) Molecular mechanisms underlying the dedifferentiation process of isolated hepatocytes and their cultures. Curr Drug Metab 7:629–660
38. Paine AJ, Andreakos E (2004) Activation of signalling pathways during hepatocyte isolation: relevance to toxicology in vitro. Toxicol In Vitro 18:187–193
39. Smith CM, Nolan CK, Edwards MA et al (2012) A comprehensive evaluation of metabolic activity and intrinsic clearance in suspensions and monolayer cultures of cryopreserved primary human hepatocytes. J Pharml Sci 101:3989–4002
40. Yajima K, Uno Y, Murayama N et al (2014) Evaluation of 23 lots of commercially available cryopreserved hepatocytes for induction assays of human cytochromes P450. Drug Metab Dispos 42:867–871
41. Liu HL, Wang YJ, Guo HT et al (2004) Cryopreservation and gel collagen culture of porcine hepatocytes. World J Gastroenterol 10:1010–1014
42. Ijima H, Ogata R, Murasawa Y et al (2010) Albumin production activity of primary rat hepatocytes is improved on type V collagen. J Biosci Bioeng 109:179–181
43. Swift B, Brouwer KL (2010) Influence of seeding density and extracellular matrix on bile acid transport and Mrp4 expression in sandwich-cultured mouse hepatocytes. Mol Pharm 7:491–500
44. Olsavsky Goyak KM, Laurenzana EM, Omiecinski CJ (2010) Hepatocyte differentiation. In: Maurel P (ed) Methods in molecular biology. Springer Science Business Media LLC, New York, pp 115–138
45. Sidhu JS, Liu F, Omiecinski CJ (2004) Phenobarbital responsiveness as a uniquely sensitive indicator of hepatocyte differentiation status: requirement of dexamethasone and extracellular matrix in establishing the functional integrity of cultured primary rat hepatocytes. Exp Cell Res 292:252–264
46. Page JL, Johnson MC, Olsavsky KM et al (2007) Gene expression profiling of extracellular matrix as an effector of human hepatocyte phenotype in primary cell culture. Toxicol Sci 97:384–397

47. Hou YT, Ijima H, Matsumoto S et al (2010) Effect of a hepatocyte growth factor/heparin-immobilized collagen system on albumin synthesis and spheroid formation by hepatocytes. J Biosci Bioeng 110:208–216
48. Ijima H, Mizumoto H, Nakazawa K et al (2009) Hepatocyte growth factor and epidermal growth factor promote spheroid formation in polyurethane foam/hepatocyte culture and improve expression and maintenance of albumin production. Biochem Eng J 47:19–26
49. Li BL, Qu Q, Zhao YP et al (2005) Expression of albumin during hepatocyte differentiation by human bone marrow stem cells. Zhonghua wai ke za zhi 43:713–715
50. Moshage HJ, de Haard HJW, Princen HMG et al (1985) The influence of glucocorticoid on albumin synthesis and its messenger RNA in rat in vivo and in hepatocyte suspension culture. Biochim Biophys Acta 824:27–33
51. Ali-Osman F, Brunner JM, Kutluk TM et al (1997) Prognostic significance of glutathione *S*-transferase π expression and subcellular localization in human gliomas. Clin Cancer Res 3:2253–2261
52. Lavon N, Benvenisty N (2005) Study of hepatocyte differentiation using embryonic stem cells. J Cell Biochem 96:1193–1202
53. Van Ommeslaeghe K, Elaut G, Brecx V et al (2003) Amide analogues of TSA: synthesis, binding mode analysis and HDAC inhibition. Bioorg Med Chem Lett 13:1861–1864
54. Vinken M, Papeleu P, Snykers S et al (2006) Involvement of cell junctions in hepatocyte culture functionality. Crit Rev Toxicol 36:299–318
55. Sa-ngiamsuntorn K, Wongkajornsilp A, Kasetsinsombat K et al (2011) Upregulation of CYP 450s expression of immortalized hepatocyte-like cells derived from mesenchymal stem cells by enzyme inducers. BMC Biotechnol 11:89
56. Maezawa K, Miyazato K, Matsunaga T et al (2008) Expression of cytochrome P450 and transcription factors during in vitro differentiation of mouse embryonic stem cells into hepatocytes. Drug Metab Pharmacokinet 23:188–195
57. Ek M, Soderdahl T, Kuppers-Munther B et al (2007) Expression of drug metabolizing enzymes in hepatocyte-like cells derived from human embryonic stem cells. Biochem Pharmacol 74:496–503
58. Tolonen A, Petsalo A, Turpeinen M et al (2007) In vitro interaction cocktail assay for nine major cytochrome P450 enzymes with 13 probe reactions and a single LC/MSMS run: analytical validation and testing with monoclonal anti-CYP antibodies. J Mass Spectrom 42:960–966
59. Floby E, Briem S, Terelius Y et al (2004) Use of a cocktail of probe substrates for drug-metabolizing enzymes for the assessment of the metabolic capacity of hepatocyte preparations. Xenobiotica 34:949–959
60. Schrem H, Klempnauer J, Borlak J (2002) Liver-enriched transcription factors in liver function and development part I: the hepatocyte nuclear factor network and liver-specific gene expression. Pharmacol Rev 54:129–158
61. Gonzalez FJ (2008) Regulation of hepatocyte nuclear factor 4α-mediated transcription. Drug Metab Pharmacokinet 23:2–7
62. Jover R, Moya M, Gomez-Lechon MJ (2009) Transcriptional regulation of cytochrome P450 genes by the nuclear receptor hepatocyte nuclear factor 4-alpha. Curr Drug Metab 10:508–519
63. Omiecinski CJ, Remmel RP, Hosagrahara VP (1999) Concise review of the cytochrome P450s and their roles in toxicology. Toxicol Sci 48:151–156
64. Turner NA, Pitot HC (1989) Dependence of the induction of cytochrome P-450 by phenobarbital in primary cultures of adult rat hepatocytes on the composition of the culture medium. Biochem Pharmacol 38: 2247–2251
65. Yoshida Y, Kimura N, Oda H et al (1996) Insulin suppresses the induction of CYP2B1 and CYP2B2 gene expression by phenobarbital in adult rat cultured hepatocytes. Biochem Biophys Res Commun 229:182–188
66. Draheim V, Reichel A, Weitschies W et al (2010) *N*-glycosylation of ABC transporters is associated with functional activity in sandwich-cultured rat hepatocytes. Eur J Pharm Sci 41:201–209
67. Yabe Y, Galetin A, Houston JB (2011) Kinetic characterization of rat hepatic uptake of 16 actively transported drugs. Drug Metab Dispos 39:1808–1814
68. Hewitt NJ, Lechon MJG, Houston JB et al (2007) Primary hepatocytes: current understanding of the regulation of metabolic enzymes and transporter proteins, and pharmaceutical practice for the use of hepatocytes in metabolism, enzyme induction, transporter, clearance, and hepatotoxicity studies. Drug Metab Rev 39:159–234
69. Inoue C, Yamamoto H, Nakamura T et al (1989) Nicotinamide prolongs survival of primary cultured hepatocytes without involving loss of hepatocyte-specific functions. J Biol Chem 264:4747–4750
70. Su T, Waxman DJ (2004) Impact of dimethyl sulfoxide on expression of nuclear receptors and drug-inducible cytochromes P450 in primary rat hepatocytes. Arch Biochem Biophys 424:226–234

71. Krause P, Markus PM, Schwartz P et al (2000) Hepatocyte-supported serum-free culture of rat liver sinusoidal endothelial cells. J Hepatol 32:718–726
72. Kim Y, Lasher CD, Milford LM et al (2010) A comparative study of genome-wide transcriptional profiles of primary hepatocytes in collagen sandwich and monolayer cultures. Tissue Eng Part C Methods 16:1449–1460
73. Kinasiewicz A, Dudzinski K, Chwojnowski A et al (2007) Threedimensional culture of hepatocytes on spongy polyethersulfone membrane developed for cell transplantation. Transplant Proc 39:2914–2916
74. Hammond AH, Fry JR (1992) Effect of serum-free medium on cytochrome P450-dependent metabolism and toxicity in rat cultured hepatocytes. Biochem Pharmacol 44:1461–1464
75. Kimura M, Moteki H, Ogihara M (2011) Inhibitory effects of dexamethasone on epidermal growth factor-induced DNA synthesis and proliferation in primary cultures of adult rat hepatocytes. Biol Pharm Bull 34:682–687
76. Gomez-Lechon MJ, Garcia MD, Castell JV (1983) Effect of glucocorticoids on the expression of gamma-glutamyltransferase and tyrosine aminotransferase in serum-free-cultured hepatocytes. Hoppe Seylers Z Physiol Chem 364:501–508
77. Dong J, Mandenius CF, Lubberstedt M et al (2008) Evaluation and optimization of hepatocyte culture media factors by design of experiments (DoE) methodology. Cytotechnology 57:251–261
78. Coecke S, Rogiers V, Bayliss M et al (1999) The use of long-term hepatocyte cultures for detecting induction of drug metabolising enzymes: the current status. ATLA 27:579–638
79. Kagaya N, Kamiyoshi A, Tagawa Y et al (2005) Suppression of cell death in primary rat hepatocytes by alpha1-acid glycoprotein. J Biosci Bioeng 99:81–83
80. Stevenson DJ, Morgan C, McLellan LI et al (2007) Reduced glutathione levels and expression of the enzymes of glutathione synthesis in cryopreserved hepatocyte monolayer cultures. Toxicol In Vitro 21:527–532
81. Matsunaga T, Maruyama M, Matsubara T et al (2012) Mechanisms of CYP3A induction by glucocorticoids in human fetal liver cells. Drug Metab Pharmacokinet 27:653–657
82. Shi D, Yang D, Yan B (2010) Dexamethasone transcriptionally increases the expression of the pregnane X receptor and synergistically enhances pyrethroid esfenvalerate in the induction of cytochrome P450 3A23. Biochem Pharmacol 80:1274–1283
83. Kanou M, Usui T, Ueyama H et al (2004) Stimulation of transcriptional expression of human UDP-glucuronosyltransferase 1A1 by dexamethasone. Mol Biol Rep 31:151–158
84. Duanmu Z, Locke D, Smigelski J et al (2002) Effects of dexamethasone on aryl (SULT1A1)- and hydroxysteroid (SULT2A1)-sulfotransferase gene expression in primary cultured human hepatocytes. Drug Metab Dispos 30:997–1004
85. LeCluyse E, Bullock P, Madan A et al (1999) Influence of extracellular matrix overlay and medium formulation on the induction of cytochrome P-450 2B enzymes in primary cultures of rat hepatocytes. Drug Metab Dispos 27:909–915
86. Hang H, Shi X, Gu G et al (2009) A simple isolation and cryopreservation method for adult human hepatocytes. Int J Artif Organs 32:720–727
87. Novicki DL, Irons GP, Strom SC et al (1982) Cryopreservation of isolated rat hepatocytes. In Vitro 18:393–399
88. Katenz E, Vondran FWR, Schwartlander R et al (2007) Cryopreservation of primary human hepatocytes: the benefit of trehalose as an additional cryoprotective agent. Liver Transpl 13:38–45
89. Mitry RR, Lehec SC, Hughes RD (2010) Cryopreservation of human hepatocytes for clinical use. In: Maurel P (ed) Methods in molecular biology. Springer Science Business Media LLC, New York, pp 107–113
90. Loretz LJ, Li AP, Flye MW et al (1989) Optimization of cryopreservation procedures for rat and human hepatocytes. Xenobiotica 19:489–498
91. Southard JH, van Gulik TM, Ametani MS et al (1990) Important components of the UW solution. Transplantation 49:251–257
92. Kunieda T, Maruyama M, Okitsu T et al (2003) Cryopreservation of primarily isolated porcine hepatocytes with UW solution. Cell Transplant 12:607–616
93. Hubel A, Conroy M, Darr TB (2000) Influence of preculture on the prefreeze and postthaw characteristics of hepatocytes. Biotechnol Bioeng 71:173–183
94. Alexandre E, Viollon-Abadie C, David P et al (2002) Cryopreservation of adult human hepatocytes obtained from resected liver biopsies. Cryobiology 44:103–113
95. Li AP, Gorycki PD, Hengstler JG et al (1999) Present status of the application of cryopreserved hepatocytes in the evaluation of xenobiotics: consensus of an international expert panel. Chem Biol Interact 121:117–123
96. Rijntjes PJM, Moshage HJ, Van Gemert PJL et al (1986) Cryopreservation of adult human

hepatocytes: the influence of deep freezing storage on the viability, cell seeding, survival, fine structures and albumin synthesis in primary cultures. J Hepatol 3:7–18
97. Smith DJ, Schulte M, Bischof JC (1998) The effect of dimethylsulfoxide on the water transport response of rat hepatocytes during freezing. J Biomech Eng 120:549–558
98. Jeong HS, Kim KH, Nam YK et al (2004) Optimization of cryoprotectants for cryopreservation of rat hepatocyte. Biotechnol Lett 26:829–833
99. Ruegg CE, Silber PM, Mughal RA et al (1997) Cytochrome-P450 induction and conjugated metabolism in primary human hepatocytes after cryopreservation. In Vitro Mol Toxicol 10:217–222
100. Solanas E, Sostres C, Serrablo A et al (2012) Incubation with dimethyl sulfoxide *prior* to cryopreservation improves functionality of thawed human primary hepatocytes. Biopreserv Biobank 10:446–453
101. Stephenne X, Najimi M, Sokal EM (2010) Hepatocyte cryopreservation: is it time to change the strategy? World J Gastroentero 16:1–14
102. Miyamoto Y, Suzuki S, Nomura K et al (2006) Improvement of hepatocyte viability after cryopreservation by supplementation of long-chain oligosaccharide in the freezing medium in rats and humans. Cell Transplant 15:911–919
103. Stevenson DJ, Morgan C, Goldie E et al (2004) Cryopreservation of viable hepatocyte monolayers in cryoprotectant media with high serum content: metabolism of testosterone and kaempherol post-cryopreservation. Cryobiology 49:97–113
104. Hamel F, Grondin M, Denizeau F et al (2006) Wheat extracts as an efficient cryoprotective agent for primary cultures of rat hepatocytes. Biotechnol Bioeng 95:661–670
105. Grondin M, Hamel F, Sarhan F et al (2008) Metabolic activity of cytochrome p450 isoforms in hepatocytes cryopreserved with wheat protein extract. Drug Metab Dispos 36:2121–2129
106. Di L, Trapa P, Obach RS et al (2012) A novel relay method for determining low-clearance values. Drug Metab Dispos 40: 1860–1865
107. Yang H, Roth CM, Ierapetritou MG (2009) A rational design approach for amino acid supplementation in hepatocyte culture. Biotechnol Bioeng 103:1176–1191

Chapter 11

Coculture and Long-Term Maintenance of Hepatocytes

Merav Cohen, Gahl Levy, and Yaakov Nahmias

Abstract

The liver is the largest internal organ in mammals, serving a wide spectrum of vital functions. Loss of liver function due to drug toxicity, progressive fatty liver disease, or viral infection is a major cause of death in the United States of America. Pharmaceutical and cosmetic toxicity screening, basic research and the development of bioartificial liver devices require long-term hepatocyte culture techniques that sustain hepatocyte morphology and function. In recent years, several techniques have been developed that can support high levels of liver-specific gene expression, metabolic function, and synthetic activity for several weeks in culture. These include the collagen double gel configuration, hepatocyte spheroids, coculture with nonparenchymal cells, and micropatterned cocultures. This chapter will cover the current status of hepatocyte culture techniques, including media formulation, oxygen supply, and heterotypic cell–cell interactions.

Key words Liver, Hepatocytes, Coculture, Culture medium, Nonparenchymal cells, Metabolism, Oxygen

1 Introduction

The liver is the largest internal organ in the human body, ascribed with over 500 functions. Among these functions are embryonic hematopoiesis, protein synthesis (e.g., albumin and fibrinogen), bile acid production and xenobiotic metabolism, as well as maintaining energetic homeostasis by regulating carbohydrate, lipid, and amino acid metabolism [1]. The liver is unique in its ability to fully regenerate after loss of as much as 80 % of its mass [2]. Despite its ability to regenerate from various injuries, loss of liver function due to drug toxicity, viral infection, and progressive alcoholic- and nonalcoholic fatty liver disease has become a major cause of death in the United States of America, reaching nearly 32,000 individuals in 2010, a number that is expected to rise, given that metabolic syndrome and obesity have reached epidemic proportions. Despite recent advances in split graft transplantation, there are roughly 16,000 people awaiting a liver transplantation in the United States of

Mathieu Vinken and Vera Rogiers (eds.), *Protocols in In Vitro Hepatocyte Research*, Methods in Molecular Biology, vol. 1250, DOI 10.1007/978-1-4939-2074-7_11, © Springer Science+Business Media New York 2015

America alone. Consequently, the availability of human liver tissue is extremely limited, resulting in high prices and limited supply of human hepatocytes crucial for pharmaceutical, clinical, and research applications. This shortage is aggravated by the inability to expand primary hepatocytes in vitro [3]. Most available human hepatocytes are therefore isolated from marginal livers (i.e., fatty or fibrotic livers) or those rejected from transplantation and subsequently sold as freshly isolated or cryopreserved. Pharmaceutical industry is a major consumer of human hepatocytes. It is currently estimated that the capitalized cost of developing a drug from lead discovery through clinical trials is $1.8 billion [4]. Animal studies are inadequate in evaluating drug toxicity due to species-specific variations [5], limiting liver drug metabolism models to the use of primary human hepatocytes for absorption, distribution, metabolism, excretion, and toxicity screening. Hepatocyte cultures are also utilized in the study of metabolism [6, 7], liver development [8], regeneration [9], viral infection [10], and inflammation [11]. They are also crucial for the development of an extracorporeal bioartificial liver system, aimed at extending the life of those waiting for transplantation, by providing critical detoxification, anti-inflammatory, metabolic and synthetic functions, as well as stimulation of regeneration in the injured liver [12].

In the past 3 decades, reliable techniques for primary hepatocyte culture have emerged [13, 14]. Albeit, in standard monolayer cultures, primary cells rapidly lose their polarized architectures, leading to the loss of liver-specific functions [15]. These cells accumulate actin stress fibers, are described as becoming fibroblast-like, lack bile canaliculi, and die within a few days [16, 17]. Therefore, maintaining culture environments that support the three-dimensional architecture and polarization of the cells is critical for retaining liver functions for long periods ex vivo. This need was addressed in later developments that enable maintenance of cell function and structure for several months in culture under several widely different configurations. Culture in sandwich configuration or spheroids relies on the assembly of three-dimensional-like structures, while cocultures with endothelial cells or fibroblasts depend on heterotypic cell–cell interactions [18]. Finally, oxygen supply and medium formulation are critical aspects in the design of the optimal hepatocyte microenvironment [19].

This chapter will cover the current knowledge of hepatocyte culture, including drawbacks, limitations, and the most common culture techniques. We will describe culture media formulation, hepatocyte culture techniques, coculture techniques, and the importance of oxygen supply.

2 Materials

2.1 General Solutions and Materials

1. 10× concentrated phosphate-buffered saline (PBS). Dissolve 80 g NaCl, 2.0 g KCl, and 14.4 g Na_2HPO_4 in 1 L ultrapure water. Adjust to pH 7.4 and store at 4 °C.
2. 1 N NaOH solution. Add 4 g NaOH to 10 mL ultrapure water.
3. Matrigel (Corning, Massachusetts).
4. Rat tail collagen type I (BD Biosciences, San Joje, California). Store at 4 °C.
5. Trypsin–ethylenediaminetetraacetic acid solution.
6. Incubator (37 °C ± 1 °C, 90 % ± 5 % humidity, 5 % ± 1 % CO_2).
7. Laminar air flow cabinet.
8. Plastic culture dishes.

2.2 Hepatocyte Culture Medium

1. Fetal bovine serum. Heat inactivate the fetal bovine serum by placing the thawed aliquot in a waterbath set at 60 °C for 30 min.
2. 200 mM L-glutamine (*see* **Note 1**).
3. 0.1 mg epidermal growth factor (EGF) from murine submaxillary gland. Prepare a solution by adding to 1 mL ultrapure water in sterile conditions. Aliquot to 100 μL and store at −20 °C.
4. 10,000 IU penicillin–10 mg streptomycin/mL solution. Aliquot to 5 mL and store at −20 °C.
5. 100 mg hydrocortisone sodium succinate. Prepare a solution by adding to 2 mL ultrapure water in sterile conditions. Store at 4 °C.
6. 100× concentrated insulin–transferrin–selenium (ITS) liquid media supplement (Sigma-Aldrich, Israel).
7. Hepatocyte culture medium. Under sterile conditions, add 0.2 mM L-glutamine, 100 μL EGF solution, 75 μL hydrocortisone solution, 5 mL liquid ITS liquid media supplement, 5 mL penicillin–streptomycin solution, and 50 mL fetal bovine serum to 440 mL of high glucose Dulbecco's Modified Eagle's Medium. Hepatocyte culture medium should be stored at 4 °C and used within a few weeks (*see* **Notes 2** and **3**).

2.3 Endothelial Cell Culture Medium

EGM™-2mv Bulletkit™ (Lonza, Switzerland). Thaw the Bullkit™ components and prepare according to the manufacturer's instruction. Endothelial cell culture medium should be stored at 4 °C and used within a few weeks.

2.4 Fibroblast Cell Culture Medium

1. Fetal bovine serum. Heat inactivate the fetal bovine serum by placing the thawed aliquot in a waterbath set at 60 °C for 30 min.
2. 200 mM L-glutamine (*see* **Note 1**).
3. 10,000 IU penicillin–10 mg streptomycin/mL solution. Aliquot to 5 mL and store at −20 °C.
4. Fibroblast culture medium. Under sterile conditions, add 0.2 mM L-glutamine, 5 mL penicillin–streptomycin solution, and 50 mL fetal bovine serum to 440 mL of high glucose Dulbecco's Modified Eagle's Medium. Fibroblast culture medium should be stored at 4 °C and used within a few weeks (*see* **Notes 2** and **3**).

2.5 Coculture Medium

1. Fetal bovine serum. Heat inactivate the fetal bovine serum by placing the thawed aliquot in a waterbath set at 60 °C for 30 min.
2. Roswell Park Memorial Institute (RPMI) medium (Sigma-Aldrich, Israel).
3. 10,000 IU penicillin–10 mg streptomycin/mL solution. Aliquot to 5 mL and store at −20 °C.
4. 100 mM dexamethasone solution in ethanol. Aliquot and store at −20 °C.
5. 100× ITS liquid media supplement (Sigma-Aldrich, Israel).
6. 1 M 4-(2-hydroxyethyl)-1-piperazineethanesulfonic acid (HEPES) (Sigma-Aldrich, Israel) solution.
7. 2 Microgram/mL fibroblast growth factor 2 (FGF2) solution (Peprotech, Israel). Reconstitute in 0.1 % bovine serum albumin in PBS. Aliquot and store at −20 °C.
8. 10 Microgram/mL vascular endothelial growth factor (VEGF) solution (Peprotech, Israel). Reconstitute in 0.1 % bovine serum albumin in PBS. Aliquot and store at −20 °C.
9. 10 Microgram/mL hepatocyte growth factor (HGF) solution (Peprotech, Israel). Reconstitute in 0.1 % bovine serum albumin in PBS. Aliquot and store at −20 °C.
10. 50 mg/mL heparin sodium salt solution. Prepare in double distilled water. Sterilize by filtration and store at 4 °C.
11. Coculture medium. Under sterile conditions, add 0.5 mL HEPES solution, 0.5 mL dexamethasone solution, 10 mL fetal bovine serum, 5 mL penicillin–streptomycin solution, 1.25 mL FGF2 solution, 250 μL VEGF solution, 250 μL HGF solution, 1 mL heparin solution, and 5 mL of ITS liquid media supplement solution to 500 mL RPMI medium. Coculture medium should be stored at 4 °C and used within a few weeks (*see* **Notes 2** and **3**).

3 Methods

3.1 Hepatocyte Culture Medium

While primary hepatocytes can be maintained in several types of culture medium, there is a clear distinction between serum-free and serum-containing medium formulations. Serum-free formulations are often based on William's E medium and are ideal for short-term cultures up to 10 days. One such formulation that yields excellent results is Hepatocyte Maintenance Medium commercially available from Lonza (Israel). Serum-containing formulations were developed for long-term cultures (i.e., several weeks) and require an adaptation period [20, 21].

3.2 Hepatocyte Mono-culture Techniques

3.2.1 Hepatocyte Sandwich Technique

A common technique for primary hepatocyte culture is to seed a monolayer of cells on a thin layer of collagen gel. Under these conditions, hepatocytes secrete albumin and urea, and show minimal cytochrome P450 activity. Unfortunately, these liver-specific functions decline within the first week, suggesting that significant survival factors are missing (Fig. 1). In 1989, Dunn and colleagues suggested to culture hepatocytes in a collagen sandwich configuration [21]. Adding a second collagen layer induces the formation of distinct apical and lateral surfaces. Cells cuboidal morphology was maintained for 42 days with albumin and urea secretion slowly rising and stabilizing after 10 days in culture. Cells that were cultured on a thin layer of collagen for 1 week, after which a second layer of collagen was added, exhibited recovery of albumin secretion levels.

1. Acquire primary hepatocytes (*see* **Note 4**).
2. Place all solutions for establishing hepatocyte sandwich cultures on ice.

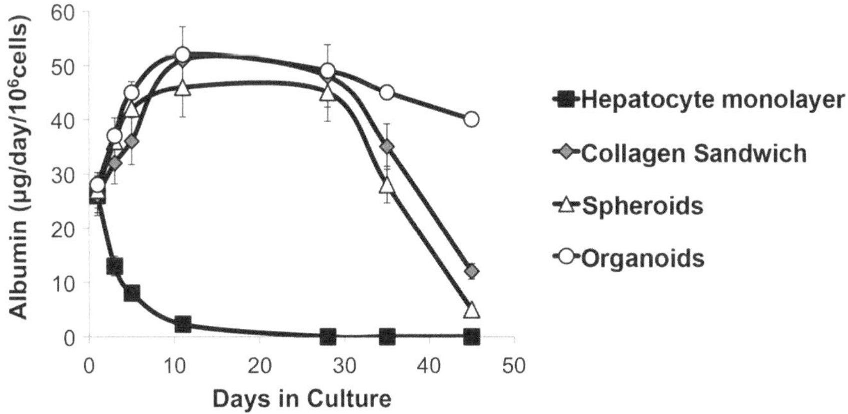

Fig. 1 Long-term albumin production in cultures of primary rat hepatocytes. Primary hepatocytes cultured alone (i.e., hepatocyte monolayer) rapidly lose albumin secretion in vitro. Cells seeded on Matrigel (i.e., spheroids) or in a collagen sandwich configuration (i.e., collagen sandwich) stabilize over 7–10 days and maintain liver-specific function for 28 days. Self-assembled hepatic organoids (i.e., organoids) maintain albumin secretion for over 50 days in culture

3. Determine the final volume of collagen solution needed. For a cell culture plate with a surface of 10 cm^2, use 0.5 mL collagen solution.
4. Determine the volume of 10× concentrated PBS by dividing the final volume calculated by 10.
5. Calculate the volume of collagen needed by dividing the final volume by the collagen concentration as specified by the manufacturer.
6. The volume of 1 N NaOH solution needed is given by the following formula: volume of collagen × 0.023 mL = mL NaOH.
7. Calculate the volume of ultrapute water requied: water volume = (Total volume of solution) − (collagen volume + 1 N NaOH volume + 10 × concantrated PBs volume).
8. Under sterile conditions and on ice, add the calculated volume of 1 N NaOH solution to the calculated volume of 10× concentrated PBS and water.
9. Add the calculated volume of collagen and mix gently.
10. Add 0.5 mL mixture into each well of a 6-well plate and distribute evenly.
11. Incubate at 37 °C and 5 % CO_2 for 30–40 min.
12. Seed 100,000 cells/cm^2 in a final volume of 1 mL/10 cm^2. This roughly translates to 1×10^6 hepatocytes in each well of a 6-well plate.
13. Incubate the cells overnight at 37 °C and 5 % CO_2.
14. Prepare fresh collagen solution.
15. Aspirate culture medium and add 0.5 mL of collagen solution into each well.
16. Incubate at 37 °C and 5 % CO_2 for 30–40 min.
17. Add 1 mL of hepatocyte culture medium to each well.
18. Incubate at 37 °C and 5 % CO_2.

3.2.2 Hepatocyte Spheroid Technique

Hepatocytes can be cultured in the form of hepatic spheroids. When cultured on soft or nonadhesive extracellular matrix, cells form spherical aggregates during the first 48 h. Cells in those aggregates maintain their morphology and liver-specific functions for over 1 month (Fig. 1). A disadvantage of this technique includes the lack of control over spheroid size and thus variation in the transport of metabolites. Another disadvantage is the formation of a necrotic core in big aggregates [22]. Cell morphology in collagen sandwich and spheroid culture configurations is different (Fig. 2). While hepatocytes entrapped in a collagen matrix show a cuboidal morphology [23], spheroids are round and form closely

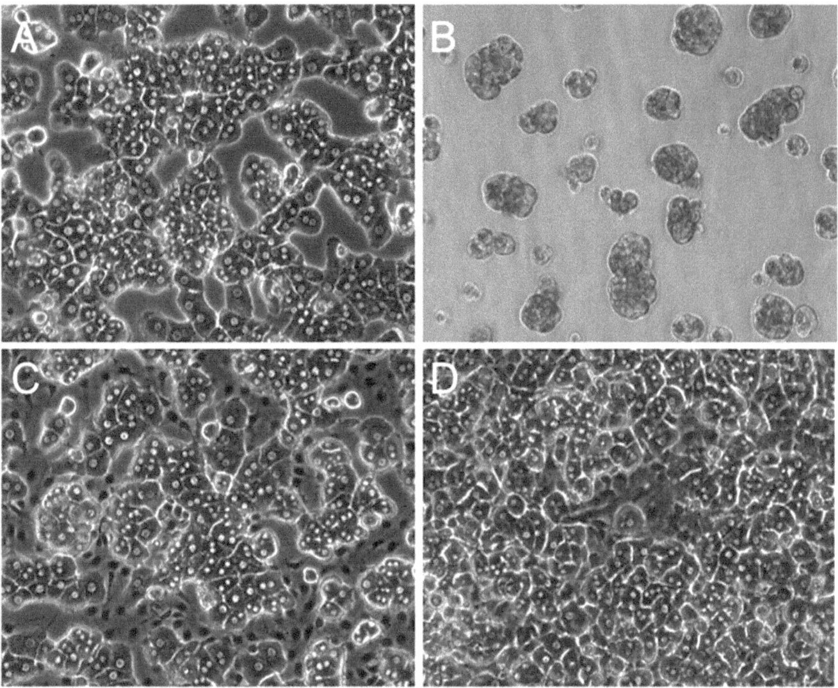

Fig. 2 Phase images of hepatocyte morphology in mono-culture and coculture. (**a**) Hepatocytes cultured in a collagen double gel configuration form plate-like structures and stabilize synthetic and enzymatic activity a week after isolation. Cells in double gel were shown to form cell–cell contacts apical bile canaliculi, express laterally localized E-cadherin [41], but not sinusoidal receptors, such as epidermal growth factor receptor [41] and low density lipoprotein receptor [10]. (**b**) Hepatocyte spheroids formed on Matrigel exhibit significant synthetic and enzymatic activity. Spheroids were shown to express E-cadherin [41], form extensive bile canaliculi [42], and show sinusoidal surface markers at the interface between cells in the spheroid [42]. (**c**) Hepatocytes cocultured with liver sinusoidal endothelial cells show both traditional polarity markers [43, 44] and express a high level of sinusoidal receptors at the interface between the hepatocytes and the liver sinusoidal endothelial cells. At least part of the interaction between hepatocytes and liver sinusoidal endothelial cells has been shown to be mediated by growth factors [28, 45]. (**d**) Hepatocytes cocultured with 3T3 fibroblasts grow in distinct clusters and exhibit hepatic cell–cell contacts, including connexin32 [46] and bile canaliculi [47], but also do not express the epidermal growth factor receptor and low density lipoprotein receptor. At least part of the interaction between hepatocytes and 3T3 fibroblasts was shown to be mediated by N-cadherin and decorin [48]

associated aggregates, which are not found in the mature liver. An assumption is that spheroids resemble the organization of liver during regeneration [1, 24]. Unfortunately, hepatocytes cultured in either configuration cannot generate large quantities of cells and show little to no proliferation capacity [25].

1. Suspend the freshly isolated hepatocytes in serum-containing hepatocyte medium at a concentration of 1.2×10^6 viable cells/mL.
2. Mix 1:1 with ice-cold Matrigel.
3. Place 80 μL of the mix in each well of a 96-well plate (i.e., 264 μL of the mix/cm^2).

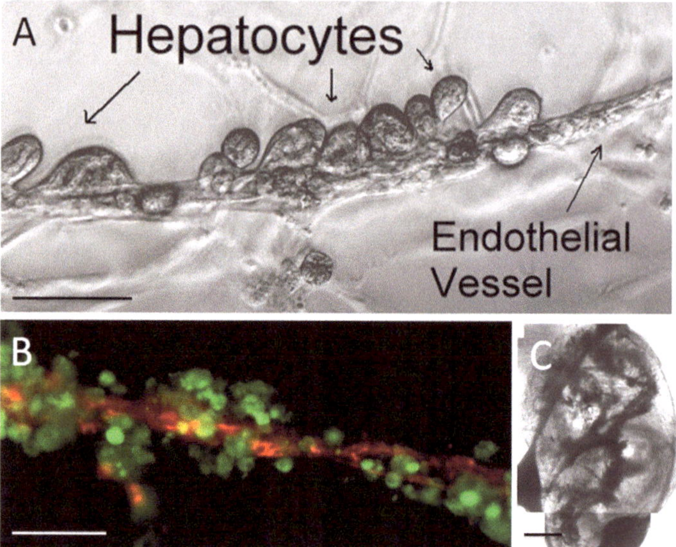

Fig. 3 Hepatic-endothelial organoid formation. (**a**) Phase contrast image of hepatocytes decorating a dermal microvascular endothelial structure formed on Matrigel. Bar = 100 μm. (**b**) Immunofluorescence staining of human hepatocytes (albumin; *green*) enveloping a dermal microvascular endothelial structure (CD31; *red*) formed on Matrigel. Bar = 100 μm. (**c**) Hepatic-endothelial organoid, 3 mm in diameter, day 28 in culture. Note the complex internal structure. Bar = 1 mm

4. Incubate at 37 °C and 5 % CO_2 for 15 min.
5. Add 100 μL of hepatocyte culture medium to each well.

3.3 Hepatocyte Coculture Technique

Cellular functions are influenced by a sum of extracellular factors, which include neighboring cells, extracellular matrix, soluble factors, and physical forces [26]. Adjacent nonparenchymal liver cells play a major role in the regulation of hepatocyte function [27]. Although the nature of the interactions between hepatocytes and endothelial cells of the liver has yet to be elucidated, recent findings suggest it is integral to the maturation of the liver [28, 29], its response to injury [28], epithelial polarization [10], and hepatitis C virus infection [10, 30]. Recently, we have shown that endothelial vascular structures, formed on Matrigel, can recruit primary hepatocytes in a HGF-dependent mechanism, to form liver sinusoid-like structures in vitro (Fig. 3) [31]. The liver-like structures that self-assemble maintain a high level of hepatic liver-specific function and gene expression for over 2 months in culture [31]. The inclusion of fibroblasts in the culture stabilizes the structures and induces proliferation, which culminates in the formation of a liver organoid that is several mm in size. We have formed liver sinusoid-like structures from both rat and human primary hepatocytes, fibroblasts, and endothelial cells. This method is uniquely suited for studying the role of tissue architecture and mechanical properties at single cell resolution and the effects of heterotypic cell interactions underlying processes, such as tissue morphogenesis [29], differentiation [32], and angiogenesis.

3.3.1 Endothelial Cell Culture

1. Acquire and culture endothelial cells (*see* **Note 5**).
2. Wash the cells twice with 5 mL PBS prior to passage.
3. Add 1 mL of 0.25 % trypsin–ethylenediaminetetraacetic acid solution. Tilt back and forth to wash the flask surface. Quickly remove and discard the solution.
4. Add 2 mL of 0.25 % trypsin–ethylenediaminetetraacetic acid solution and incubate at 37 °C for 2 min.
5. Add 10 mL of warm EGM™-2mv medium to the tissue culture flask. Slightly tap the flask bottom to release cells.
6. Remove the 12 mL cell suspension. Aspirate to create a single cell suspension. Split 1:3 into new gelatin-coated tissue culture flasks.

3.3.2 Fibroblast Cell Culture

1. Acquire and culture fibroblasts (*see* **Note 6**).
2. Wash the cells twice with 5 mL PBS prior to passage.
3. Add 1 mL of 0.25 % trypsin–ethylenediaminetetraacetic acid solution. Tilt back and forth to wash the flask surface. Quickly remove and discard the solution.
4. Add 2 mL of 0.25 % trypsin–ethylenediaminetetraacetic acid solution and incubate at 37 °C for 2 min.
5. Add 10 mL of warm fibroblast cell culture medium to the tissue culture flask. Slightly tap the flask bottom to release cells.
6. Remove the 12 mL cell suspension. Aspirate to create a single cell suspension. Split 1:3 into new tissue culture flasks.

3.3.3 Hepatocyte-Endothelial Cell-Fibroblast Coculture

1. Suspend primary hepatocytes in coculture medium at a concentration of 3.6×10^6 viable cells/mL (*see* **Note 7**).
2. Suspend the endothelial cells in coculture medium at a concentration of 3.6×10^6 viable cells/mL.
3. Suspend the fibroblasts in coculture medium at a concentration of 3.6×10^5 viable cells/mL.
4. Mix the cell suspensions at 1:1:1 ratio.
5. Mix 1:1 with ice-cold Matrigel.
6. Place 80 μL of the mix in each well of a 96-well plate (i.e., 264 μL of the mix/cm^2).
7. Incubate at 37 °C and 5 % CO_2 for 15 min.
8. Add 100 μL of coculture medium to each well.

3.4 Oxygen Supply

One of the limiting factors in hepatic tissue engineering is oxygen and nutrient transport. Oxygen consumption rate of hepatocytes can be as high as 0.9 nmol/s/10^6 cells [33, 34]. In vivo, in order to supply cells with this amount of oxygen, the liver is connected to a highly oxygenated arterial network, in addition to the portal circulation, delivering close to 1.29 nmol oxygen/s/10^6 cells. The low solubility of oxygen in aqueous media is compensated in vivo by the presence of oxygen-binding hemoglobin [35]. In vitro, oxygen is supplied by diffusion from the air–liquid interface, severely limiting the cell density that can be seeded in a given culture area [25]. It has been previously proposed that high oxygen tensions stimulate production of free radicals that can damage cultured cells [36, 37]. However, our group demonstrated that hepatic damage at high oxygen tensions is a result of serum adaptation at the early stages of culture. Removing serum completely from the hepatocyte culture medium, while increasing oxygen tension to 95 %, caused a dramatic threefold increase in albumin synthesis and 74 % increase in cytochrome P450 activity [20]. Gene expression, functional polarization, and drug metabolism were similarly enhanced in both rat and human primary hepatocytes. Remarkably, these oxygenated cocultures showed an ability to predict in vivo hepatic clearance rates of both rapid and slow clearing drugs, such as carbamazepine and antipyrine, with a correlation coefficient of 0.92.

4 Notes

1. L-Glutamine rapidly degrades at temperatures above 4 °C and should be added ex tempore.
2. Human cells require ascorbic acid, which is absent from the formulation indicated. Add ascorbic acid (Sigma-Aldrich, Israel) when using human hepatocytes. Prepare a 100 mM solution by adding 1.76 g acetic acid to 100 mL ultrapure water. Aliquot and store at −20 °C. Add 5 mL solution to 500 mL hepatocyte culture medium.
3. Phenol red is known to be a weak estrogen [38]. Change to phenol red-free basal medium if this is an experimental concern.
4. Primary hepatocytes do not proliferate and should be freshly isolated. Several companies offer high-quality human hepatocytes in a collagen-coated 6-well plate format, including BD Biosciences San Joje, California and Cambrex Lonza (Switzerland). Adherent hepatocytes in a 6-well format can be easily released by trypsin–ethylenediaminetetraacetic acid or collagenase H treatment during the first 24 h after isolation. Rat hepatocytes are typically acquired by means of a 2-step collagenase perfusion technique [39, 40].

5. Endothelial cells for establishing cocultures include dermal microvascular endothelial Lonza (Switzerland), umbilical vein endothelial cells Lonza (Switzerland), or liver endothelial cells (ScienCell, Carlsbad, California) on gelatin-coated tissue culture flasks in EGM™-2mv culture medium. Endothelial vascular network formation is strongly dependent on the quality and age of the endothelial cells used. Microvascular and umbilical vein endothelial cells should be used prior to passage 6–8. Liver endothelial cells (i.e., nonsinusoidal) should be used before passage 3. Cells should be split when 80 % confluent and culture medium should be changed every 48 h.
6. Fibroblasts for establishing cocultures include 3T3-J2, MEFs, or T10½ fibroblasts in tissue culture flasks in fibroblast cell culture medium. Cells should be split when 80 % confluent. Culture medium should be changed every 48 h.
7. Use hepatocytes within 1–2 h following their isolation.

Acknowledgements

The authors wish to thank Dr. Maria Shulman for technical advice. This work was funded by a European Research Council Starting Grant TMIHCV (project number 242699), a Marie Curie Reintegration Grant microLiverMaturation (project number 248417), the Israel-Japan Ministry of Science (award number 9645), the British Council BIRAX Regenerative Medicine award (number 33BX12HGYN) and HeMiBio, a jointly funded consortium by the European Commission and Cosmetics Europe as part of the SEURAT-1 cluster (project number HEALTH-F5-2010-266777).

References

1. Taub R (2004) Liver regeneration: from myth to mechanism. Nat Rev Mol Cell Biol 5:836–847
2. Nahmias Y, Berthiaume F, Yarmush ML (2007) Integration of technologies for hepatic tissue engineering. Adv Biochem Eng Biotechnol 103:309–329
3. Block GD, Locker J, Bowen WC et al (1996) Population expansion, clonal growth, and specific differentiation patterns in primary cultures of hepatocytes induced by HGF/SF, EGF and TGF alpha in a chemically defined (HGM) medium. J Cell Biol 132:1133–1149
4. Paul SM, Mytelka DS, Dunwiddie CT et al (2010) How to improve R&D productivity: the pharmaceutical industry's grand challenge. Nat Rev Drug Discov 9:203–214
5. Pritchard JF, Jurima-Romet M, Reimer ML et al (2003) Making better drugs: decision gates in non-clinical drug development. Nat Rev Drug Discov 2:542–553
6. Rodriguez JV, Pizarro MD, Scandizzi AL et al (2008) Construction and performance of a minibioreactor suitable as experimental bioartificial liver. Artif Organs 32:323–328
7. Hewitt NJ, Lechon MJ, Houston JB et al (2007) Primary hepatocytes: current understanding of the regulation of metabolic enzymes and transporter proteins, and pharmaceutical practice for the use of hepatocytes in metabolism, enzyme induction, transporter, clearance, and hepatotoxicity studies. Drug Metab Rev 39:159–234

8. Kamiya A, Kinoshita T, Ito Y et al (1999) Fetal liver development requires a paracrine action of oncostatin M through the gp130 signal transducer. EMBO J 18:2127–2136
9. Michalopoulos GK, Khan Z (2005) Liver regeneration, growth factors, and amphiregulin. Gastroenterology 128:503–506
10. Nahmias Y, Casali M, Barbe L et al (2006) Liver endothelial cells promote LDL-R expression and the uptake of HCV-like particles in primary rat and human hepatocytes. Hepatology 43:257–265
11. Richards CD, Brown TJ, Shoyab M et al (1992) Recombinant oncostatin M stimulates the production of acute phase proteins in HepG2 cells and rat primary hepatocytes in vitro. J Immunol 148:1731–1736
12. Nyberg SL (2012) Bridging the gap: advances in artificial liver support. Liver Transpl 18(Suppl 2):S10–S14
13. Dunn JC, Tompkins RG, Yarmush ML (1991) Long-term in vitro function of adult hepatocytes in a collagen sandwich configuration. Biotechnol Prog 7:237–245
14. Koide N, Shinji T, Tanabe T et al (1989) Continued high albumin production by multicellular spheroids of adult rat hepatocytes formed in the presence of liver-derived proteoglycans. Biochem Biophys Res Commun 161:385–391
15. Elliott NT, Yuan F (2011) A review of three-dimensional in vitro tissue models for drug discovery and transport studies. J Pharm Sci 100:59–74
16. Guguen-Guillouzo C, Clement B, Baffet G et al (1983) Maintenance and reversibility of active albumin secretion by adult rat hepatocytes co-cultured with another liver epithelial cell type. Exp Cell Res 143:47–54
17. Bissell DM, Hammaker LE, Meyer UA (1973) Parenchymal cells from adult rat liver in non-proliferating monolayer culture: functional studies. J Cell Biol 59:722–734
18. Bhatia SN, Balis UJ, Yarmush ML et al (1999) Effect of cell-cell interactions in preservation of cellular phenotype: cocultivation of hepatocytes and nonparenchymal cells. FASEB J 13: 1883–1900
19. Nishikawa M, Kojima N, Komori K et al (2008) Enhanced maintenance and functions of rat hepatocytes induced by combination of on-site oxygenation and coculture with fibroblasts. J Biotechnol 133:253–260
20. Kidambi S, Yarmush RS, Novik E et al (2009) Oxygen-mediated enhancement of primary hepatocyte metabolism, functional polarization, gene expression, and drug clearance. Proc Natl Acad Sci U S A 106:15714–15719
21. Dunn JC, Yarmush ML, Koebe HG et al (1989) Hepatocyte function and extracellular matrix geometry: long-term culture in a sandwich configuration. FASEB J 3:174–177
22. Zakim D, Boyer T (1996) Hepatology: a textbook of liver disease, vol 1. W.B. Saunders Company, Philadelphia
23. Berthiaume F, Moghe PV, Toner M et al (1996) Effect of extracellular matrix topology on cell structure, function, and physiological responsiveness: hepatocytes cultured in a sandwich configuration. FASEB J 10:1471–1484
24. Michalopoulos GK, DeFrances MC (1997) Liver regeneration. Science 276:60–66
25. Yarmush ML, Toner M, Dunn JC et al (1992) Hepatic tissue engineering. Development of critical technologies. Ann N Y Acad Sci 665:238–252
26. Bhatia SN, Yarmush ML, Toner M (1997) Controlling cell interactions by micropatterning in co-cultures: hepatocytes and 3T3 fibroblasts. J Biomed Mater Res 34:189–199
27. Kmiec Z (2001) Cooperation of liver cells in health and disease. Adv Anat Embryol Cell Biol 161:III-XIII, 1–151
28. LeCouter J, Moritz DR, Li B et al (2003) Angiogenesis-independent endothelial protection of liver: role of VEGFR-1. Science 299: 890–893
29. Cleaver O, Melton DA (2003) Endothelial signaling during development. Nat Med 9: 661–668
30. Gardner JP, Durso RJ, Arrigale RR et al (2003) L-SIGN (CD 209L) is a liver-specific capture receptor for hepatitis C virus. PNAS 100: 4498–4503
31. Nahmias Y, Schwartz RE, Hu WS et al (2006) Endothelium-mediated hepatocyte recruitment in the establishment of liver-like tissue in vitro. Tissue Eng 12:1627–1638
32. Lammert E, Cleaver O, Melton D (2001) Induction of pancreatic differentiation by signals from blood vessels. Science 294:564–567
33. Balis UJ, Behnia K, Dwarakanath B et al (1999) Oxygen consumption characteristics of porcine hepatocytes. Metab Eng 1:49–62
34. Foy BD, Rotem A, Toner M et al (1994) A device to measure the oxygen uptake rate of attached cells: importance in bioartificial organ design. Cell Transplant 3:515–527
35. Fisher RJ, Peattie RA (2007) Controlling tissue microenvironments: biomimetics, transport phenomena, and reacting systems. Adv Biochem Eng Biotechnol 103:1–73

36. Fariss MW (1990) Oxygen toxicity: unique cytoprotective properties of vitamin E succinate in hepatocytes. Free Radic Biol Med 9:333–343
37. Martin H, Sarsat JP, Lerche-Langrand C et al (2002) Morphological and biochemical integrity of human liver slices in long-term culture: effects of oxygen tension. Cell Biol Toxicol 18:73–85
38. Berthois Y, Katzenellenbogen JA, Katzenellenbogen BS (1986) Phenol red in tissue culture media is a weak estrogen: implications concerning the study of estrogen-responsive cells in culture. Proc Natl Acad Sci U S A 83: 2496–2500
39. Seglen PO (1976) Preparation of isolated rat liver cells. Methods Cell Biol 13:29–83
40. Berry MN, Grivell AR, Grivell MB et al (1997) Isolated hepatocytes: past, present and future. Cell Biol Toxicol 13:223–233
41. Moghe PV, Berthiaume F, Ezzell RM et al (1996) Culture matrix configuration and composition in the maintenance of hepatocyte polarity and function. Biomaterials 17: 373–385
42. Abu-Absi SF, Friend JR, Hansen LK et al (2002) Structural polarity and functional bile canaliculi in rat hepatocyte spheroids. Exp Cell Res 274:56–67
43. Goulet F, Normand C, Morin O (1988) Cellular interactions promote tissue-specific function, biomatrix deposition and junctional communication of primary cultured hepatocytes. Hepatology 8:1010–1018
44. Morin O, Normand C (1986) Long-term maintenance of hepatocyte functional activity in co-culture: requirements for sinusoidal endothelial cells and dexamethasone. J Cell Physiol 129:103–110
45. Davidson AJ, Zon LI (2003) Biomedicine: love, honor, and protect (your liver). Science 299:835–837
46. Sugimachi K, Sosef MN, Baust JM et al (2004) Long-term function of cryopreserved rat hepatocytes in a coculture system. Cell Transplant 13:187–195
47. Bhandari RN, Riccalton LA, Lewis AL et al (2001) Liver tissue engineering: a role for coculture systems in modifying hepatocyte function and viability. Tissue Eng 7:345–357
48. Khetani SR, Szulgit G, Rio JAD et al (2004) Exploring interactions between rat hepatocytes and nonparenchymal cells using gene expression profiling. Hepatology 40:545–554

Chapter 12

Primary Hepatocytes in Sandwich Culture

Janneke Keemink, Marlies Oorts, and Pieter Annaert

Abstract

Hepatocytes in sandwich configuration constitute of primary hepatocytes cultured between two layers of extracellular matrix. Sandwich-cultured hepatocytes maintain expression of liver-specific proteins and gradually form intact bile canaliculi with functional biliary excretion of endogenous compounds and xenobiotics. Both freshly isolated and cryopreserved hepatocytes can be used to establish sandwich cultures. Therefore, this preclinical model has become an invaluable in vitro tool to evaluate hepatobiliary drug transport, metabolism, hepatotoxicity, and drug interactions. In this chapter, commonly used procedures to cultivate primary hepatocytes from human and rat in sandwich configuration are described.

Key words Primary hepatocytes, Sandwich culture, In vitro model, Cryopreserved hepatocytes, Hepatic drug disposition, Hepatotoxicity

1 Introduction

Primary hepatocytes are extensively used to assess hepatobiliary drug disposition and drug-induced hepatotoxicity. However, rapid deterioration of liver-specific functions after isolation and loss of cell polarity limits the applicability in suspension and conventional culture conditions. Therefore, in 1989, Dunn and group cultured rat hepatocytes between two layers of collagen, enabling their long-term maintenance [1]. Since then, a number of procedures have been described to culture primary hepatocytes from different preclinical species and human in this configuration, also known as sandwich-cultured hepatocytes [2, 3]. Sandwich-cultured hepatocytes maintain polygonal morphology, albumin secretion, as well as overall transcriptional activity [1]. Moreover, sandwich-cultured hepatocytes gradually repolarize and form extensive biliary networks. In addition, they show increased expression of relevant metabolic enzymes and transporters as compared to conventionally cultured hepatocytes (i.e., on a single layer of collagen). For these reasons, sandwich-cultured hepatocytes constitute a preferred comprehensive in vitro model to investigate mechanisms underlying

Mathieu Vinken and Vera Rogiers (eds.), *Protocols in In Vitro Hepatocyte Research*, Methods in Molecular Biology, vol. 1250, DOI 10.1007/978-1-4939-2074-7_12, © Springer Science+Business Media New York 2015

Table 1
Applications of sandwich-cultured hepatocytes

Applications of SCH	Examples
Clearance prediction	
Metabolism	In vitro intrinsic clearance of slowly metabolized tolbutamide [4]
Transporter mediated uptake	Characterization of digoxin uptake [24]
Biliary excretion	
Accumulation method	Prediction of biliary clearance of angiotensin II receptor blockers and HMG-CoA reductase inhibitors [7]
Efflux method	Basolateral versus biliary clearance of rosuvastatin [25]
Imaging	Biliary excretion of a fluorescent bile acid derivative [26]
Hepatobiliary disposition pathways	Evaluation of mechanisms underlying systemic exposure to furamidine and CPD-0801 [27]
Drug–drug interactions	
Induction	Omeprazole, phenobarbital, and rifampicine induce several enzymes and transporters [28]
Inhibition	Ritonavir inhibits cytochrome P450 3A4 and reduces midazolam metabolism [29]
Mechanisms underlying hepatotoxicity	
Reactive metabolite formation	Elucidation of the role of metabolites of valproic acid in hepatotoxicity [12]
Drug-induced cholestasis	Incubation of SCH with compounds and bile acid mixture can identify cholestatic compounds [10]
Other	Evaluation of mechanisms underlying toxicity of substrates of equilibrative and concentrative nucleoside transporters [30]

Sandwich-cultured hepatocytes are used to evaluate different aspects of hepatic drug disposition and hepatotoxicity

hepatobiliary drug disposition. This includes hepatic drug uptake, metabolism, basolateral efflux, and biliary excretion (Table 1) [4–7]. Importantly, sandwich-cultured hepatocytes have frequently been utilized to generate intrinsic biliary clearance values which, upon scaling to whole liver, correlate well with in vivo drug clearance values in both human and rat [7, 8]. Furthermore, sandwich-cultured hepatocytes allow prediction of inhibition and induction potential of compounds involved in drug interactions (Table 1) [9]. Finally, sandwich-cultured hepatocytes have been successfully used to determine potential hepatotoxicity of xenobiotics, including the underlying mechanisms (e.g., drug-induced cholestasis) (Table 1) [10–12].

Availability of freshly isolated human hepatocytes is restricted and unpredictable. Moreover, frequent hepatocyte isolation from animals should be limited from an ethical perspective. Therefore, preparation of sandwich cultures from cryopreserved hepatocytes allows rational use of hepatocytes. These cultures generally result in transporter and enzyme activities comparable to activities in cultures established with freshly isolated hepatocytes [2, 13]. Additionally, optimal use of hepatocytes can be obtained with multiwell plates. Both rat and human hepatocytes have been cultured in culture plates of the 96-well format [14, 15]. Sandwich-cultured hepatocytes thus represent a medium-throughput to high-throughput model which allows the assessment of all aspects of hepatic drug disposition and drug-induced hepatotoxicity in early stages of drug development.

2 Materials

In all procedures, sterile ultrapure water (i.e., prepared by purifying deionized water to attain a resistivity of 18.2 MΩ cm at 25 °C) should be used. All procedures should be performed under sterile conditions in a laminar airflow cabinet.

2.1 Collagen Extraction from Rat Tails

1. (Frozen) rat tails.
2. 0.5 M acetic acid solution. Dilute 14.7 mL glacial acetic acid in 400 mL sterile ultrapure water. Add sterile ultrapure water to 500 mL. Work in sterile conditions.
3. 0.1 M acetic acid solution. Dilute the 0.5 M acetic acid solution 5 times with sterile ultrapure water. Work in sterile conditions.
4. 25 % NaCl in 0.5 M acetic acid solution. Dissolve 125 g NaCl in 400 mL ultrapure water. Add 14.7 mL glacial acetic acid. Add ultrapure water to 500 mL. Pass through a 0.2 μm filter. Work in sterile conditions.

2.2 Coating of Cell Culture Plates and Overlay of Hepatocytes

All components used for the coating and overlay should be kept on ice and should be used immediately to prevent gelation of matrix components. The required volume of coating and overlay can be calculated according to the plate size and number of wells (Table 2).

Table 2
Volume of collagen solutions used to coat different sizes of cell culture plates for hepatocytes

	Collagen volume		PBS volume
Plate format	Gelled (μL)	Rigid (mL)	(mL)
60-mm dish	250	3	3
6-Well plate	100	1.5	1.5
12-Well plate	70	1	1
24-Well plate	50	0.5	0.5
48-Well plate		0.25	0.25
96-Well plate		0.1	0.1

PBS is used to hydrate and wash collagen coated plates after coating

1. The most common extracellular matrix mixtures for sandwich-cultured hepatocytes are gelled collagen, rigid collagen, and Matrigel or Geltrex:
   - 1.5 mg/mL gelled collagen. Prepare a 0.2 N NaOH solution in deionized water and sterile filter this using a 0.2 μm filter. This solution can be kept under sterile conditions at 4 °C for 6 months. In a 50 mL tube, add 40 % of rat tail collagen I freshly prepared or commercially obtained (BD Biosciences, Belgium), 30 % sterile deionized water and 10 % 10× concentrated Dulbecco's Modified Eagle Medium. Neutralize the mixture by adding 20 % sterile 0.2 N NaOH solution. The NaOH solution should be added dropwise under constant shaking until the yellow color turns fuchsia, which corresponds with pH 7.4.
   - 50 μg/mL rigid collagen. Prepare a 0.02 N acetic acid solution in deionized water and sterile filter this using a 0.2 μm filter. In a 50 mL tube, dilute rat tail collagen I with the sterile acetic acid solution to a final concentration of 50 μg/mL.
   - 0.25 mg/mL Matrigel® or Geltrex® working solution. Prepare by diluting Matrigel (BD Biosciences, Belgium) or Geltrex (Life technologies, United Kingdom) in ice-cold culture medium.
2. Phosphate-buffered saline (PBS). 137 mM NaCl, 2.7 mM KCl, 10 mM $Na_2HPO_4 \cdot 2H_2O$, 1.8 mM KH_2PO_4 in deionized water. Adjust to pH 7.4 and store for maximum 6 months at 4 °C.
3. Sterile cell culture plates (Tables 2 and 3).

Table 3
Reference values for cell seeding and culture media volumes used for rat and human hepatocytes (*see* Note 18)

Plate format	Cell density ($\times 10^6$ cells/mL)	Seeding volume (mL)	Culture medium volume (mL)
60-mm dish	1	3	3
6-Well plate	1	1.5	1.5
12-Well plate	1	0.8	1
24-Well plate	1	0.4	0.5
48-Well plate	1	0.2	0.25
96-Well plate	0.4	0.1	0.1

4. 1 mL syringe.
5. 30G × ½′ needles.
6. Cell scraper.

2.3 Thawing of Cryopreserved Hepatocytes

1. Thawing medium. Dulbecco's Modified Eagle Medium containing 10 % fetal bovine serum, 2 mM L-glutamine, 100 IU/mL penicillin, 100 μg/mL streptomycin, 4 μg/mL insulin, and 1 μM dexamethasone.
2. Seeding medium for cryopreserved hepatocytes. Williams' E medium containing 10 % fetal bovine serum, 2 mM L-glutamine, 100 IU/mL penicillin, 100 μg/mL streptomycin, 4 μg/mL insulin, and 1 μM dexamethasone (*see* **Notes 1** and **2**).
3. 10× concentrated PBS.
4. Percoll® (Sigma-Aldrich, Belgium).
5. Trypan blue.
6. Counting chamber (Marienfeld, Germany).

2.4 Culturing Hepatocytes in Sandwich Configuration

1. Freshly isolated or cryopreserved hepatocytes can be used (*see* **Notes 3–5**).
2. Seeding medium for freshly isolated hepatocytes. Williams' E medium containing 5 % fetal bovine serum, 2 mM L-glutamine, 100 IU/mL penicillin, 100 μg/mL streptomycin, 4 μg/mL insulin, and 1 μM dexamethasone (*see* **Notes 1** and **2**).
3. Seeding medium for cryopreserved hepatocytes. Williams' E medium containing 10 % fetal bovine serum, 2 mM L-glutamine, 100 IU/mL penicillin, 100 μg/mL streptomycin, 4 μg/mL insulin, and 1 μM dexamethasone (*see* **Notes 1** and **2**).
4. Culture medium. Williams' E medium containing 2 mM L-glutamine, 100 IU/mL penicillin, 100 μg/mL streptomycin, 1 % insulin–transferrin–selenium + premix (VWR, Belgium), and 0.1 μM dexamethasone (*see* **Notes 2** and **6**).
5. Trypan blue.
6. Counting chamber (Marienfeld, Germany).
7. 10× concentrated PBS.
8. 1 mL syringe.
9. 30G × ½′ needles.
10. Coated cell culture plates (Table 2).
11. VisiCam 3.0 digital camera and operation and acquisition software Visicam Image Analyser 7 (VWR, Belgium).
12. Olympus IX70 inverted tissue culture microscope (Olympus Optical C, Germany).
13. Incubator (37 °C ± 1 °C, 90 % ± 5 % humidity, 5 % ± 1 % CO_2).
14. Laminar air flow cabinet.

3 Methods

3.1 Collagen Extraction from Rat Tails

3.1.1 Day 1

1. Put the (frozen) rat tails in 70 % ethanol for about 30 min (i.e., 10–15 tails for ±10 g collagen) and bring into the laminar airflow cabinet.
2. Strip off the skin using tweezers, break off the tail starting from the end and cut off the collagen fibers.
3. Place the fibers in 0.5 M acetic acid solution. Use 200 mL/2 g dry collagen.
4. Let the collagen dissolve in a sterile bottle for at least 48 h in a cold room (i.e., 4 °C) on a magnetic stirrer.

3.1.2 Day 3

1. Centrifuge the dissolved collagen at 887 × *g* and 4 °C for 60 min.
2. Keep the supernatants and precipitate the clear solution by adding 1/9 volume of 0.5 M acetic acid containing 25 % NaCl in small portions.
3. Stir for 24 h at 4 °C.

3.1.3 Day 4

1. Centrifuge the solution at 887 × *g* and 4 °C for 60 min.
2. Keep the supernatants and add 1/8 volume of 0.5 M acetic acid containing 25 % NaCl. This time the solution will become very cloudy and white.
3. Stir for 24 h at 4 °C.

3.1.4 Day 5

1. Centrifuge the solution at 887 × *g* and 4 °C for 60 min.
2. Keep the pellet and add 0.1 M acetic acid solution (i.e., about 50 mL per pellet) to dissolve.
3. Adjust the concentration to 0.15–0.20 % with 0.1 M acetic acid.
4. The collagen concentration can be measured with a spectrophotometer at 230 nm after diluting the collagen solution (i.e., 1/100) and using the 0.1 M acetic acid solution as blank. Calculate the concentration with $C = 0.621 \times$ absorption value, expressed in mg/mL or compare to commercial collagen.
5. Dilute to 3.75 mg/mL.

3.2 Coating of Cell Culture Plates and Overlay of Hepatocytes

1. Gelled collagen. Add the appropriate volume of 1.5 mg/mL gelled collagen to the desired plate (Table 2). Immediately spread the collagen gently over the cell surface with the use of a cell scraper. Avoid air bubbles. Transfer the plate into a 37 °C and 5 % CO_2 incubator for at least 2 h, and preferably overnight. Hydrate the collagen with the appropriate volume of sterile and warmed (i.e., 37 °C) PBS (Table 2) for at least 2 h before seeding the hepatocytes. The use of gelled collagen is not recommended or feasible for 48- and 96-well plates (*see* **Note 7**).

2. Rigid collagen. Add the appropriate volume of rigid collagen mixture to the desired plate (Table 2). Transfer the plate into a 37 °C and 5 % CO_2 incubator for at least 1 h and preferably overnight. Wash the wells 3 times with PBS. Plates can be used immediately or dried and stored at 2–8 °C for up to 1 week under sterile conditions.

3.3 Thawing of Cryopreserved Hepatocytes

1. Add 16 mL of isotonic Percoll (i.e., 90 % Percoll and 10 % 10× concentrated PBS) to a 50 mL centrifuge tube and add 25 mL thawing medium. Warm this to 37 °C and keep in the water bath until the cells are thawed.
2. Carefully remove the vial from the shipping container or freezer.
3. Immediately immerse the vial into a 37 °C water bath. Shake gently until the ice crystals are entirely melted. Proceed immediately to the next step.
4. Inside the laminar airflow cabinet, open the vial and pour the cells into the centrifuge tube with the mixture of prewarmed thawing medium and isotonic Percoll. Rinse the vial with prewarmed thawing medium and empty it into the centrifuge tube.
5. Dilute to 50 mL with thawing medium at 37 °C.
6. Close the lid and invert the tube slowly about 2–3 times to resuspend the cells.
7. Centrifuge at 168 × *g* and room temperature for 20 min.
8. Pour off the supernatants and resuspend the cell pellet in 20 mL of thawing medium at 37 °C.
9. For human hepatocytes, centrifuge the cell suspension at 100 × *g* and room temperature for 5 min. For rat hepatocytes, centrifuge the cell suspension at 50 × *g* and room temperature for 3 min.
10. Discard the supernatants and resuspend in seeding medium for cryopreserved hepatocytes at 37 °C (i.e., about 3 mL per 10 million cells).
11. Determine the total cell count and the number of viable cells using the trypan blue exclusion method. The viability should be at least 70 % (*see* **Note 8**).

3.4 Culturing Hepatocytes in Sandwich Configuration

3.4.1 Culturing Human Hepatocytes in Sandwich Configuration

1. For freshly isolated hepatocytes, use the cell suspension obtained after isolation. Cells are suspended in the medium, such as Williams' E medium, used during isolation. For a total of at least 100 million cells, hepatocytes should be divided over two 50 mL tubes. Centrifuge at 100 × *g* and room temperature for 5 min (*see* **Note 9**). Aspirate the supernatants and resuspend the hepatocytes in a small volume of seeding medium at 4 °C with about 3 mL per 10 million cells (*see* **Note 10**). Count the cell number and viability of the hepatocytes

with the trypan blue exclusion method. The viability should be at least 85 % (*see* **Note 8**).

2. For cryopreserved hepatocytes, continue from Subheading 3.3.
3. Dilute the cell suspension to a density of 1×10^6 cells/mL 0.4×10^6 cells/ml (see Table 3) with seeding medium at 4 °C.
4. Aspirate PBS from the coated plates and add the appropriate volume of cell suspension (Table 3) (*see* **Note 11**).
5. Transfer the plate to a 37 °C and 5 % CO_2 incubator and gently move the plate horizontally in orthogonal directions to evenly distribute the cells at the bottom of each well (*see* **Note 12**).
6. Allow the cells to attach for 24 h (*see* **Note 13**).
7. Firmly shake the plate to resuspend unadhered cells. Immediately aspirate the medium and add the appropriate volume the Matrigel®-supplemented or Geltrex®-supplemented culture medium onto the plated cells (Table 3) (*see* **Note 14**).
8. After 24 h, replace the culture medium.
9. Replace the medium daily (*see* **Note 15**).
10. Human sandwich-cultured hepatocytes are kept for minimum 6 days. After 5 days of culture, bile canaliculi are formed (Fig. 1) (*see* **Note 16**).

3.4.2 Culturing Rat Hepatocytes in Sandwich Configuration

1. For freshly isolated hepatocytes, divide the cells obtained after isolation over two 50 mL tubes. Cells are suspended in the medium, such as Williams' E medium, used during isolation. Centrifuge at $50 \times g$ and room temperature for 3 min (*see* **Note 9**). Aspirate the supernatant and resuspend the hepatocytes in a small volume of seeding medium at 4 °C with about 3 mL per 10 million cells (*see* **Note10**). Count the cell number and viability of the hepatocytes with the trypan blue exclusion method. The viability should be at least 85 % (*see* **Note 8**).
2. For cryopreserved hepatocytes, continue from Subheading 3.3.
3. If the viability is less than 85 %, perform a Percoll centrifugation to increase the viability. To this end, centrifuge the cell suspension at $50 \times g$ and room temperature for 3 min. Aspirate the supernatant and resuspend each pellet gently in 25 mL of seeding medium at 4 °C. Then add 25 mL of isotonic Percoll. Centrifuge at $70 \times g$ and 4 °C for 5 min. Aspirate the supernatant and resuspend the pellet in 50 mL seeding medium at 4 °C. Centrifuge at $50 \times g$ and room temperature for 3 min, aspirate the supernatant and add seeding medium at 4 °C with about 3 mL per 10 million cells (*see* **Note 10**). Count the number and viability of the hepatocytes with the trypan blue exclusion method (*see* **Note 8**).
4. Dilute the cell suspension with seeding medium at 4 °C to the appropriate cell density (Table 3).

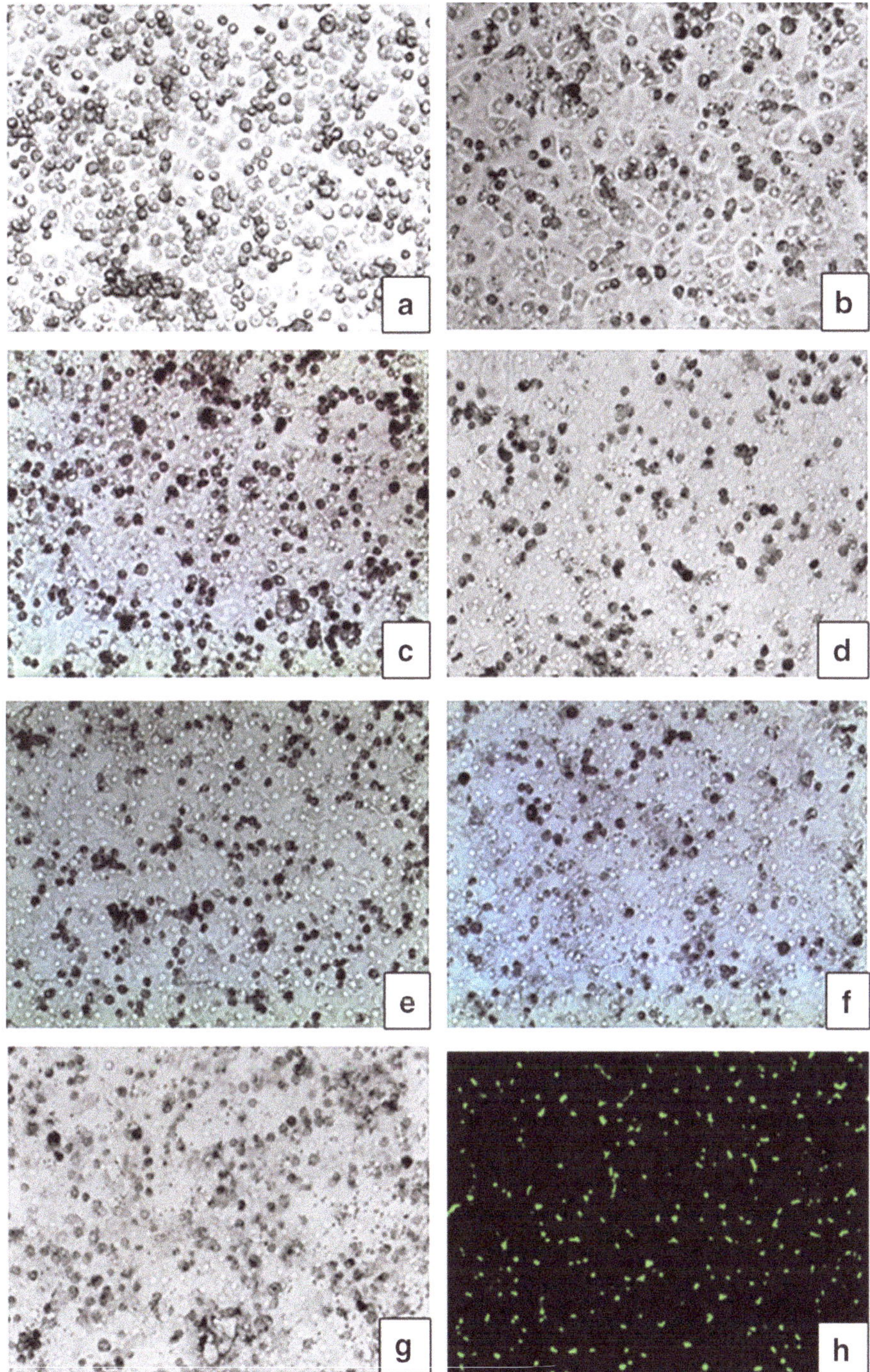

Fig. 1 Sandwich cultures established with human hepatocytes in function of culture time. Panel (**a**–**g**): light microscopy images of human sandwich-cultured hepatocytes at different days in culture (day 0–6). Panel (**h**) fluorescence microscopy image illustrating excretion of the multidrug resistance protein 2 probe 5(6)-carboxy-2′,7′-dichlorofluorescein in bile canaliculi of human sandwich-cultured hepatocytes at day 5 (100× magnification)

5. Aspirate PBS from the coated plates and add cell suspension in the indicated seeding volume (Table 3) (*see* **Note 11**).
6. Transfer the plate to a 37 °C and 5 % CO_2 incubator and gently move the plate horizontally in orthogonal directions to evenly distribute the cells at the bottom of each well (*see* **Note 12**).
7. Allow the cells to attach for 1–2 h.
8. Prepare an appropriate amount of 1.5 mg/mL gelled collagen to be used for the overlay (*see* **Note 17**).
9. Firmly shake the plate to resuspend nonadhered cells. Aspirate the medium and add the appropriate volume (Table 2) of gelled collagen onto the plated cells (*see* **Notes 14** and **15**). Move and tilt the plate in a figure 8-pattern to evenly distribute the collagen over the well.
10. Transfer the plates to a 37 °C and 5 % CO_2 incubator.
11. Allow collagen gelation for 45 min before adding a volume of prewarmed (i.e., 37 °C) seeding medium onto the sandwich-cultured hepatocytes (Table 3).
12. After 24 h, remove the seeding medium and add culture medium onto the sandwich-cultured hepatocytes.
13. Replace the medium daily (*see* **Note 15**).
14. Rat sandwich-cultured hepatocytes are usually kept for 4 days. After 3 days of culture, bile canaliculi are formed (Fig. 2) (*see* **Note 16**).

4 Notes

1. Culture media are supplemented with fetal bovine serum during the first 24 h of culture to promote attachment of hepatocytes to collagen and improve overall hepatocyte morphology. However, prolonged addition causes loss of the differentiated phenotype [16].
2. Dulbecco's Modified Eagle Medium, Modified Chee's medium, or hepatocyte maintenance medium has also been used in seeding and culture medium [2].
3. Freshly isolated hepatocytes, which are obtained following a 2-step collagenase perfusion, can be cultured in sandwich configuration [17].
4. Cryopreserved hepatocytes can be used after thawing. The use of cryopreserved hepatocytes supports a sensible use of liver tissue. The attachment capacity of cryopreserved hepatocytes is strongly batch dependent [18].
5. After isolation or thawing of cryopreserved hepatocytes, cells should be cultured as quickly as possible to minimize viability loss. Low viability will negatively affect cell attachment.

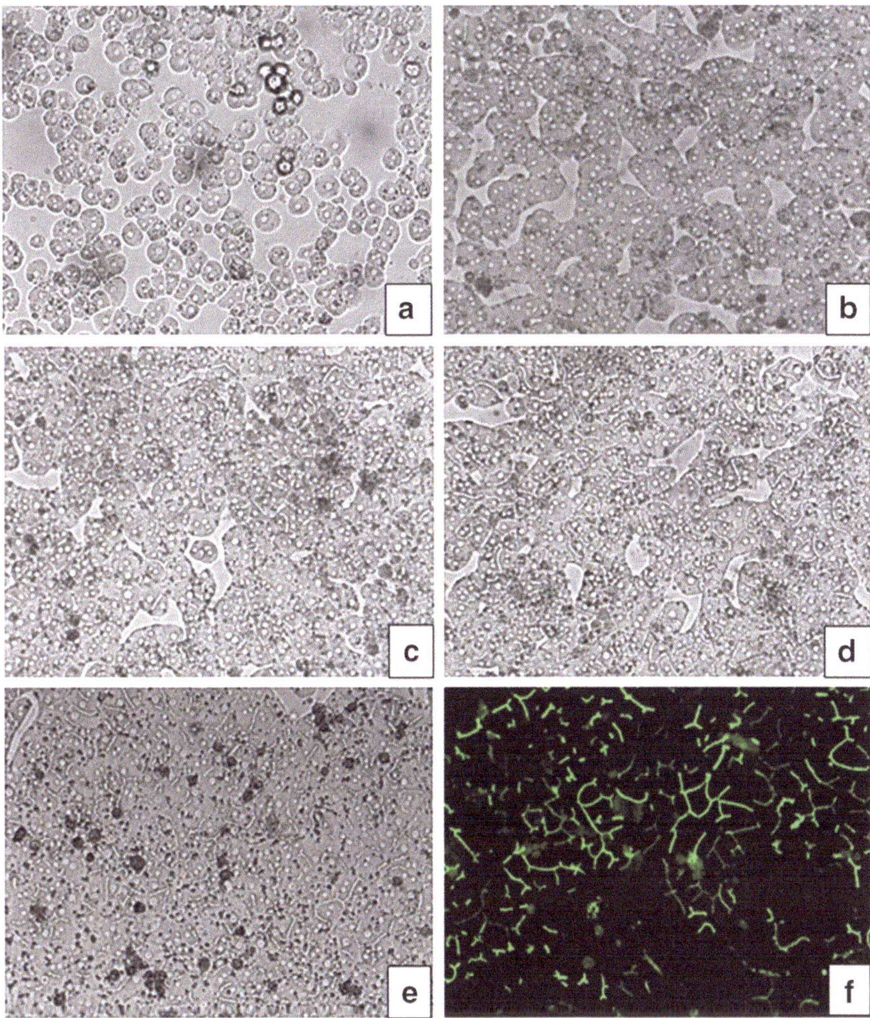

Fig. 2 Sandwich cultures established with rat hepatocytes in function of culture time. Panel (**a**–**e**) light microscopy images of rat sandwich-cultured hepatocytes at different days in culture (day 0–4). Panel (**f**) fluorescence microscopy image illustrating excretion of the multidrug resistance protein 2 probe 5(6)-carboxy-2′,7′-dichlorofluorescein in bile canaliculi of rat sandwich-cultured hepatocytes at day 5 (100× magnification)

6. In sandwich-cultured hepatocytes, cytochrome P450 enzyme activity declines as a function of culture time. Cytochrome P450 inducers, such as phenobarbital (e.g., 1 mM), dexamethasone (e.g., 10 μM), and beta-naphthoflavone (e.g., 5 μM), can be added to the culture medium to improve the metabolic competence which results in a more close resemblance to the in vivo situation [19].
7. Gelled collagen is not applicable for the coating of 48-well and 96-well format cell culture plates, since it is difficult to evenly distribute over the well surface. Hence, rigid collagen is mostly used for those well plates. However, coating with gelled

collagen results in improved cell morphology compared to the use of rigid collagen [20]. Therefore, after coating with rigid collagen, a layer of gelled collagen can be added on top. Due to the first layer, the gelled collagen evenly distributes over the wells without the need for a cell scraper. Hepatocytes seeded on this double layer show a better morphology compared to hepatocytes seeded on a single layer of rigid collagen.

8. An automatic cell counter can be used to calculate cell density and viability.
9. Cell suspensions should be centrifuged in 50 mL tubes even when a small volume is needed. Resuspending the hepatocytes in a smaller tube is difficult and requires more mechanical force, causing damage to the hepatocytes.
10. A pellet of hepatocytes should be resuspended in a small volume of medium (i.e., a volume comparable to the pellet volume) to avoid clot formation. When cells are suspended in the small volume the suspension can be diluted to the required volume.
11. After seeding 1 well, allow the cells to distribute and look under the microscope to check whether the number of cells seeded is adequate. Underseeding will result in poor cell–cell contacts and hepatocyte flattening, which can be corrected by the addition of cell suspension to the wells. Overseeding interferes with cell attachment and should be solved by diluting the initial suspension.
12. Circular swirling after cell seeding will cause the hepatocytes to accumulate in the center of the well and thus causes excessive cell death due to anoxia and poor attachment.
13. Gently moving the plate horizontally in orthogonal directions 2 h after seeding will redistribute the hepatocytes evenly in the wells.
14. Care should be taken when aspirating the medium to avoid damage onto the overlay. If necessary, aspiration force should be decreased.
15. It is recommended to work with 1 plate at a time to reduce the time in which the cells are left without medium.
16. Sandwich-cultured hepatocytes can only be maintained for a limited period of time. With the use of a perfusion system, the liver-specific phenotype (e.g., albumin secretion and metabolism) can be extended [21].
17. Matrigel or Geltrex is superior to gelled collagen for the overlay of human hepatocytes [22]. Although these extracellular matrices can be used for rat hepatocytes as well, it is not recommended. Gelled collagen is the preferred choice, since it is cheaper, results in cultures with the same quality, and is more animal friendly.

18. Except for human and rat hepatocytes, mouse, pig, dog, and monkey hepatocytes have been used as well. Hepatocytes of these species could be cultured in a similar way. However, lower seeding densities should be used for mouse hepatocytes [23].

References

1. Dunn JC, Yarmush ML, Koebe HD et al (1989) Hepatocyte function and extracellular matrix geometry: long-term culture in a sandwich configuration. FASEB J 3:174–177
2. Bi Y, Kazolias D, Duignan DB (2006) Use of cryopreserved human hepatocytes in sandwich culture to measure hepatobiliary transport. Drug Metab Dispos 34:1658–1665
3. Ye Z, Van Pelt J, Camus S et al (2010) Species-specific interaction of HIV protease inhibitors with accumulation of cholyl-glycylamido-fluorescein (CGamF) in sandwich-cultured hepatocytes. J Pharm Sci 99:2886–2898
4. Treijtel N, Barendregt A, Freidig AP et al (2004) Modeling the in vitro intrinsic clearance of the slowly metabolized compound tolbutamide determined in sandwich-cultured rat hepatocytes. Drug Metab Dispos 32: 884–891
5. Kotani N, Maeda K, Watanabe T et al (2011) Culture period-dependent changes in the uptake of transporter substrates in sandwich-cultured rat and human hepatocytes. Drug Metab Dispos 39:1503–1510
6. Jacobsen JK, Jensen B, Skonberg C et al (2011) Time-course activities of Oct1, Mrp3, and cytochrome P450s in cultures of cryopreserved rat hepatocytes. Eur J Pharm Sci 44:427–436
7. Abe K, Bridges AS, Brouwer KLR (2009) Use of sandwich-cultured human hepatocytes to predict biliary clearance of angiotensin II receptor blockers and HMG-CoA reductase inhibitors. Drug Metab Dispos 37:447–452
8. Abe K, Bridges AS, Yue W et al (2008) In vitro biliary clearance of angiotensin II receptor blockers and 3-hydroxy-3-methylglutaryl-coenzyme A reductase inhibitors in sandwich-cultured rat hepatocytes: comparison with in vivo biliary clearance. J Pharmacol Exp Ther 326:983–990
9. Bi Y, Kimoto E, Sevidal S et al (2012) In vitro evaluation of hepatic transporter-mediated clinical drug-drug interactions: hepatocyte model optimization and retrospective investigation. Drug Metab Dispos 40:1085–1092
10. Chatterjee S, Richert L, Augustijns P et al (2014) Hepatocyte-based in vitro model for assessment of drug-induced cholestasis. Toxicol Appl Pharmacol 274:124–136
11. De Bruyn T, Chatterjee S, Fattah S et al (2013) Sandwich-cultured hepatocytes: utility for in vitro exploration of hepatobiliary drug disposition and drug-induced hepatotoxicity. Expert Opin Drug Metab Toxicol 9:589–616
12. Kiang TKL, Teng XW, Karagiozov S et al (2010) Role of oxidative metabolism in the effect of valproic acid on markers of cell viability, necrosis, and oxidative stress in sandwich-cultured rat hepatocytes. Toxicol Sci 118:501–509
13. Keemink J, De Bruyn T, Augustijns P et al. (2014) Effect of cryopreservation on enzyme and transporter activities in suspended and sandwich cultured rat hepatocytes. Submitted
14. Halladay JS, Wong S, Khojasteh SC et al (2012) An all-inclusive 96-well cytochrome P450 induction method: measuring enzyme activity, mRNA levels, protein levels, and cytotoxicity from one well using cryopreserved human hepatocytes. J Pharmacol Toxicol Methods 66:270–275
15. Mingoia RT, Nabb DL, Yang CH et al (2007) Primary culture of rat hepatocytes in 96-well plates: effects of extracellular matrix configuration on cytochrome P450 enzyme activity and inducibility, and its application in in vitro cytotoxicity screening. Toxicol In Vitro 21: 165–173
16. Jasmund I, Schwientek S, Acikgoz A et al (2007) The influence of medium composition and matrix on long-term cultivation of primary porcine and human hepatocytes. Biomol Eng 24:59–69
17. Annaert P, Turncliff RZ, Booth CL et al (2001) P-glycoprotein-mediated in vitro biliary excretion in sandwich-cultured rat hepatocytes. Drug Metab Dispos 29:1277–1283
18. Lu JN, Wang CC, Young TH (2011) The behaviors of long-term cryopreserved human hepatocytes on different biomaterials. Artif Organs 35:65–72
19. Kienhuis AS, Wortelboer HM, Maas WJ et al (2007) A sandwich-cultured rat hepatocyte system with increased metabolic competence evaluated by gene expression profiling. Toxicol In Vitro 21:892–901
20. LeCluyse EL, Bullock PL, Parkinson A (1996) Strategies for restoration and maintenance of normal hepatic structure and function in

long-term cultures of rat hepatocytes. Adv Drug Deliv Rev 22:133–186

21. Xia L, Ng S, Han R et al (2009) Laminar-flow immediate-overlay hepatocyte sandwich perfusion system for drug hepatotoxicity testing. Biomaterials 30:5927–5936
22. LeCluyse EL (2001) Human hepatocyte culture systems for the in vitro evaluation of cytochrome P450 expression and regulation. Eur J Pharm Sci 13:343–368
23. Swift B, Brouwer KLR (2010) Influence of seeding density and extracellular matrix on bile Acid transport and mrp4 expression in sandwich-cultured mouse hepatocytes. Mol Pharm 7:491–500
24. Kimoto E, Chupka J, Xiao Y et al (2011) Characterization of digoxin uptake in sandwich-cultured human hepatocytes. Drug Metab Dispos 39:47–53
25. Pfeifer ND, Yang K, Brouwer KLR (2013) Hepatic basolateral efflux contributes significantly to rosuvastatin disposition I: characterization of basolateral versus biliary clearance using a novel protocol in sandwich-cultured hepatocytes. J Pharmacol Exp Ther 347: 727–736
26. De Bruyn T, Sempels W, Snoeys J et al (2014) Confocal imaging with a fluorescent bile acid analogue closely mimicking hepatic taurocholate disposition. J Pharm Sci 103:1872–1881
27. Yan GZ, Brouwer KLR, Pollack GM et al (2011) Mechanisms underlying differences in systemic exposure of structurally similar active metabolites: comparison of two preclinical hepatic models. J Pharmacol Exp Ther 337:503–512
28. Schaefer O, Ohtsuki S, Kawakami H et al (2012) Absolute quantification and differential expression of drug transporters, cytochrome P450 enzymes, and UDP-glucuronosyltransferases in cultured primary human hepatocytes. Drug Metab Dispos 40:93–103
29. Kirby BJ, Collier AC, Kharasch ED et al (2011) Complex drug interactions of HIV protease inhibitors 1: inactivation, induction, and inhibition of cytochrome P450 3A by ritonavir or nelfinavir. Drug Metab Dispos 39:1070–1078
30. Govindarajan R, Endres CJ, Whittington D et al (2008) Expression and hepatobiliary transport characteristics of the concentrative and equilibrative nucleoside transporters in sandwich-cultured human hepatocytes. Am J Physiol Gastrointest Liver Physiol 295:570–580

Chapter 13

Establishing Liver Bioreactors for In Vitro Research

Sofia P. Rebelo, Rita Costa, Marcos F.Q. Sousa,
Catarina Brito, and Paula M. Alves

Abstract

In vitro systems that can effectively model liver function for long periods of time are fundamental tools for preclinical research. Nevertheless, the adoption of in vitro research tools at the earliest stages of drug development has been hampered by the lack of culture systems that offer the robustness, scalability, and flexibility necessary to meet industry's demands. Bioreactor-based technologies, such as stirred tank bioreactors, constitute a feasible approach to aggregate hepatic cells and maintain long-term three-dimensional cultures. These three-dimensional cultures sustain the polarity, differentiated phenotype, and metabolic performance of human hepatocytes. Culture in computer-controlled stirred tank bioreactors allows the maintenance of physiological conditions, such as pH, dissolved oxygen, and temperature, with minimal fluctuations. Moreover, by operating in perfusion mode, gradients of soluble factors and metabolic by-products can be established, aiming at resembling the in vivo microenvironment. This chapter provides a protocol for the aggregation and culture of hepatocyte spheroids in stirred tank bioreactors by applying perfusion mode for the long-term culture of human hepatocytes. This in vitro culture system is compatible with feeding high-throughput screening platforms for the assessment of drug elimination pathways, being a useful tool for toxicology research and drug development in the preclinical phase.

Key words Human hepatocytes, Three-dimensional, Stirred tank bioreactor, Perfusion, Long-term toxicity studies

1 Introduction

The development of in vitro hepatic models is required for several applications, namely to understand liver physiology and to model hepatic diseases, for bioartificial liver support and to assess drug detoxification mechanisms and pathways. Liver-based in vitro models thus constitute powerful tools for toxicology research and contribute to speed up the drug development process. Primary cultures of human hepatocytes are considered as the gold standard for xenobiotic metabolism profiling. However, they present major drawbacks, such as phenotypic instability and short lifespan when cultured in two-dimensional systems [1]. Three-dimensional (3D) approaches for the culture of hepatocytes have

Mathieu Vinken and Vera Rogiers (eds.), *Protocols in In Vitro Hepatocyte Research*, Methods in Molecular Biology, vol. 1250, DOI 10.1007/978-1-4939-2074-7_13, © Springer Science+Business Media New York 2015

significantly improved culture duration and functionality, particularly phase I and II biotransformation enzyme activity and albumin synthesis [1, 2], thereby enabling long-term assessment of hepatic functions. This is mainly due to the maintenance of the polygonal shape and the basolateral polarity of hepatocytes, which contribute to prevent hepatocyte dedifferentiation and to sustain the hepatic phenotype. A plethora of culture strategies have been successfully applied to maintain three-dimensionality, such as scaffolds [3], the collagen sandwich configuration [4], and cell spheroids [5–7]. The 3D configuration of cell spheroids allows the maximization of cell–cell interactions, thus resulting in highly polarized hepatocytes with expression of polarity proteins and functional bile canaliculi [7]. Cell aggregation into spheroids may be promoted by different methods, including self-assembly in matrices or low-adhesion conditions (e.g., matrigel matrix and ultralow adhesion plates), hanging-drop platforms, or by using stirred conditions to promote aggregation, such as rotary shakers or stirred tank bioreactors (STBs) [1].

Culture in computer-controlled bioreactors provides tight control and online monitoring of culture parameters, such as pH, partial pressure of oxygen (pO_2), and temperature, resulting in reproducible conditions for cell cultivation. A number of bioreactor configurations have been described for the cultivation of hepatocytes. Most of them are based on the hollow fiber format, such as the modular extracorporeal liver support [8, 9] and the bioartificial liver device [10] systems for extracorporeal liver support or miniaturized systems, with potential for pharmacological applications [11]. An extensive review on bioreactor configurations can be found in [12].

This chapter addresses the aggregation and long-term culture of hepatic spheroids in STBs. Previous reports from our group have described the culture of mouse [5, 6, 13, 14] or human [7] liver spheroids in STBs, maintaining high cell viability throughout the culture time and stable phenotype, including phase I biotransformation activity up to 4 weeks in culture.

The optimal conditions to promote aggregation are somehow empirical and depend to a major extent on the cell characteristics. Aggregation of hepatocytes is induced by appropriate vessel hydrodynamic characteristics to promote cell–cell contacts with minimal detrimental impact on cell collisions and shear stress. Several parameters, such as type and size of paddle, impeller, and stirring rate, can affect aggregate formation and maintenance during culture. Compromises that allow for minimal shear stress without diffusion problems within the aggregates have to be made, all which minimize the formation of necrotic centers in the spheroids. In contrast to hollow fiber bioreactors, STBs are compatible with nondestructive sampling, allowing characterization of cultured cells throughout the culture period and the possibility of using the

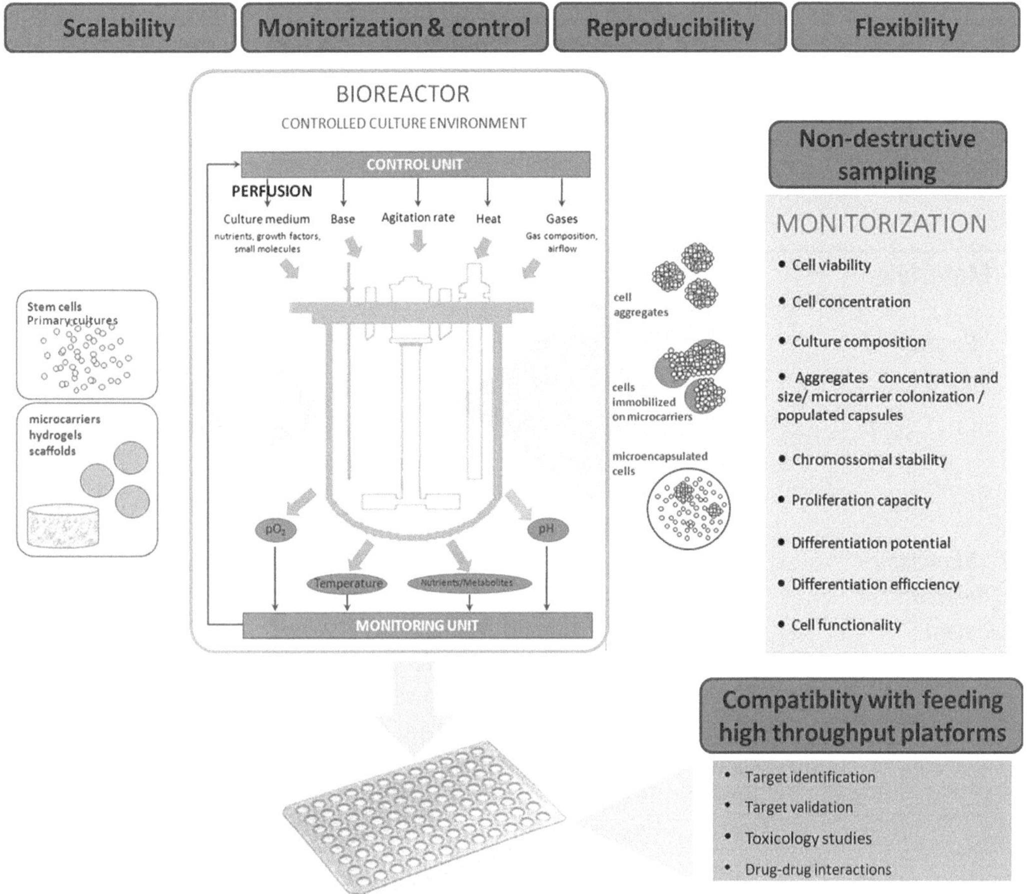

Fig. 1 Schematic representation of STBs. Major features of STBs are depicted, including scalability, monitoring and control, reproducibility, flexibility, nondestructive sampling, and compatibility with feeding of high-throughput platforms

STBs as a feeder system to perform assays in high-throughput platforms. Furthermore, STBs provide a scalable solution for the production of the large cell numbers required in industry and clinical settings and enable robust bioprocesses (Fig. 1).

Dynamic flow conditions have been applied to hepatic cultures, either in monolayer or other configurations, with successful outcomes in viability and metabolic performance [6, 7, 15]. Perfusion cultures offer attractive alternatives to batch and fed-batch cultures, as they minimize environmental variations. In perfusion operation mode, fresh medium is fed to a bioreactor containing cells that are retained by a retention system, with gradual replacement of culture medium. The continuous removal of secreted metabolites and growth factors, which may benefit specific functions, and the removal of toxic components resemble the in vivo blood circulation, reconstituting a more physiological situation.

The protocol for the establishment and long-term maintenance of spheroid cultures of human hepatocytes in computer-controlled STBs, with perfusion operation mode, might be applied to other hepatocyte cell sources, such as hepatic cell lines or hepatocyte-like cells derived from pluripotent stem cells and extended to other liver cell types to establish cocultures.

2 Materials

2.1 General Equipment

1. Bright-field and fluorescence microscope (i.e., vertical and inverted) for culture monitoring.
2. Incubator (37 °C ± 1 °C, 90 % ± 5 % humidity, 5 % ± 1 % CO_2).
3. YSI 7100 multiparameter bioanalytical system (Life Sciences Advances Inc., United States of America) or equivalent.

2.2 Bioreactor and Related Equipment

A protocol for the BIOSTAT Qplus system is described in this chapter, yet it should be noted that this procedure can be equally adapted to other commercially available STBs (e.g., DASGIP systems) or single-use vessels that can be connected to bioreactor control units. The choice of the bioreactor has to take into account minimal working volumes required (i.e., type of experiments, sampling, duration of the experiments, minimal volume for probes, and costs) and any limitations on cell source availability. A minimal *inoculum* concentration is required to start and obtain viable cell cultures.

1. Bioreactor system. 500 mL bioreactor vessels and control unit from BIOSTAT Qplus (Sartorius Stedim Biotech S.A., Germany) or equivalent.
2. Equipment to assemble the STB. Silicone tubing, 0.22 μm filters, luer connections, pinchers, tweezers, Schott flask for culture feeding or harvesting.
3. Equipment for perfusion. Dosing pumps (Sartorius Stedim Biotech S.A., Germany) or equivalent, balances for gravimetric control of perfusion operation mode, Schott flasks for addition and removal of culture medium.
4. Equipment for sampling. Mobile biosafety cabinet or alternative equipment (i.e., sterile sampling devices) to ensure maintenance of bioreactor sterility throughout and after sampling, 10 and 50 mL syringes.
5. Material for silanization of glass vessels. Dichlorodimethylsilane, toluene, and potassium hydroxide.

2.3 Biological Material

In this protocol, a procedure for aggregation and long-term culture of hepatocytes isolated from human liver tissue is provided. To inoculate an STB of 500 mL, a minimum of 50 up to 250 million viable cells are necessary, which largely depends on the quality of the liver source and on the success of the hepatocyte isolation process. This protocol might be adapted for other hepatic sources, such as hepatic cell lines, yet aggregation dynamics may differ substantially.

2.4 Cell Culture Media and Materials

1. Hepatocyte culture medium. William's E medium supplemented with 1 % GlutaMAX (Sigma, United States of America), 1 % penicillin/streptavidin, and the hepatocyte culture medium SingleQuots kit (Lonza, Switzerland).
2. Fetal bovine serum.
3. BioCoat multiwell collagen coated plates (BD Biosciences, United States of America).

2.5 Culture Characterization

1. Phosphate-buffered saline (PBS). 137 mM NaCl, 2.7 mM KCl, 10 mM $Na_2HPO_4{\cdot}2H_2O$, 1.8 mM KH_2PO_4 in deionized water. Adjust to pH 7.4 and store for maximum 6 months at 4 °C.
2. 0.1 % trypan blue solution. Dilute a stock solution of trypan blue 10 % in PBS to a concentration of 0.1 %. Filter the solution through a 0.22 μm filter. The solution may be stored for long periods at room temperature.
3. Fuchs-Rosenthal hemocytometer.
4. 0.05 % trypsin–ethylenediaminetetraacetic acid (Invitrogen, United States of America).
5. Fluorescein diacetate (Sigma, United States of America).
6. TO-PRO®3 (Invitrogen, United States of America).
7. Lysis reagent. Add 0.1 % citric acid and 0.1 % Triton X-100 to double distilled water. This solution may be stored for long periods at room temperature.
8. 0.1 % crystal violet solution in ultrapure water. This solution may be stored for long periods at room temperature.
9. 24-Well plates.
10. 1.5 mL Eppendorf tubes.

3 Methods

3.1 STB Preparation

Preparing the STB in advance is critical to ensure the success of the bioprocess will be maintained for long-term without major issues. Vessel preparation, tubing assembly and, in particular, electrode calibration constitute crucial steps and must be initiated at least 4 days before setting up the culture. To set the STB operations, follow the instructions below according to the process workflow presented in Fig. 2.

1. Start the data acquisition mode in the control unit.
2. Calibrate the pH probe with standard buffer solutions to pH 7.0 and 9.21 according to the manufacturer's instructions.

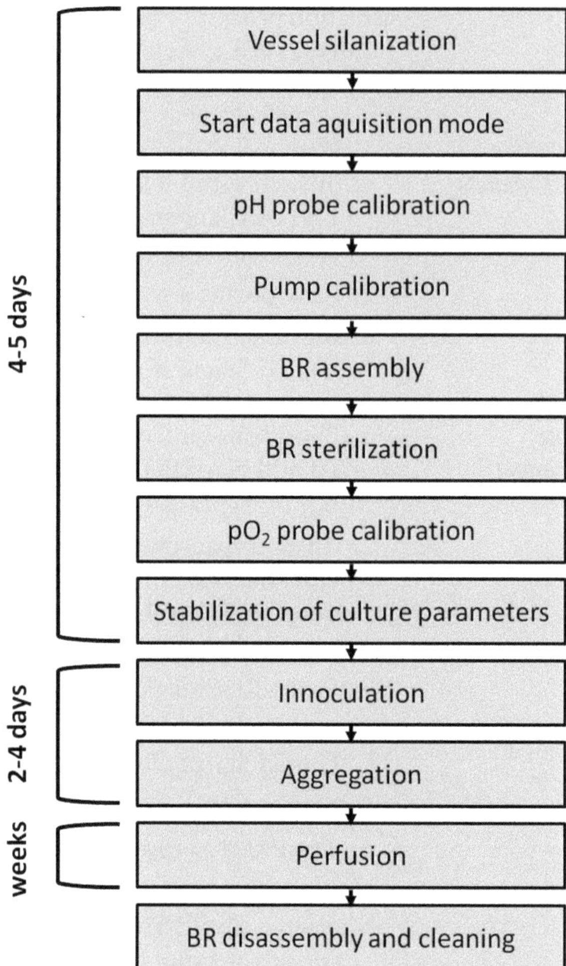

Fig. 2 Workflow for preparation of STBs for long-term perfusion cultures of human hepatocytes

3. Prepare the STB vessel by silanizing the glass vessel (*see* **Note 1**).
4. The 500 mL BIOSTAT Qplus system has multiple upper cap ports for insertion of probes that allow online monitoring of culture parameters and connections for feeding, sampling and harvesting the culture. Assemble the STB by inserting the following components as represented in Fig. 3, including the 3-blade segment impeller, pH, pO_2 and temperature probes, exhaust cooler, connection for aeration, connection for medium addition (i.e., inlet), connection for medium withdrawal (i.e., outlet), connection for inoculation and sampling.
5. For sampling connections, insert a luer connector in the extremity of the tube.
6. For perfusion, add a silicone tube to the medium harvesting connection, ensuring that it reaches the bottom of the vessel.

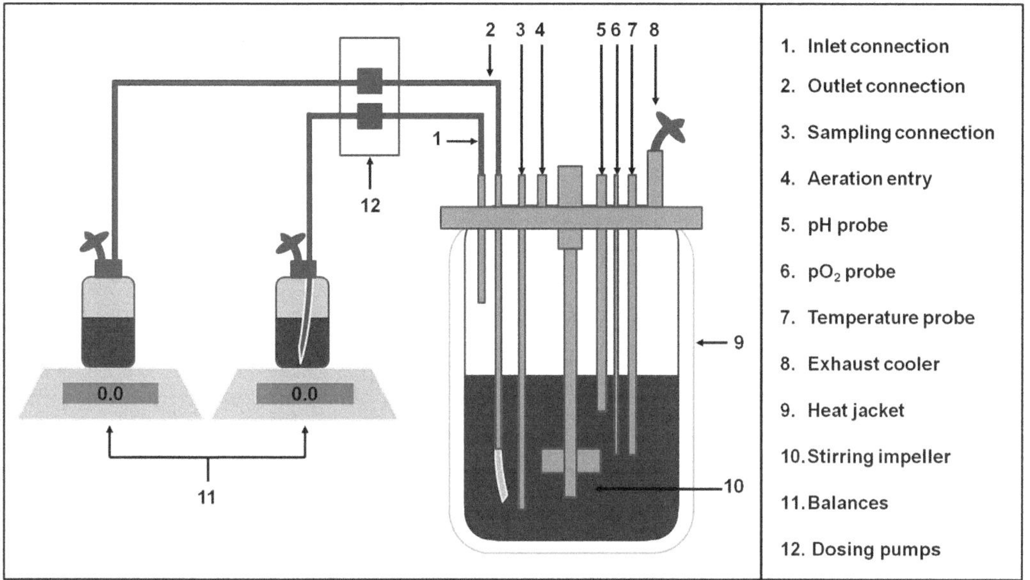

Fig. 3 Schematic representation of STBs prepared for perfusion operation mode. The representation is not to scale

7. Prepare 1 flask for inlet and 1 for outlet.
8. Ensure that the inlet and outlet perfusion tubing fit the pumps and that the tube diameter is appropriate for the flow rates applied.
9. Perform pump calibration according to the manufacturer's instructions (*see* **Note 2**).
10. Add ultrapure water to the vessel up to the level of the probes.
11. Check the rubber rings, tighten all fittings, and perform the hold-up test to certify that the STB is airtight.
12. Sterilize the STB and connecting bottles for 30–45 min at 121 °C (*see* **Note 3**).
13. Check the STB assembly and connections after sterilization, thereby looking for damaged tubes and leakages.
14. Connect the STB to the control unit with inlet and outlet connections to the jacket and cooler, temperature, pO_2 and pH probes, and the motor and aeration system to the gas supply system.
15. Fill the jacket with ultrapure water and start temperature control whenever the temperature is below 45–50 °C.
16. Start pO_2 calibration by supplying nitrogen gas (N_2) to set the baseline oxygen level (i.e., 0 %), calibrate and switch to air saturation (i.e., 100 % air with 21 % O_2) afterwards.

Table 1
Culture parameters for aggregation and long-term culture of spheroids of human hepatocytes

pH	pO_2 (%)	Temperature (°C)	Working volume (mL)	Stirring rate (rpm)	Dilution rate (day^{-1})	Inoculum concentration (cell/mL)
7.4	30	37	200–500	70–120	0.2	$2–5 \times 10^5$

17. Whenever the pO_2 and temperature readings are stable, finish the calibration of the pO_2 probe.
18. Replace the ultrapure water with culture medium containing 10 % fetal bovine serum until reaching 80 % of culture working volume (i.e., 200–250 mL).
19. Select the culture parameters (Table 1) and switch on the unit controllers.
20. Wait for stabilization of culture parameters (i.e., temperature, pH, and pO_2) before cell inoculation.

3.2 Bioreactor Inoculation

Before cell inoculation, the culture parameters set in the STB (i.e., temperature, pH, and pO_2) must be stable. All handlings must ensure sterile conditions and minimize interference with the culture.

1. After isolation and purification of human hepatocytes from liver tissue, determine the viable cell concentration by the trypan blue exclusion method (*see* **Note 4**).
2. Since the number of viable cells depends on the quality of the liver source and on the isolation and purification procedures, determination of the total number of viable cells at the end of the isolation process is necessary to establish the final working volume and *inoculum* cell concentration within the range described in Table 1.
3. Once the culture parameters are set, determine the volume of cell suspension necessary for inoculation in the STB in order to attain the established *inoculum* and the volume to seed two-dimensional collagen-coated plates, which will be used as culture controls (*see* **Note 6**).
4. Centrifuge the required volumes of cell suspension for 2 min at $180 \times g$ and 4 °C.
5. Remove the supernatant and gently ressuspend the cell pellet in the proper volume of hepatocyte culture medium containing 10 % fetal bovine serum, to bring the STB to its working volume (i.e., typically 20 % of the working volume).
6. Proceed with inoculation using the bioreactor sampling tube. Remove residual cell suspension from the tubing by washing with an extra volume of approximately 10 mL.

3.3 Bioreactor Sampling

Nondestructive sampling is an important characteristic of STB, as it provides the possibility of monitoring the culture throughout time. Representative sampling will ensure adequate culture monitoring. For daily culture monitoring, collect at least 3 mL of sample and distribute as described below.

3.3.1 Morphological Characterization and Microscopic Assessment of Cell Viability

1. Use a cut pipette tip to transfer 0.5–1 mL sample to a 24-well plate.
2. Use fluorescent dyes, such as 10 μg/mL fluorescein diacetate and 1 μM TO-PRO®3, to visualize live and dead cells, respectively.
3. Once small aggregates are detected in the cultures, determine their average diameter by measuring the diameter of at least 20–30 aggregates per sample.
4. Count the number of aggregates per sample under the microscope to estimate the number of aggregates/mL culture.

3.3.2 Cell Quantification and Metabolite Analysis

1. Distribute 2 × 1 mL sample in 1.5 mL Eppendorf tubes. Due to sampling heterogeneity, at least duplicates for cell quantification should be taken.
2. Centrifuge for 5 min at 300 × *g* and 4 °C.
3. Use the cell pellet for cell quantification based on the trypan blue exclusion method, nuclei counting, protein, or DNA quantification (*see* **Note 5**).
4. Use the supernatant for metabolite analysis. An additional step of centrifugation may be necessary for cell debris removal at 1,000 × *g* and 4 °C. The supernatants may be stored at −20 °C for further analysis.
5. Use the YSI 7100 multiparameter bioanalytical system to quantify the concentration of metabolites in the supernatants according to the manufacturer's instructions.

3.4 Monitoring of Aggregation Dynamics

The aggregation period typically lasts 3–4 days to yield compact cell aggregates, but is highly dependent on several factors (*see* **Note 7**). Thus, during the initial steps of aggregation, it is crucial to monitor the culture frequently (i.e., 3–6 h). Culture monitoring may be performed at larger intervals (i.e., 24–48 h) once the aggregates are compact (Fig. 4).

1. Determine the cell concentration by the trypan blue exclusion method (*see* **Note 4**). If necessary, trypsinize the cell aggregates (*see* **Note 5**).
2. For monitoring of hepatocyte aggregation, use a bright-field microscope to visualize the formation of duplets and triplets of cells (Fig. 4) and, as culture proceeds, visualize multicellular aggregates (Fig. 4).

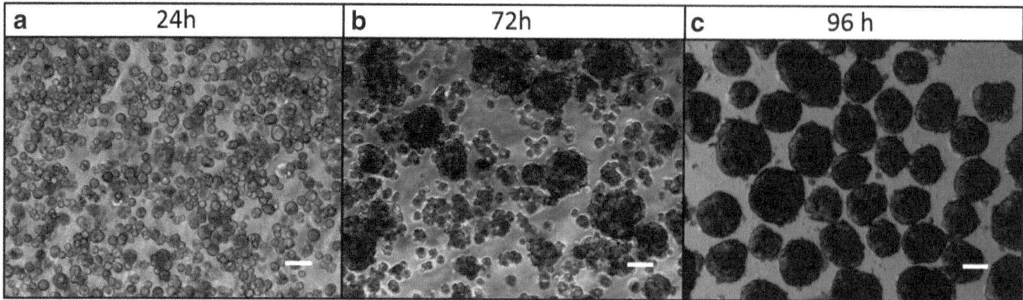

Fig. 4 Primary cultures of human hepatocytes in different time points post-inoculation. (**a**) 24 h post-inoculation: duplets and triplets of cells are formed. (**b**) 72 h post-inoculation: cell aggregates with irregular shape and loose morphology can be observed. (**c**) 96 h post-inoculation: compact cell aggregates are formed. Scale bars represent 50 μm

3. After determination of the average aggregate diameter at each sampling point, as described in section 3.3.1, gradually increase the stirring rate if the average diameter increases until 80 to 100 μm.

3.5 Establishment of Perfusion

Perfusion may be initiated whenever compact cell aggregates are formed. Spheroids with an average diameter higher than 80 μm are typically formed after 3–4 days (Fig. 4). Perfusion systems imply retention of biological components inside the STB through multiple strategies, such as cell retention devices. In the present protocol, the strategy used to retain cell aggregates in the STB is gravity through sedimentation of cell aggregates. For this purpose, it is estimated the time of sedimentation of cell aggregates to determine whether it is compatible with the flow rates used in the outlet. Single cells and cell debris will be washed out from the STB. The culture may be maintained for several weeks under perfusion conditions.

1. Fill the inlet flask with hepatocyte culture medium without fetal bovine serum in sterile conditions (*see* **Note 8**).
2. Determine the flow rate as a function of the dilution rate and working volume in the STB (*see* **Note 9**).
3. Initiate the pumps manually to fill up the inlet and outlet tubes before initiation of perfusion. Be sure to use low flow rates to fill the outlet tube to prevent cell wash-out from the STB.
4. Initiate perfusion by setting the previously determined flow rates to the pumps.
5. Use gravimetric measurements to corroborate the volumes added or withdrawn from the STB and adjust the flow rate of the pumps if necessary.
6. Adjust the perfusion flow rates throughout the culture time, as the culture working volume changes with time due to sampling.

7. To quantify the secretion or metabolization of culture components, such as albumin and glucose, a general mass balance equation for a continuous system must be applied, as represented in the following equation:

$$q = \frac{\frac{\Delta C}{\Delta t} - D(\text{Cin} - \text{Cout})}{[X]v\ \text{average}}$$

With parameters:

q = specific synthesis/consumption rate

$\frac{\Delta C}{\Delta t}$ = rate change of the metabolite in the supernatant throughout time

D = dilution rate

Cin = inlet concentration of the metabolite

Cout = outlet concentration of the metabolite

$[X]v$ average = average cell concentration during Δt

3.6 Bioreactor Disassembly

1. Switch off all the controllers.
2. Disconnect the cables and tubing from the control unit.
3. Prepare the STB for sterilization.
4. Clean and store the pO_2 and pH electrodes according to the manufacturer's instructions.
5. Clean the STB vessel and tubing with 70 % ethanol, certifying that there are no culture residues, particularly in the impeller and tubing, and let dry.

4 Notes

1. The silanization process makes the glass surfaces less hydrophilic, which is important when working with adherent cells (i.e., to prevent adhesion to the glass vessels). Washing and silanization solutions are commercially available. A washing solution for this procedure may be prepared by dissolving 56 g of potassium hydroxide in 76 mL of distilled water with constant agitation in ice, since it is an exothermic reaction. Then, fill to 1,000 mL with 96 % ethanol and keep the solution at room temperature in an amber glass bottle for 14 days before use. All the solutions used in the silanization protocol can be reused and should be stored in an amber glass container. After washing with the washing solution, the vessel must dry to air. When dry, use a small volume of dichlorodimethylsilane, ensuring that it covers the entire vessel inner surface. Repeat this procedure using toluene and let the vessel dry.

2. Precise pump calibration will ensure that the perfusion runs properly. Although it may be performed according to the manufacturer's instructions, gravimetric control might be used to calibrate the pumps. To do so, use 2 balances, silicone tubing with equal diameter as the one applied in the perfusion and 2 liquid containers. Measure a fixed volume of water (e.g., 100 mL), put it in 1 of the balances and initiate the pump until the tube is filled. Set the balances to 0, initiate the pumps again, and determine the volume pumped in a fixed period of time (e.g., 30 min). Divide the volume in g/h and insert the flow rate measured in mL/h into the pump software if applicable.
3. Before autoclaving the bioreactor, cover tubing, filter ends, probes, and the stirrer system with aluminum foil, and tape them with steam indicator tape. Open the clamp from the exhaust cooler to allow pressure release during sterilization.
4. The trypan blue exclusion method for determination of the viable cell concentration makes use of a vital dye that selectively enters cells whose membrane is damaged, thus enabling the distinction between live (i.e., bright) and dead (i.e., blue) cells. For sample analysis and cell counting under the brightfield microscope, a hemocytometer (i.e., Fuchs-Rosenthal or equivalent) might be used.
5. Cell quantification may be performed by using different methods, according to the aggregation status and desired read-out or application. During the initial period of aggregation, it is possible to trypsinize the cell aggregates for 5 min at 37 °C and to count viable cells in single cell suspension using the trypan blue exclusion method. However, as aggregation proceeds and aggregates become compact, this method is not applicable. Thus, cell quantification may be performed by nuclei counting using a cell lysis reagent and crystal violet to visualize nuclei under the microscope. Alternative methods, such as protein or DNA quantification, may be performed.
6. To seed two-dimensional plates, collagen-coated plates are used and the preferred cell density is 2×10^5 cells/cm^2 (i.e., 0.4×10^6 cells are used for 24-well plates). Medium should be exchanged 12 h after seeding. The two-dimensional condition will provide information on the adhesive properties of the isolated hepatocytes, thus is used as control for the spheroid culture.
7. Aggregation of primary cultures of human hepatocytes is highly dependent on the applied isolation procedure and purification methods. Factors, such as collagenase incubation time, type of collagenase, and percoll purification, strongly affect the aggregative potential and determine whether it occurs slowly or rapidly. Thus, the *inoculum* might be adjusted according to the aforementioned factors. The culture must be

frequently monitored during the first steps of aggregation to ensure that the stirring rate is adjusted according to the aggregation dynamics.

8. After aggregation, animal serum-free medium is preferred to pursue the culture, since serum might interfere with cytochrome P450 enzyme activity.
9. The determination of the flow rate used for perfusion will depend on the daily dilution rate and the working volume of the STB. As an example, for a daily dilution rate of 0.2/day and a volume of 400 mL in the STB, the volume to be replaced in the STB would be 80 mL/day (i.e., 0.2/day × 400 mL = 80 mL/day). Set the flow rate in the pump to the value determined and confirm that the outlet flow rate is low enough to allow sedimentation of cell aggregates, otherwise cell wash-out from the STB may occur.

Acknowledgements

This work was financially supported by the grants PTDC/EBB-BIO/112786/2009 and SFRH/BD/70264/2010 provided by Fundação para a Ciência e Tecnologia, Portugal. The authors would like to acknowledge Janne Jensen and Peter Bjorquist from Cellartis and Rui Tostões, Sofia Leite, and Margarida Serra for implementation of the aggregation of primary cultures of human hepatocytes at iBET. The authors also acknowledge João Clemente for the implementation and support on the perfusion operation mode in the bioreactors.

References

1. Lecluyse EL, Witek RP, Andersen ME et al (2012) Organotypic liver culture models: meeting current challenges in toxicity testing. Crit Rev Toxicol 42:501–548
2. Godoy P, Hewitt NJ, Albrecht U et al (2013) Recent advances in 2D and 3D in vitro systems using primary hepatocytes, alternative hepatocyte sources and non-parenchymal liver cells and their use in investigating mechanisms of hepatotoxicity, cell signaling and ADME. Arch Toxicol 87:1315–1530
3. Vasanthan KS, Subramanian A, Krishnan UM et al (2012) Role of biomaterials, therapeutic molecules and cells for hepatic tissue engineering. Biotechnol Adv 30:742–752
4. Dunn JC, Tompkins RG, Yarmush ML (1991) Long-term in vitro function of adult hepatocytes in a collagen sandwich configuration. Biotechnol Prog 7:237–245
5. Miranda JP, Leite SB, Muller-Vieira U et al (2009) Towards an extended functional hepatocyte in vitro culture. Tissue Eng Part C Methods 15:157–167
6. Tostoes RM, Leite SB, Miranda JP et al (2011) Perfusion of 3D encapsulated hepatocytes: a synergistic effect enhancing long-term functionality in bioreactors. Biotechnol Bioeng 108:41–49
7. Tostoes RM, Leite SB, Serra M et al (2012) Human liver cell spheroids in extended perfusion bioreactor culture for repeated-dose drug testing. Hepatology 55:1227–1236
8. Hoffmann SA, Muller-Vieira U, Biemel K et al (2012) Analysis of drug metabolism activities

in a miniaturized liver cell bioreactor for use in pharmacological studies. Biotechnol Bioeng 109:3172–3181

9. Zeilinger K, Holland G, Sauer IM et al (2004) Time course of primary liver cell reorganization in three-dimensional high-density bioreactors for extracorporeal liver support: an immunohistochemical and ultrastructural study. Tissue Eng 10:1113–1124
10. Flendrig LM, la Soe JW, Jorning GG et al (1997) In vitro evaluation of a novel bioreactor based on an integral oxygenator and a spirally wound nonwoven polyester matrix for hepatocyte culture as small aggregates. J Hepatol 26:1379–1392
11. Novik E, Maguire TJ, Chao P et al (2010) A microfluidic hepatic coculture platform for cell-based drug metabolism studies. Biochem Pharmacol 79:1036–1044
12. Ebrahimkhani MR, Neiman JA, Raredon MS et al (2014) Bioreactor technologies to support liver function in vitro. Adv Drug Deliv Rev 69–70:132–157
13. Leite SB, Teixeira AP, Miranda JP et al (2011) Merging bioreactor technology with 3D hepatocyte-fibroblast culturing approaches: improved in vitro models for toxicological applications. Toxicol In Vitro 25:825–832
14. Miranda JP, Rodrigues A, Tostoes RM et al (2010) Extending hepatocyte functionality for drug-testing applications using high-viscosity alginate-encapsulated three-dimensional cultures in bioreactors. Tissue Eng Part C Methods 16:1223–1232
15. De Bartolo L, Bader A (2001) Review of a flat membrane bioreactor as a bioartificial liver. Ann Transplant 6:40–46

Chapter 14

Epigenetic Modifications as Antidedifferentiation Strategy for Primary Hepatocytes in Culture

Jennifer Bolleyn, Joanna Fraczek, Vera Rogiers, and Tamara Vanhaecke

Abstract

A well-known problem of cultured primary hepatocytes is their rapid dedifferentiation. During the last years, several strategies to counteract this phenomenon have been developed, of which changing the in vitro environment is the most popular one. However, mimicking the in vivo setting in vitro by adding soluble media additives or the restoration of both cell–cell and cell–extracellular matrix contacts is not sufficient and only delays the dedifferentiation process instead of counteracting it. In this chapter, new strategies to prevent the deterioration of the liver-specific phenotype of primary hepatocytes in culture by targeting the (epi)genetic mechanisms that drive hepatocellular gene expression are described.

Key words Histone acetylation, DNA methylation, HDAC inhibition, DNMT inhibition, TSA, Decitabine, Primary hepatocyte

1 Introduction

During the drug development process, new chemical compounds need to be tested for their toxicological features and in particular their potential liver toxicity. As the functionality of hepatocytes in the liver plays a key role in the pharmacokinetic behavior and consequent potential toxicity of pharmaceuticals, cultures of primary hepatocytes are currently considered as the golden standard to evaluate drug toxicity in vitro [1]. Indeed, these cells preserve several hepatic functions, including cytochrome P450 (CYP) activity and albumin secretion, although only for a limited period of time [2]. Due to this progressive deterioration caused by changes in mRNA and protein levels of biotransformation enzymes and drug transporters, primary hepatocyte cultures are not suitable for long-term studies [3–7]. In fact, hepatocytes are triggered to proliferate towards a less differentiated state by the loss of their cell architecture and ischemia–perfusion injury induced during the isolation procedure [4, 8, 9]. Moreover, proliferative and inflammatory

Mathieu Vinken and Vera Rogiers (eds.), *Protocols in In Vitro Hepatocyte Research*, Methods in Molecular Biology, vol. 1250, DOI 10.1007/978-1-4939-2074-7_14, © Springer Science+Business Media New York 2015

signaling cascades are initiated, forcing the hepatocytes to leave their quiescent in vivo G_0 state and to reenter the cell cycle in the G_1 phase [4, 8–10]. This proliferation trigger has, however, a detrimental outcome on the expression of phase I and II biotransformation enzymes and liver-enriched transcription factors [3, 5]. The first approaches to counteract this dedifferentiation process consisted of mimicking the in vivo hepatocyte micro-environment in vitro [2, 7]. As such, addition to the culture media of soluble differentiation-promoting factors, including hormones, growth factors, vitamins, amino acids, and other additives, was introduced. Also, the restoration of cell–cell and cell–extracellular matrix contacts was found to be imperative to preserve and restore hepatic polarization together with an in vivo-like morphology. Notwithstanding, these antidedifferentiation strategies do not completely prevent hepatic dedifferentiation, but merely delay this process [7, 11]. More recent approaches aim at preventing dedifferentiation by targeting (epi)genetic mechanisms that drive hepatocellular gene expression [12].

2 Epigenetic Mechanisms in the Regulation of Gene Expression

Gene transcription is initiated through binding of RNA polymerase together with several nuclear transcription factors to their corresponding response elements in gene promoters. The accessibility of the latter, however, is determined by the chromatin status that in turn is dictated by epigenetic mechanisms, including reversible histone modifications and DNA methylation [13].

2.1 Histone (De)acetylation

The nucleosome, being the structural unit of the eukaryotic genome, is composed of an octamer of histone proteins and 147 base pairs of DNA [14]. The *N*-terminal tails of the core histone proteins contain amino acid residues that are subject to posttranslational modifications, including the addition of methyl, acetyl, and phosphate groups, as well as adenosine diphosphate ribose. Furthermore, covalent binding of proteins, such as ubiquitin and small ubiquitin modifier, and of the amino acid citrulline may occur. These covalent modifications alter the interactions between histones and DNA and consequently affect the gene transcriptional status. Currently, acetylation is among the best studied posttranslational histone modifications. Deacetylation of lysine residues, catalyzed by histone deacetylases (HDACs), tightens the negatively charged DNA strands more strongly to the positively charged histone cores by increasing their electrostatic attraction. This hampers the accessibility of the RNA polymerase and transcription factors to gene promoters and thus leads to transcriptional repression [15–17]. In contrast, a more relaxed chromatin configuration is

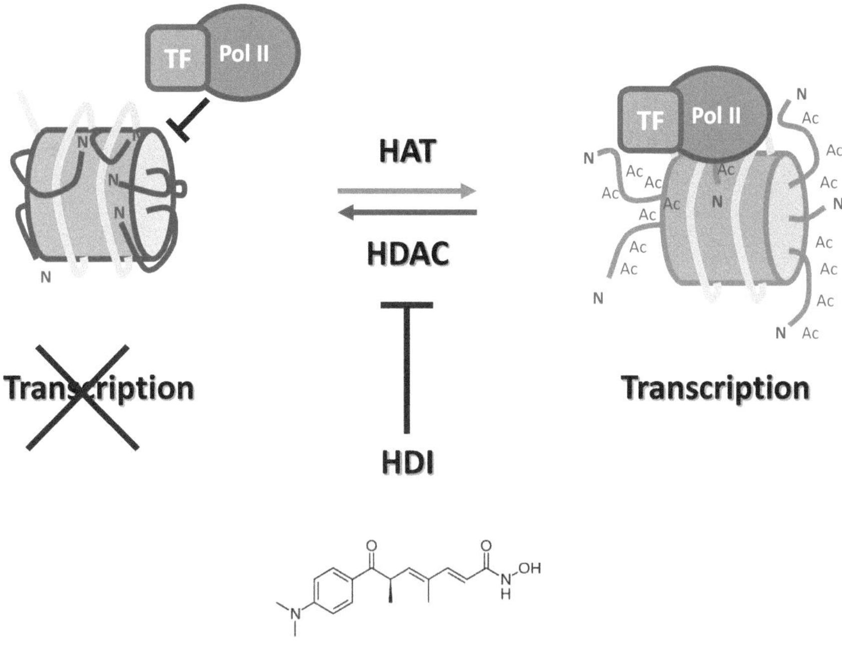

Fig. 1 Mechanism of histone (de)acetylation. Histone deacetylases (HDACs) tighten the DNA strand on the histone cores, thereby creating a more compact structure, which is associated with transcriptional repression. In contrast, by introducing an acetyl group by histone acetyltransferases (HATs), transcription is activated through a more relaxed chromatin structure. (*HAT* histone acetyltransferase, *HDAC* histone deacetylase, *HDI* histone deacetylase inhibitor, *Pol II* polymerase II, *TF* transcription factor)

formed when acetyl groups are introduced on the lysine residues by histone acetyltransferases (HATs), resulting in more readily transcription factor binding and thus transcriptional activation (Fig. 1) [16, 18, 19].

2.2 DNA Methylation

DNA methylation implies the covalent addition of a methyl group by means of DNA methyltransferases (DNMTs) on the 5-carbon of cytosine bases usually located in CpG dinucleotide sites [17]. Because methylation of cytosines efficiently inhibits transcription factor binding, it represents a stable, heritable, and reversible mark that is generally associated with transcriptional repression (Fig. 2). In addition, the presence of 5-methylcytosine subsequently attracts methylated DNA-binding domain proteins, which on their turn attract corepressor molecules, such as histone-modifying enzymes [12, 13, 17, 20, 21]. Thus, DNA methylation also alters the level of chromatin compaction resulting in gene silencing [22, 23].

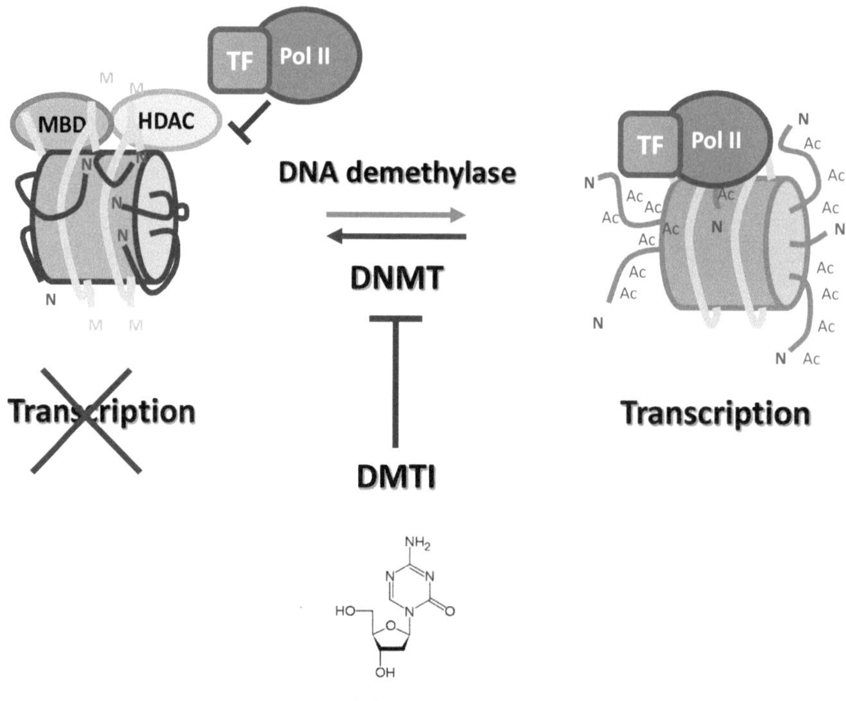

Fig. 2 Mechanism of DNA methylation. DNA methylation will covalently modify cytosine bases found in CpG dinucleotides by adding a methyl group by means of DNA methyltransferases. The presence of 5-methylcytosine subsequently attracts the methylated DNA-binding domain proteins, which attract corepressor molecules such as histone-modifying enzymes. Thus, DNA methylation will alter the level of chromatin compaction resulting in gene silencing (*DMTI* DNA methyltransferase inhibitor, *DNMT* DNA methyltransferase, *HDAC* histone deacetylase, *MBD* methylated DNA-binding domain proteins, *Pol II* polymerase II, *TF* transcription factor)

3 Epigenetic Modifications as Antidedifferentiation Strategy for Hepatic In Vitro Systems

In several pathologies, including cancer, differential expression and function of HDACs and DNMTs have been suggested to be a major mechanism of disease. Consequently, several novel therapies are based on the modulation of these enzymes [15, 24–28]. By adding HDAC inhibitors (HDIs) and DNMT inhibitors (DMTIs) to a plethora of in vitro hepatic cancer cell models, the induction of differentiation together with an antiproliferative effect has indeed been described in numerous reports [13, 29]. In this context, Trichostatin A (TSA), the HDI prototype, was shown to induce the expression of CCAAT-enhancer-binding protein alpha and both of the phase I biotransformation enzymes CYP3A4 and alcohol dehydrogenase 1B in the human hepatoma HepG2 cell line [30–32]. Other hepatic functions, including ammonia removal and albumin secretory capacity, were also augmented in response to TSA exposure in both HepG2 and Huh7 hepatoma cells.

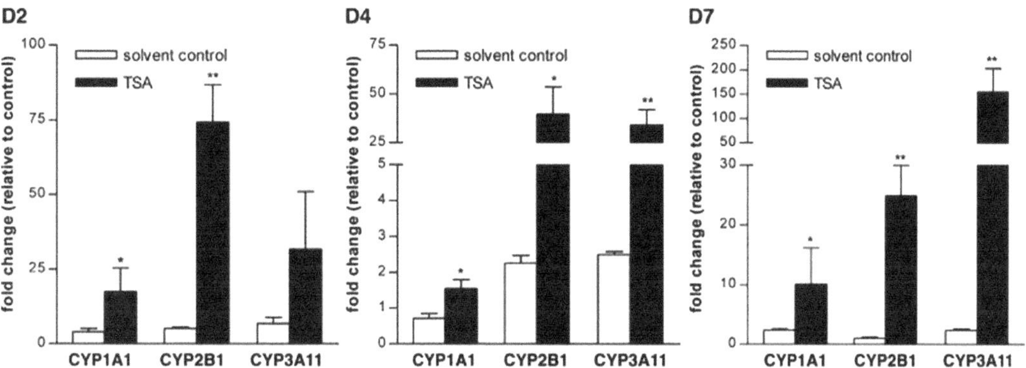

Fig. 3 Effects of TSA on *CYP* gene expression. Hepatocytes were cultured and remained either unexposed (i.e., control) or were exposed to 0.083 % ethanol (i.e., solvent control) or 25 μM TSA for 7 days. After 2 (D2), 4 (D4), and 7 days (D7), samples were taken and quantitative reverse transcriptase polymerase chain reaction analysis was performed using 18S as an endogenous control. Data are expressed as fold change relative to the corresponding control. Values represent mean ± standard deviation ($n = 3$) ($^{*}p < 0.05$ and $^{**}p < 0.01$ compared to solvent control values). Reproduced with permission from Ref. 35

Simultaneously, upregulated expression of apolipoprotein CIII, human blood coagulation factor, and glutamine synthetase were observed [33]. When inhibiting DNMT activity, upregulation of several phase I and phase II biotransformation genes could be demonstrated [32, 34].

With respect to primary hepatocytes in culture, it has been shown that treatment with epigenetic modifiers, such as TSA, could hinder the reentry of cells in the cell cycle, the latter being triggered during 2-step collagenase isolation of hepatocytes [35, 36]. In fact, the time schedule of adding TSA to the culture medium appears crucial to arrest hepatocytes in a particular stage of the cell cycle. Indeed, when TSA is added to the perfusion medium during the isolation procedure, epidermal growth factor-stimulated hepatocyte proliferation is blocked in the early G_1 phase of the cell cycle. This is evidenced by the absence of the proto-oncogene *c*-jun and cyclin D1, indicating a state of proliferative quiescence. On the other hand, when cells are cultivated in the presence of TSA from the time of cell seeding onwards, the point of arrest is located in the early S phase. This is proven by the absence of the $S/G_2/M$-phase marker cyclin-dependent kinase 1 [35]. When a more metabolically stable TSA derivative, namely, Ω-carboxypenthyl para-dimethylaminobenzamide hydroxamate (4-Me_2N-BAVAH), is used during the culture experiments, a G_1 cell cycle arrest could be induced. This is the case when added to the cell culture medium from the time of plating onwards [36]. Both mRNA and protein levels of CYP1A1, CYP2B1/2 and CYP3A11/2 are elevated together with their enzymatic activities (Fig. 3) [37, 38]. Furthermore, the protein expression of some key hepatic transcription factors, such as hepatic nuclear factor 4α

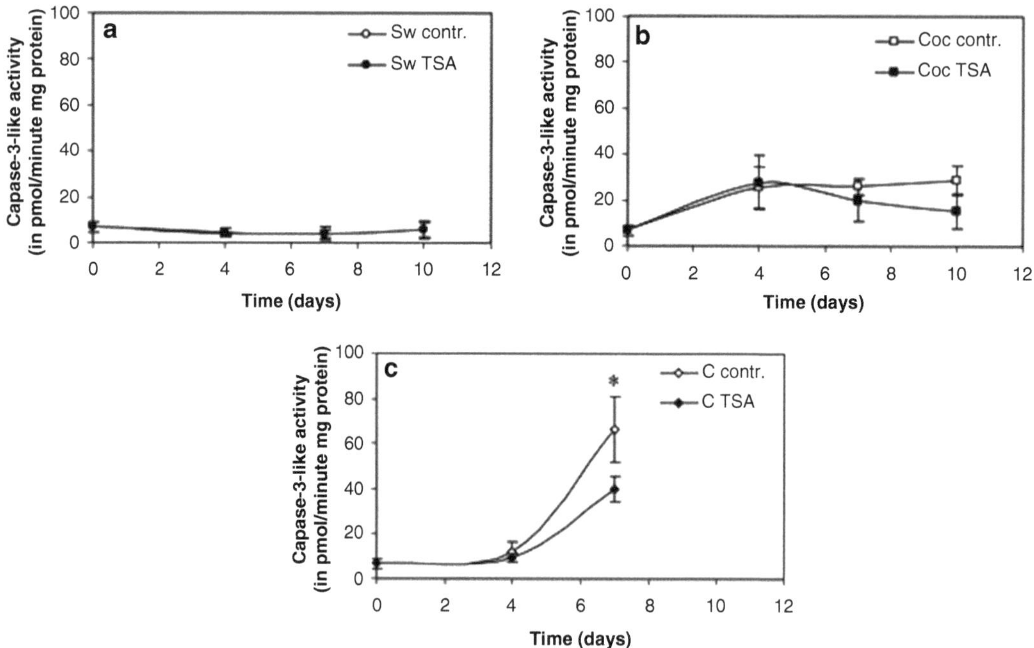

Fig. 4 Measurement of caspase-3-like activity. Caspase-3-like activity measured in rat hepatocytes either cultured in the absence or presence of 1 μM TSA in a collagen I gel sandwich configuration (**a**), as a coculture with rat liver epithelial cells (**b**) or as a conventional monolayer culture (**c**). Rat hepatocytes were cultured, harvested at the indicated time points and aspartylglutamylalanylaspartic acid-ase activity was measured. The activities are expressed as mean ± standard deviation ($n = 3$). Statistical significance between control and TSA-treated hepatocyte cultures was tested by a paired Student's *t*-test ($^*p < 0.05$) (*C* conventional monolayer, *Coc* coculture, *contr* control, *Sw* collagen I gel sandwich culture). Reproduced with permission from Ref. 39

and CCAAT-enhancer-binding protein, is also upregulated. The same holds true for albumin secretion capacity and gap junctional intracellular communication [35, 37, 39, 40]. Also, the effect of HDIs on the occurrence of spontaneous apoptosis in primary hepatocyte cultures has been studied. Upon exposure to TSA or structurally related HDIs, reduced caspase-3 activation and downregulation of the pro-apoptotic B-cell lymphoma 2 (Bcl-2)-like proteins Bid and Bax were observed in rat hepatocyte cultures (Fig. 4). Simultaneously, an increase in the protein level of the antiapoptotic Bcl-2 family member BclxL was noticed. These findings, together with a prolonged liveability of the cells in culture, clearly indicate that the inhibition of HDACs delays the onset of apoptosis, which spontaneously occurs during primary hepatocyte cultivation [35, 36, 41, 42].

In parallel to the above, it was shown that inhibition of DNMTs displays similar positive effects on the maintenance of the phenotype of primary hepatocytes in culture. When the DMTI decitabine was added to the culture medium of the cells, either alone or in combination with an HDI, a positive impact on their albumin secretory capacity and CYP1A1 protein expression was found.

Furthermore, DNMT inhibition also caused a concentration-dependent inhibition of DNA synthesis in epidermal growth factor-induced hepatocyte monolayers. Moreover, it could be shown that combined exposure to HDIs and DMTIs exhibits a synergistic effect on the acquisition and maintenance of the differentiated hepatic phenotype in vitro [13, 43]. Taken together, these data support the hypothesis that the modulation of epigenetic mechanisms is an interesting approach to restore and maintain the in vivo-like phenotype of primary hepatocytes in culture.

4 Conclusion

The hepatic models that are currently available to screen for hepatotoxicity do not succeed to reproduce the in vivo hepatic functionality for more than a few days. Mimicking the in vivo micro-environment of hepatocytes by adding soluble media additives and reintroducing cell–cell and cell–extracellular matrix contacts could already significantly improve the morphology and functionality of primary hepatocytes in culture. Nevertheless, these methodologies focus on reducing the consequences of hepatocellular deterioration and not on affecting the underlying molecular mechanisms. New epigenetic approaches rather target hepatocellular gene expression and represent a powerful tool to tackle dedifferentiation of primary hepatocytes in culture. Although these innovative approaches hold great promise in ameliorating hepatic functions in vitro, we are only standing at the verge to discover the different levels and mechanisms that are needed to further improve hepatocyte functioning.

References

1. Burt AD, Day CP (2002) Pathophysiology of the liver. In: MacSween RNM, Burt AD, Portmann BC, Ishak KG, Scheuer PJ, Anthony PP (eds) Pathology of the liver. Churchill Livingstone, New York, NY, pp 67–105
2. Papeleu P, Elaut G, Rogiers V et al (2002) Cell cultures as *in vitro* tools for biotransformation studies. In: Pandalai SG (ed) Recent research developments in drug metabolism and disposition drug. Transworld Research Network, Kerala, India, pp 199–234
3. Baker TK, Carfagna MA, Gao H et al (2001) Temporal gene expression analysis of monolayer cultured rat hepatocytes. Chem Res Toxicol 14:1218–1231
4. Elaut G, Henkens T, Papeleu P et al (2006) Molecular mechanisms underlying the dedifferentiation process of isolated hepatocytes and their cultures. Curr Drug Metab 7:629–660
5. Beigel J, Fella K, Kramer PJ et al (2008) Genomics and proteomics analysis of cultured primary rat hepatocytes. Toxicol In Vitro 22:171–181
6. Rowe C, Goldring CEP, Kitteringham NR et al (2010) Network analysis of primary hepatocyte dedifferentiation using a shotgun proteomics approach. J Proteome Res 9:2658–2668
7. LeCluyse EL, Witek RP, Andersen ME et al (2012) Organotypic liver culture models: meeting current challenges in toxicity testing. Crit Rev Toxicol 42:501–548
8. Paine AJ, Andreakos E (2004) Activation of signaling pathways during hepatocyte isolation: relevance to toxicology *in vitro*. Toxicol In Vitro 18:187–193
9. Vinken M, Papeleu P, Snykers S et al (2006) Involvement of cell junctions in hepatocyte culture functionality. Crit Rev Toxicol 36:299–318

10. Vinken M, Maes M, Oliveira A et al (2014) Primary hepatocytes and their cultures in liver apoptosis research. Arch Toxicol 88:199–212
11. Shulman M, Nahmias Y (2013) Long-term culture and co-culture of primary rat and human hepatocytes. Methods Mol Biol 945:287–302
12. Godoy P, Hewitt NJ, Albrecht U et al (2013) Recent advances in 2D and 3D *in vitro* systems using primary hepatocytes, alternative hepatocyte sources and non-parenchymal liver cells and their use in investigating mechanisms of hepatotoxicity, cell signaling and ADME. Arch Toxicol 87:1315–1530
13. Fraczek J, Bolleyn J, Vanhaecke T et al (2013) Primary hepatocyte cultures for pharmaco-toxicological studies: at the busy crossroad of various anti-dedifferentiation strategies. Arch Toxicol 87:577–610
14. Choudhuri S (2011) From Waddington's epigenetic landscape to small noncoding RNA: some important milestones in the history of epigenetics research. Toxicol Mech Methods 21:252–274
15. Witt O, Deubzer HE, Milde T et al (2009) HDAC family: what are the cancer relevant targets? Cancer Lett 277:8–21
16. Riggs MG, Whittaker RG, Neumann JR et al (1977) n-Butyrate causes histone modification in HeLa and Friend erythroleukaemia cells. Nature 268:462–464
17. Guil S, Esteller M (2009) DNA methylomes, histone codes and miRNAs: tying it all together. Int J Biochem Cell Biol 41:87–95
18. Hebbes TR, Thorne AW, Crane-Robinson C (1988) A direct link between core histone acetylation and transcriptionally active chromatin. EMBO J 7:1395–1402
19. Durrin LK, Mann RK, Kayne PS et al (1991) Yeast histone H4 *N*-terminal sequence is required for promoter activation *in vivo*. Cell 65:1023–1031
20. Meehan RR, Lewis JD, McKay S et al (1989) Identification of a mammalian protein that binds specifically to DNA containing methylated CpGs. Cell 58:499–507
21. Nan X, Meehan RR, Bird A (1993) Dissection of the methyl-CpG binding domain from the chromosomal protein MeCP2. Nucleic Acids Res 21:4886–4892
22. Nan X, Ng HH, Johnson CA et al (1998) Transcriptional repression by the methyl-CpG-binding protein MeCP2 involves a histone deacetylase complex. Nature 393:386–389
23. Fuks F, Hurd PJ, Wolf D et al (2003) The methyl-CpG-binding protein MeCP2 links DNA methylation to histone methylation. J Biol Chem 278:4035–4040
24. Batty N, Malouf GG, Issa JPJ (2009) Histone deacetylase inhibitors as anti-neoplastic agents. Cancer Lett 280:192–200
25. Fandy TE (2009) Development of DNA methyltransferase inhibitors for the treatment of neoplastic diseases. Curr Med Chem 16:2075–2085
26. Kristensen L, Nielsen H, Hansen L (2009) Epigenetics and cancer treatment. Eur J Pharmacol 625:131–142
27. Prince HM, Bishton MJ, Harrison SJ (2009) Clinical studies of histone deacetylase inhibitors. Clin Cancer Res 15:3958–3969
28. Spiegel S, Milstien S, Grant S (2012) Endogenous modulators and pharmacological inhibitors of histone deacetylases in cancer therapy. Oncogene 31:537–551
29. Gore S (2009) *In vitro* basis for treatment with hypomethylating agents and histone deacetylase inhibitors: can epigenetic changes be used to monitor treatment? Leuk Res 33:S2–S6
30. Kim JY, Ahn MR, Kim DK et al (2004) Histone deacetylase inhibitor stimulate CYP3A4 proximal promoter activity in HepG2 cells. Arch Pharm Res 27:407–414
31. Dannenberg LO, Chen HJ, Tian H et al (2006) Differential regulation of the alcohol dehydrogenase 1B (ADH1B) and ADH1C genes by DNA methylation and histone deacetylation. Alcohol Clin Exp Res 30:928–937
32. Dannenberg L, Edenberg H (2006) Epigenetics of gene expression in human hepatoma cells: expression profiling the response to inhibition of DNA methylation and histone deacetylation. BMC Genomics 7:181
33. Yamashita YI, Shimada M, Harimoto N et al (2003) Histone deacetylase inhibitor trichostatin A induces cell-cycle arrest/apoptosis and hepatocyte differentiation in human hepatoma cells. Int J Cancer 103:572–576
34. Bakker J, Lin X, Nelson WG (2002) Methyl-CpG binding domain protein 2 represses transcription from hypermethylated pi-class glutathione *S*-transferase gene promoters in hepatocellular carcinoma cells. J Biol Chem 277:22573–22580
35. Papeleu P, Loyer P, Vanhaecke T et al (2003) Trichostatin A induces differential cell cycle arrests but does not induce apoptosis in primary cultures of mitogen-stimulated rat hepatocytes. J Hepatol 39:374–382
36. Papeleu P, Wullaert A, Elaut G et al (2007) Inhibition of NF-kappaB activation by the histone deacetylase inhibitor 4-Me_2N-BAVAH induces an early G1 cell cycle arrest in primary hepatocytes. Cell Prolif 40:640–655
37. Henkens T, Papeleu P, Elaut G et al (2007) Trichostatin A, a critical factor in maintaining

the functional differentiation of primary cultured rat hepatocytes. Toxicol Appl Pharmacol 218:64–71

38. Snykers S, Henkens T, De Rop E et al (2009) Role of epigenetics in liver-specific gene transcription, hepatocyte differentiation and stem cell reprogrammation. J Hepatol 51:187–211
39. Henkens T, Vinken M, Lukaszuk A et al (2006) Differential effects of hydroxamate histone deacetylase inhibitors on cellular functionality and gap junctions in primary cultures of mitogenstimulated hepatocytes. Toxicol Lett 178:37–43
40. Vinken M, Henkens T, Vanhaecke T et al (2006) Trichostatin A enhances gap junctional intercellular communication in primary cultures of adult rat hepatocytes. Toxicol Sci 91: 484–492
41. Vanhaecke T, Henkens T, Kass GEN et al (2004) Effect of the histone deacetylase inhibitor Trichostatin A on spontaneous apoptosis in various types of adult rat hepatocyte cultures. Biochem Pharmacol 68:753–760
42. Fraczek J, Deleu S, Lukaszuk A et al (2009) Screening of amide analogues of Trichostatin A in cultures of primary rat hepatocytes: search for potent and safe HDAC inhibitors. Invest New Drugs 27:338–346
43. Fraczek J, Vinken M, Tourwe D et al (2011) Synergetic effects of DNA demethylation and histone deacetylase inhibition in primary rat hepatocytes. Invest New Drugs 30:1715–1724

Chapter 15

Transfection of Primary Hepatocytes with Liver-Enriched Transcription Factors Using Adenoviral Vectors

Marta Benet, Ramiro Jover, and Roque Bort

Abstract

Primary cultured hepatocytes are probably the best model to study endogenous metabolic pathways, toxicity, or drug metabolism. Many of these studies require expression of ectopic genes. It would be desirable to use a method of transfection that allows dose–response studies, high efficiency of transfection, and the possibility to express several genes at the same time. Adenoviral vectors fulfill these requirements, becoming a valuable tool for primary hepatocyte transfection. Moreover, they are easy to generate and do not require a high level of biocontainment. In the present chapter, we describe the generation, cloning, amplification, and purification of an adenoviral vector capable of infecting primary cultured hepatocytes. This recombinant adenovirus induces robust expression of the protein of interest in hepatocytes within a wide range of doses.

Key words Primary hepatocyte culture, Transcription factor, Adenoviral vector

1 Introduction

Primary cultured hepatocytes are cells isolated from liver and maintained in culture. They have a number of features that renders them the most appropriate hepatocyte model. Cultured hepatocytes indeed have some intrinsic advantages that result in the closest representation of the hepatic in vivo situation. As intact cells, they retain active uptake and excretion mechanisms as well as metabolism to a level comparable to hepatocytes in the liver [1]. Integrated metabolism pathways (i.e., both phase I and phase II biotransformation enzymes), physiological cofactor-enzyme levels, and active gene expression are reasonably well maintained, resulting in an in vitro model that has been successfully used to investigate both metabolism and toxic effects in liver cells [2–4]. Hepatoma cell lines offer some advantages, such as high availability and an unlimited life span. However, currently available hepatoma cell lines are not a good alternative to cultured hepatocytes, as they show an extensively dedifferentiated hepatic phenotype.

Mathieu Vinken and Vera Rogiers (eds.), *Protocols in In Vitro Hepatocyte Research*, Methods in Molecular Biology, vol. 1250, DOI 10.1007/978-1-4939-2074-7_15, © Springer Science+Business Media New York 2015

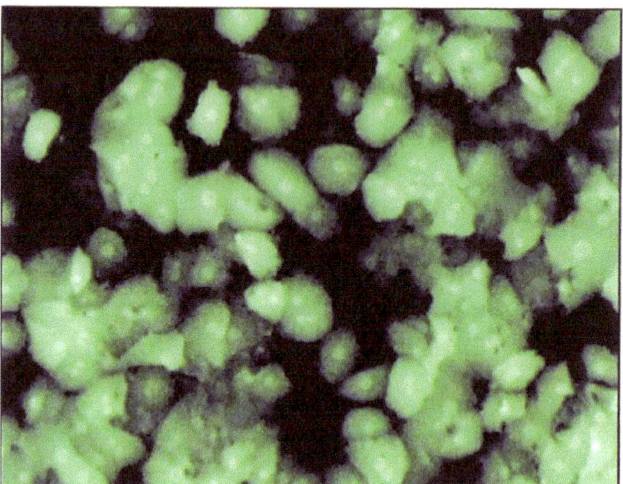

Fig. 1 GFP expression in primary cultured hepatocytes infected with an adenoviral vector expressing GFP. Human hepatocytes cultured for 24 h were transduced with 10 MOI Ad-GFP. After 48 h, images were taken using a fluorescence microscopy. Reproduced in modified form with permission [13]

Therefore, the effect of a transfected transcription factor in these cells may not have in vivo relevance. Thus, it is of interest to develop a simple technique allowing dose-dependent expression of multiple genes in primary cultured hepatocytes.

Primary cultured hepatocytes are generally difficult to transfect, limiting their use as an in vitro research model. The most common methods for cell transfection, such as liposome-mediated or electroporation and nucleofection, perform poorly in primary cultured hepatocytes with efficiencies 1–2 orders of magnitude less than the hepatocellular carcinoma derived cell line HepG2 [5, 6]. Moreover, hepatocytes in culture do not proliferate, limiting the generation of a homogeneous cultured population of transfected hepatocytes by cell sorting or other purification methods. Therefore, a successful transfection method suitable for primary cultured hepatocytes should yield high transfection efficiency with low toxicity.

Replication-defective viruses have been satisfactorily used in vivo and in vitro to deliver genes into hepatocytes [7–10]. When compared to other viruses, adenoviral vectors show several advantages. Indeed, they enable high transfection efficiency (Fig. 1), allow dose–response studies and expression of multiple genes, and do not integrate into the host genomic DNA.

In the current chapter, we present a detailed protocol to obtain a pure and high-titer viral stock of an adenoviral vector. We also show an example where we studied the regulation of cytochrome P450 (CYP) 3A4 mRNA by CCAAT/enhancer-binding protein beta liver-enriched inhibitory protein (C/EBPβ-LIP) in primary cultured human hepatocytes.

2 Materials

2.1 Basic Cell Culture

1. HEK293 complete medium. Dulbecco's Modified Eagle's Medium with 4.5 g/L glucose, 50 IU/mL penicillin, 50 μg/mL streptomycin, and 10 % fetal bovine serum.
2. Phosphate-buffered saline (PBS). 137 mM NaCl, 2.7 mM KCl, 10 mM $Na_2HPO_4{\cdot}2H_2O$, 1.8 mM KH_2PO_4 in deionized water. Adjust to pH 7.4, sterilize by passing through a 0.22 μm filter and store for maximum 6 months at 4 °C. *Prior* to use, PBS should be placed for 30 min at room temperature.
3. 0.25 % trypsin–ethylenediaminetetraacetic acid with phenol-red.
4. Laminar air flow cabinet.
5. Incubator (37 ± 1 °C, 90 ± 5 % humidity, 5 ± 1 % CO_2/air).

2.2 Development of an Adenoviral Vector (*See* Note 1)

1. 6 and 10 cm cell culture dishes.
2. Shuttle plasmid pAC/CMVpLpA containing the cDNA of interest at 1 μg/mL.
3. 1 μg/mL pJM17 vector (i.e., a dl309 adenovirus-5 genome backbone).
4. FuGENE HD transfection reagent (Promega, Spain).
5. Fluorescence microscopy.

2.3 Cloning of an Adenoviral Vector

1. 2.5 % agar in distilled water. Dissolve 2.5 g of low melting point agarose in 100 mL of distilled water. Microwave until a clear solution is obtained (*see* **Note 2**).
2. 2× concentrated Modified Eagle Medium complete medium. 2× concentrated Modified Eagle Medium (Gibco/BRL, United States of America) containing 2 % penicillin–streptomycin and 4 % fetal bovine serum.
3. HEK293 agarose/medium. Mix equal volumes of 2× concentrated Modified Eagle Medium and 2.5 % agarose solution. Both ingredients should be at 42 °C before mixing to avoid solidification of the mixture.
4. Thermostated bath.

2.4 Amplification and Titration

1. 50 mL polypropylene tube.
2. Vivapure AdenoPack kit (Sartorius, Germany).

2.5 Infection of Primary Cultured Hepatocytes

1. Hepatocyte seeding medium. Ham's F12/William's E medium (i.e., 1/1 ratio) supplemented with 2 % newborn calf serum, 100 U/mL penicillin, 50 μg/mL streptomycin, 0.1 % bovine serum albumin, 10 nM insulin, 25 μg/mL transferrin, 0.1 μM sodium selenite, 65.5 μM ethanolamine, 7.2 μM lin-

oleic acid, 17.5 mM glucose, 6.14 mM ascorbic acid, and 0.64 mM *N*-omega-nitro-l-arginine methyl ester. Prepare in a laminar air flow cabinet and store for maximum 7 days at 4 °C. *Prior* to use, the hepatocyte seeding medium should be placed for 30 min in a thermostated bath at 37 °C.

2. 3.5 μg/cm^2 fibronectin-coated plastic dishes.
3. Hepatocyte culture medium. Hepatocyte seeding medium without serum and supplemented with 100 nM dexamethasone and 10 nM insulin.
4. High titer adenovirus stock.

3 Methods

3.1 Development of an Adenoviral Vector

1. The day before transfection, evenly seed 2×10^6 HEK293 cells in 6 cm diameter cell culture plates. Use two plates for each adenovirus.
2. Change the HEK293 complete medium 2 h before transfection. Cells should be 80 % confluent before use.
3. Prepare the FuGENE HD:DNA complexes (Table 1) (*see* **Note 3**). Add HEK293 medium without antibiotics and serum to a sterile 1.5 mL tube. Add FuGENE HD transfection reagent directly into the medium without touching the sides of the tube. Mix 1 s and incubate at room temperature for 5 min. Add the DNA, mix 1 s, and incubate at room temperature for 20–25 min.
4. Add 200 μL of the corresponding FuGENE HD:DNA complex to each 6 cm diameter cell culture plate without removing the cell medium.
5. Swirl the plates to distribute the complex evenly.
6. Transfer to an incubator at 37 °C and 5 % CO_2.
7. Check for transfection efficiency with the positive control plate using a fluorescence microscopy.
8. Change the medium every 2 days (*see* **Note 4**).

Table 1
Preparation of the FUGENE HD:DNA complexes

Plates	Vector	Medium (μL)	FUGENE (μL)	pJM17 (μL)	DNA (μL)
Negative control		400	/	/	/
Positive control (1 μg/μl)	pIRES2EGFP	384	12	/	4
Adenovirus (1 μg/μl)	pAC-derived	384	12	2	2
Adenovirus (1 μg/μl)	pAC-derived	384	12	2.6	1.4

9. Cells normally start to detach and lyze between 7 and 15 days after transfection. When this happens, do not change the medium and wait until 80 % of the cells are detached (*see* **Note 5**).
10. Collect the medium and the cells using a cell scraper.
11. Perform two freeze and thawing cycles to lyze any remaining intact cells.
12. Centrifuge for 10 min at 8,000×*g* and 4 °C to remove cell debris.
13. Keep supernatant (i.e., 500 μL aliquots) at −80 °C. This is the first viral stock.
14. Before going into cloning, the adenoviral vector should be tested (*see* **Note 6**).

3.2 Cloning of an Adenoviral Vector (*See* Note 7)

1. The day before transfection, seed ten plates of HEK293 cells containing 2×10^6 cells per 6 cm diameter cell culture plate.
2. Check if the HEK293 cells are approximately 80 % confluent before use.
3. Prepare serial dilutions of the viral stock in 1 mL HEK293 complete medium. Typically, dilutions between 10^{-3} and 10^{-11} are prepared.
4. Remove the medium from the plates and add 1 mL of each dilution. A control plate with only medium should be also included.
5. Transfer the plates to the incubator for 2 h at 37 °C and 5 % CO_2.
6. In the meantime, prepare 65 mL of HEK293 agarose/medium and keep at 42 °C until use.
7. Aspirate the virus-containing media and wash the monolayers twice with PBS.
8. Slowly add 6 mL of the warm HEK293 agarose/medium and allow the mixture to solidify at room temperature up to 10 min before returning the plates to the incubator.
9. Viral plaques should be visible between 7 and 15 days after plating.
10. Select a plate where viral plaques (i.e., viral clones) are clearly separated to avoid cross-contamination between clones.
11. To collect adenoviral plaques, use an automatic 20–200 μL volume pipette with a shortened sterile pipette tip and remove an agarose plug by directly punching over the plaque. Pick between 10 and 20 plaques.
12. Place each agarose plug in a separate sterile 1.5 mL microcentrifuge tube containing 1 mL of HEK293 complete medium.

13. Elute the virus from the agarose by several freeze and thaw cycles of the medium containing the viral plug. Centrifuge for 10 min at 8,000 × g and 4 °C.
14. Keep each supernatant at −80 °C or use it immediately as described below.
15. Select 1 or 2 clones of the adenoviral vector. For this purpose, we routinely infect HeLa cells seeded in 24-well plates with 500 μL of the eluted virus of each independent clone (*see* **Note 6**).

3.3 Amplification and Titration

3.3.1 Amplification (See **Note 8***)*

1. The day before transfection, seed five plates of HEK293 cells containing 4.5×10^6 cells per 10 cm diameter cell culture plate.
2. Check if the HEK293 cells are approximately 80 % confluent before use.
3. Add 125 μL of the selected adenoviral clone in 30 mL of HEK293 complete medium.
4. Remove the medium from the plates and add 6 mL of the medium containing the adenoviral vector. Transfer to an incubator at 37 °C and 5 % CO_2.
5. Cells normally start to detach lyze between 24 and 48 h. Do not change the medium and wait until 90 % of the cells are detached.
6. Collect the media and the cells using a cell scraper into 50 mL tubes.
7. Keep the supernatant at −80 °C until purification or start purification using the Vivapure AdenoPack 20 kit following the manufacturer's recommendation (*see* **Note 9**).

3.3.2 Titration

1. The day before transfection, seed ten plates of HEK293 cells containing 2×10^6 cells per 6 cm diameter cell culture plate.
2. Check if the HEK293 cells are approximately 80 % confluent before use.
3. Prepare serial dilutions of the viral stock in 1 mL HEK293 complete medium. Typically, dilutions between 10^{-3} and 10^{-11} are prepared.
4. Remove the medium from the plates and add 1 mL of each dilution. A control plate with only medium should be also included.
5. Transfer the plates to the incubator for 2 h at 37 °C and 5 % CO_2.
6. In the meantime, prepare 65 mL of HEK293 agarose/medium and keep at 42 °C until use.
7. Aspirate the virus-containing media and wash the monolayers twice with PBS.

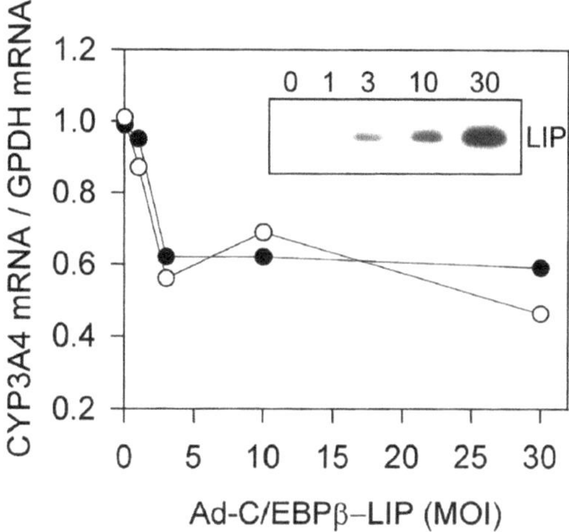

Fig. 2 Adenoviral mediated expression of C/EBPβ-LIP in primary cultured hepatocytes. Human hepatocytes cultured for 24 h were transduced with 1–30 MOI Ad-LIP. The *inset* depicts dose-dependent expression of LIP protein by immunoblot. In parallel, CYP3A4 mRNA levels were measured in control and infected cells. CYP3A4 is downregulated by C/EBPβ-LIP. Reproduced in modified form with permission [13]

8. Select a plate where viral plaques (i.e., viral clones) are clearly separated and their number remains constant after 24 h. To calculate the adenoviral titer in plaque forming units per milliliter (PFU/mL), multiply the number of plaques by the dilution factor. Once you know the adenovirus titer, the multiplicity of infection (MOI) for a specific infection is calculated using the formula MOI = [(μL adenovirus × PFU/μL)/cell number].

3.4 Infection of Primary Cultured Hepatocytes

1. Use freshly isolated primary human hepatocytes [11].
2. Seed the primary hepatocytes on fibronectin-coated plastic dishes and place the cell cultures in an incubator at 37 °C and 5 % CO_2 for 2 h.
3. Remove the hepatocyte seeding medium and replace by hepatocyte culture medium. Place the cell cultures back into an incubator at 37 °C and 5 % CO_2 for 24 h.
4. Thaw the adenovirus stock on ice. Replace the hepatocyte culture medium and add adenovirus at MOI ranging from 5 to 50 PFU/cell. Transfer the cells to an incubator at 37 °C and 5 % CO_2.
5. Change the medium next day.
6. After 48–72 h of infection, the cells are ready to use.
7. In the example shown (Fig. 2), we found that CYP3A4 mRNA is downregulated following transfection of C/EBPβ-LIP in primary cultured hepatocytes.

4 Notes

1. There are different commercially available kits to generate replication defective recombinant adenoviral vectors. Our system is based upon the method described in [12].
2. It is not necessary to prepare a fresh 2.5 % agarose solution in distilled water. We normally prepare 500 mL and microwave it each time we need it. However, it is essential to cool it down till 42 °C before mixing with the medium, which also needs to be handled at 42 °C.
3. A detailed protocol of the use of FuGENE HD transfection reagent can be found in the product manual. The protocol described in this chapter works very well for HEK293 cells.
4. After cotransfection with pJM17 and shuttle vector, cells are maintained without passaging until cell detachment and lysis occurs (i.e., 10–21 days). This is a critical period due to excessive cell growth. To reduce cell growth, the medium is prepared with 3 % fetal bovine serum instead of 10 %. Even then, cells become very confluent and the medium tends to become very acidic (i.e., yellow). It is important to maintain the pH at approximately 7.4 with sodium bicarbonate or changing the medium very often. Only when this is impossible, you can make a 1:4 passage. Do not use *N*-[2-hydroxyethyl]piperazine-*N′*[2-ethaansulfonzuur], as this is toxic to HEK293 cells.
5. Some adenoviruses will generate and lyze the cells as early as 7 days, while others will require 3 weeks. It is important to maintain the control plate in parallel to confirm that cell lysis is induced by the adenovirus.
6. After the first lysis, we recommend preliminary validation of the virus before starting further procedures. We routinely reinfect HEK293 cells with the first lysis and check for cell lysis within 72 h. In parallel, the presence of the expressed protein can be detected by immunoblot analysis in HeLa cells or other cell types. Other methods can be performed depending on the protein expressed, but it should be as simple and reliable as possible.
7. The first viral stock is composed of a mixture of different adenoviral vectors due to aberrant recombination of pJM17 and shuttle vector. Thus, it is important to clone this initial viral stock to obtain a pure adenoviral vector originating from a single successful recombination (i.e., discrete plaque). One of the most common aberrant adenoviruses is an empty adenovirus. This is generated by losing the cloned cDNA from the shuttle vector during recombination.
8. The cell lysate obtained after adenoviral cloning has a very low titer. For this reason, it is desirable to amplify, purify, and

concentrate to generate a pure high-titer recombinant adenoviral stock (i.e., 1 or more orders of magnitude higher).

9. This kit generally provides pure adenovirus with a titer in the range of 10^8–10^{11} PFU/mL.

Acknowledgements

This work was financially supported by the European Commission (project CARCINOGENOMICS) and SAF2010-15376.

References

1. Jigorel E, Le Vee M, Boursier-Neyret C et al (2005) Functional expression of sinusoidal drug transporters in primary human and rat hepatocytes. Drug Metab Dispos 33:1418–1422
2. Castell JV, Gomez-Lechon MJ, David M et al (1990) Acute-phase response of human hepatocytes: regulation of acute-phase protein synthesis by interleukin-6. Hepatology 12:1179–1186
3. Donato MT, Gomez-Lechon MJ, Castell JV (1990) Effect of xenobiotics on monooxygenase activities in cultured human hepatocytes. Biochem Pharmacol 39:1321–1326
4. Donato MT, Guillen MI, Jover R et al (1997) Nitric oxide-mediated inhibition of cytochrome P450 by interferon-gamma in human hepatocytes. J Pharmacol Exp Ther 281: 484–490
5. Tur-Kaspa R, Teicher L, Levine BJ et al (1986) Use of electroporation to introduce biologically active foreign genes into primary rat hepatocytes. Mol Cell Biol 6:716–718
6. Ourlin JC, Vilarem MJ, Daujat M et al (1997) Lipid-mediated transfection of normal adult human hepatocytes in primary culture. Anal Biochem 247:34–44
7. Guzman C, Benet M, Pisonero-Vaquero S et al (2013) The human liver fatty acid binding protein (FABP1) gene is activated by FOXA1 and PPARalpha and repressed by C/EBPalpha: implications in FABP1 downregulation in nonalcoholic fatty liver disease. Biochim Biophys Acta 1831:803–818
8. Herz J, Gerard RD (1993) Adenovirus-mediated transfer of low density lipoprotein receptor gene acutely accelerates cholesterol clearance in normal mice. Proc Natl Acad Sci U S A 90:2812–2816
9. Kay MA, Landen CN, Rothenberg SR et al (1994) *In vivo* hepatic gene therapy: complete albeit transient correction of factor IX deficiency in hemophilia B dogs. Proc Natl Acad Sci U S A 91:2353–2357
10. Moya M, Benet M, Guzman C et al (2012) Foxa1 reduces lipid accumulation in human hepatocytes and is downregulated in nonalcoholic fatty liver. Plos ONE 7:e30014
11. Gomez-Lechon MJ, Lopez P, Donato T et al (1990) Culture of human hepatocytes from small surgical liver biopsies: biochemical characterization and comparison with *in vivo*. In Vitro Cell Dev Biol 26:67–74
12. Gomez-Foix AM, Coats WS, Baque S et al (1992) Adenovirus-mediated transfer of the muscle glycogen phosphorylase gene into hepatocytes confers altered regulation of glycogen metabolism. J Biol Chem 267:25129–25134
13. Jover R, Bort R, Gomez-Lechon MJ et al (2002) Downregulation of human CYP3A4 by the inflammatory signal interleukin-6: molecular mechanism and transcription factors involved. FASEB J 16:1799–1801

Part II

In Vitro Methods to Probe Hepatocyte Functionality and Toxicity

Chapter 16

Transcriptomics of Hepatocytes Treated with Toxicants for Investigating Molecular Mechanisms Underlying Hepatotoxicity

Vaibhav Shinde, Regina Stöber, Harshal Nemade, Isaia Sotiriadou, Jürgen Hescheler, Jan Hengstler, and Agapios Sachinidis

Abstract

Transcriptomics is a powerful tool for high-throughput gene expression profiling. Transcriptome microarray experiments conducted with RNA isolated from hepatocytes after exposure to toxicants enable a deep insight into the molecular mechanisms of hepatotoxicity. This understanding, along with structure–activity relationships underlying hepatotoxicity, will provide a novel strategy to design cost-effective and safer therapeutics. Transcriptomics studies conducted with established hepatotoxic drugs in various in vitro and in vivo hepatotoxicity test systems have contributed to the elucidation of the mechanistic basis of liver insults, which were later on substantiated at the proteomics and metabolomics levels. The present chapter is focused on comprehensive transcriptomics of cultured primary hepatocytes treated with chemicals by applying Affymetrix microarray technology. It also describes the detailed protocol for culturing of hepatocytes, their exposure to toxicants as well as sample collection, including RNA isolation, RNA target preparation and finally the hybridization to gene chips for microarray expression analysis.

Key words Hepatotoxicity, Transcriptomics, Sandwich-cultured hepatocytes, Microarrays

1 Introduction

The liver is the primary organ for detoxification of chemicals. Indeed, the close association of the liver with the gastrointestinal tract and systemic circulation results in its vast exposure to exogenous ingested or injected chemical compounds, such as drugs. This exposure sometimes directly causes drug-induced liver injury or idiosyncratic hepatotoxicity. Hepatotoxicity has been the main reason for postmarket withdrawal of drugs, including alpidem, bendazac, clomacron, clometacin, exifone, sitaxentan, ximelagatran, bromfenac and troglitazone [1–3]. As the liver comprises of hepatocytes by 60 % of the total cell number and by 80 % of the total volume, research on hepatocytes has been primarily focused on the development of in vitro cell-based assays for hepatotoxicity

Mathieu Vinken and Vera Rogiers (eds.), *Protocols in In Vitro Hepatocyte Research*, Methods in Molecular Biology, vol. 1250, DOI 10.1007/978-1-4939-2074-7_16, © Springer Science+Business Media New York 2015

predictions of novel drugs as well as environmental chemicals [4–7]. The presently developed assays contribute to the analysis of cytotoxicity as well as specific biochemical effects of hepatocytes, but are not sufficient to address the mechanistic aspects of hepatotoxicity. The emergence of the toxicogenomics field has paved a way to gain insight into the mechanistic liver toxicity aspects at the molecular level. Transcriptomics studies on microarray platforms typically reveal the expression of thousands of genes perturbed by drugs at any time point and thus provide high-throughput data to understand the mechanistic basis at the gene expression level, which can be further validated by proteomics and metabolomics techniques. The fundamental concept of microarrays is based on the ability of nucleic acids to make complementary strands with DNA. There are various types of commercially available microarrays from different companies, including photolithographic technology-based in situ synthesized oligonucleotide arrays (Affymetrix), ink-jet technology-based in situ synthesized oligonucleotide arrays (Agilent Technologies), and spotted oligonucleotide arrays (Codelink, Amersham). Using these microarrays, toxicological studies carried out with different drugs have been conducted in various biological systems [8–10]. For example, microarray analysis of human and rat hepatocytes exposed to acetaminophen revealed repression of mitochondrial function, oxidoreductase activity, and energy-consuming pathways [11]. Another study using human primary hepatocytes contributed to the understanding of the hepatotoxicity mechanisms of trovafloxacin and chlorpromazine [12, 13]. Microarray-based studies also enabled the characterization of the toxicological response of primary human hepatocytes, human hepatoma HepG2 and HepaRG cells to phenobarbital, beta-naphthoflavone, and rifampicin [14]. To further shed light onto toxicological mechanisms, various hepatotoxic drugs or chemicals, such as clofibrate, omeprazole, ethionine, thioacetamide, benzbromarone, propylthouracil and bromobenzene, have been administered to rodents and microarray studies have been conducted on their liver samples. The obtained drug-specific differentially expressed genes have been considered as a starting point for the development of mechanism-based biomarkers for hepatotoxicity [15–20]. Such studies have collectively highlighted the importance of transcriptomics in toxicology.

The protocol presented in this chapter is focused on routine Affymetrix microarray-based gene expression studies. Photolithography technology is used to synthesize 25 base pair oligonucleotide probes on the silicon component inside Affymetrix GeneChips. Each gene has been designated by a perfect match probe set of 10–20 probes. Each perfect match probe hereby has a mismatch probe, which differs by 1 base pair in the middle of the oligonucleotide. The biotin-labeled RNA is hybridized to the array chips and the fluorescence signal intensities obtained from the mismatch probe set is corrected for perfect match probe sets to measure gene expression [21]. The RNA target preparation for hybridization on GeneChips is based on the T7 in vitro transcription technology (IVT). In this method, the first strand of complementary DNA (cDNA) is synthesized from total RNA by reverse transcription and is consequently converted into double-stranded DNA that is used as a template to synthesize amplified RNA (aRNA). The latter is then biotin-labeled, purified, and fragmented for hybridization onto the array chips. This chapter also describes methods used for culturing primary hepatocytes and drug exposure as well as for sample collection. Furthermore, the chapter includes a full description of the methods related to RNA isolation, using a commercially available RNA isolation kit, and to target RNA preparation for hybridization on GeneChip for microarray analysis.

2 Materials

2.1 Hepatocyte Cultivation

1. Hepatocyte cultivation medium. William's E medium (PAN Biotech, Germany) with 100 U/mL penicillin, 0.1 mg/mL streptomycin, 50 μg/mL gentamicin, 20 mM L-glutamine, 100 nM dexamethasone, 2 ng/mL insulin-transferrin-selenium (Sigma, Germany).
2. Sera plus (PAN Biotech, Germany).
3. Dulbecco's Modified Eagle Medium (Biozol Diagnostic, Germany).
4. Lyophilized collagen from rat tail (Roche Diagnostics, Germany).
5. Acetic acid.
6. Incubator (37 ± 1 °C, 90 ± 5 % humidity, 5 ± 1 % CO_2).
7. Laminar air flow cabinet.
8. Thermostated bath (37 °C).

2.2 RNA Isolation

1. RNeasy mini kit (QIAGEN, Germany).
2. RNase-free DNase set (QIAGEN, Germany).
3. TRIzol (Life technologies, Germany).
4. 70 % ethanol in nuclease-free water. Store at −20 °C.
5. Chloroform.
6. Nuclease-free tubes.
7. 1 mL syringe with 24 G needle.

2.3 RNA Target Preparation

1. GeneChip Human Genome U133 Plus 2.0 Array (Affymetrix, USA) (Fig. 1).
2. GeneChip 3′ IVT express kit (Affymetrix, USA). This includes aRNA binding buffer concentrate, RNA binding beads, aRNA wash solution concentrate, aRNA elution solution, nuclease-free water, 5× concentrated array fragmentation buffer,

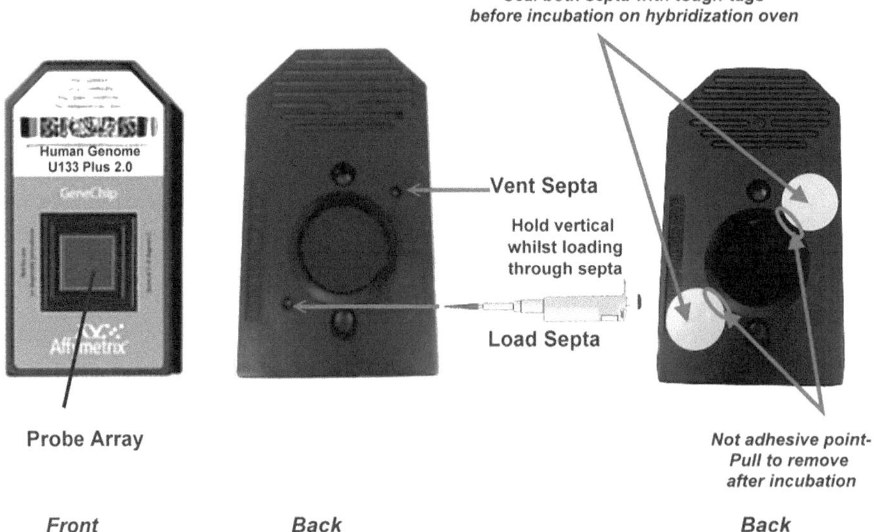

Fig. 1 GeneChip Human Genome U133 Plus 2.0 Array. This array includes the expression of about 47,400 transcripts, of which 38,500 are human genes (courtesy of Affymetrix, Inc., Santa Clara, California, USA)

polymerase chain reaction (PCR) tubes, U-bottom plate, first-strand enzyme mix, first-strand buffer mix, second-strand enzyme mix, second-strand buffer mix, IVT enzyme mix, IVT labeling buffer, IVT biotin label, control RNA, nuclease-free water, poly-A control stock, poly-A control dilution buffer, 20× concentrated hybridization controls and control oligo B2.

3. GeneChip hybridization, wash and stain kit (Affymetrix, USA).
4. Ethanol.
5. Thermal cycler with heated lid.
6. Magnetic stand for 96-well plates (Ambion, Germany).
7. Orbital shaker for 96-well plates.
8. Nanodrop ND-1000 device (NanoDrop Technologies, Germany).
9. Lab-line titer plate shaker.
10. U-bottom plates.
11. Heat block.

2.4 Microarray Components (Fig. 2)

1. GeneChip hybridization oven 645 (Affymetrix, USA).
2. GeneChip fluidics station-450 (Affymetrix, USA).
3. GeneChip scanner 3000-7G (Affymetrix, USA).

2.5 Software

1. Affymetrix GeneChip operating software.
2. Partek Genomic Suite 6.25.
3. Online tools for functional annotation, such as database for annotation, visualization and integrated discovery.

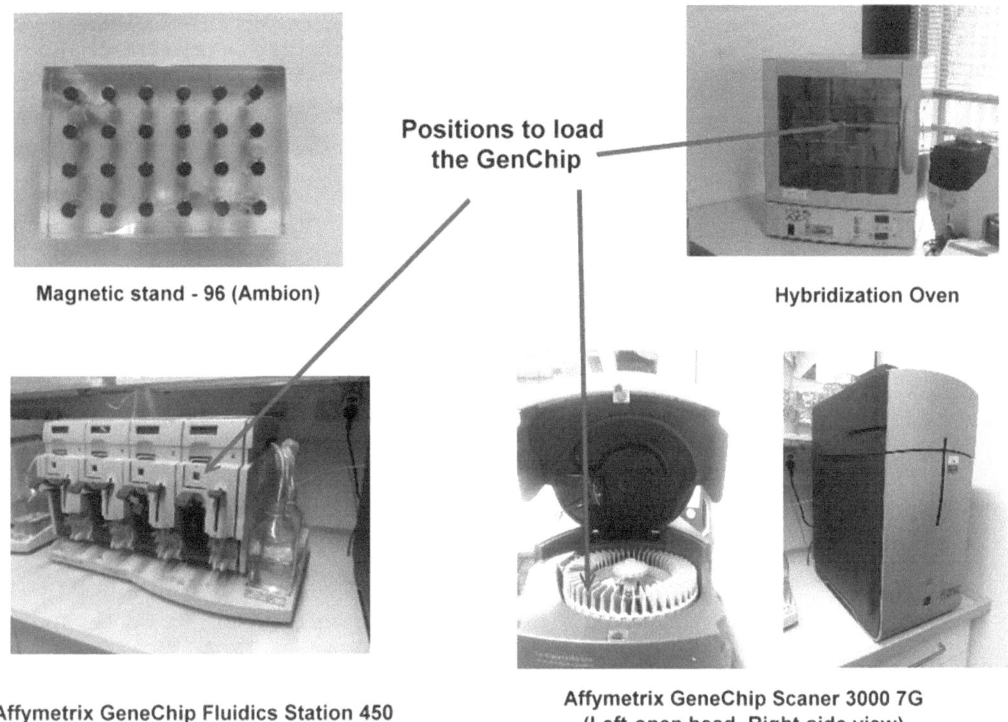

Fig. 2 Microarray components. These instruments are routinely used for microarray studies

3 Methods

3.1 Hepatocyte Cultivation

All of the steps should be performed under sterile conditions. Collagen solutions should be constantly kept on ice to prevent gelification.

1. In a laminar airflow cabinet, add 9 mL of chilled 0.2 % acetic acid to the bottle of lyophilized collagen type I, resulting in a final concentration of 1.11 mg/mL. Let dissolve for at least 3 h and preferably overnight in the fridge until no collagen clumps are visible in the solution. Designate this solution as acidic collagen solution.
2. In a 50 mL sterile tube, mix 1 volume of chilled 10× concentrated Dulbecco's Modified Eagle Medium and 9 volumes of acidic collagen solution.
3. Neutralize the collagen solution by adding dropwise chilled 1 M NaOH and constantly shaking until the yellow color turns into fuchsia. Keep the neutralized collagen solution on ice to prevent gelification.
4. Use the neutralized collagen solution to coat the cell culture plates (i.e. 700 μL for 6 cm diameter culture dishes, 350 μL for 6-well plates, 300 μL for 12-well plates, and 200 μL for

24-well plates). While coating, distribute the collagen solution evenly by quick movements of a cell scraper and gently shaking the plate up and down and side to side. Avoid the formation of air bubbles. Always prepare the neutralized collagen solution ex tempore.

5. Transfer the cell culture plates with smoothly distributed gelled collagen into an incubator at 37 °C and 5 % CO_2 for 45 min.
6. After gel polymerization, place the cell culture plates under the hood for 10 min for adjustment to room temperature (*see* **Note 1**).
7. Prepare a hepatocyte cell suspension in hepatocyte culturing medium with 10 % sera plus (i.e. 2×10^6 cells in 3 mL for 6 cm diameter culture plate, 1×10^6 cells in 2 mL for 6-well plates, 0.6×10^6 cells in 1 mL for 12-well plates and 0.3×10^6 cells in 0.5 mL for 24-well plates). Add gently the appropriate amount of cell suspension to the side of each well.
8. Gently shake the plate side to side for even distribution of cells. Transfer the culture plates in an incubator at 37 °C and 5 % CO_2 for 3 h to allow cell attachment.
9. For coating with the second layer of collagen, gently aspirate the medium from each well. Wash gently three times with pre-warmed William's E medium and add the appropriate amount of freshly prepared neutralized collagen solution onto the plated cells. Gently shake the plate for even distribution of the neutralized collagen solution over the surface. Designate these plates as collagen gel sandwich plates.
10. Transfer the completed collagen gel sandwich plates for 45 min in an incubator at 37 °C and 5 % CO_2 for collagen gelification.
11. After gel polymerization, add gently an appropriate amount of pre-warmed hepatocyte culturing medium.
12. Incubate the cell culture plates in an incubator for 20 h at 37 °C and 5 % CO_2. Refresh the medium every day.

3.2 Hepatocyte Treatment

1. Prepare the hepatotoxic compound solution by dissolving the compounds of interest in suitable solvents. Make further dilutions in hepatocyte culturing medium in such a way that the final vehicle concentration is maximum 0.01 %. Select the final concentration of hepatotoxicant based on relevant in vivo doses (e.g. C_{max}) or in vitro concentrations (e.g. IC_{20}).
2. After 20 h of incubation at 37 °C and 5 % CO_2, aspirate the medium from the cell culture plates and add hepatocyte culturing medium containing compounds or vehicle. Transfer the cell culture plates to an incubator at 37 °C and 5 % CO_2.
3. Collect the samples after 24 h of exposure.

3.3 Sample Collection

1. Transfer the cell culture plates on ice and gently aspirate the medium.
2. Add 1 mL TRIzol reagent and thoroughly disrupt the cells in the collagen matrix with a cell scraper (*see* **Note 2**).
3. Transfer TRIzol containing cells into sterile 2 mL nuclease-free tubes and sonicate the sample for 30 s on ice with pulse 5 s and break 2 s.

3.4 Sample Storage and Transportation

1. Store the TRIzol containing cells at −80 °C until RNA isolation.
2. Ship the TRIzol containing cells or RNA in a big styrofoam box filled with dry ice.

3.5 RNA Isolation (Fig. 3)

Use the RNeasy mini kit for RNA isolation. Carry out all procedures at room temperature unless otherwise specified.

1. Thaw the samples on ice. Triturate the samples on the wall of the tubes using a 24 G needle and a 1 mL syringe (*see* **Note 2**).
2. Add 200 μL chloroform, mix and centrifuge the tubes at 12,000 × *g* and 4 °C for 15 min.
3. Collect the uppermost layer without disturbing the middle or lower layer in 1.5 mL nuclease-free tubes and add equal volume of chilled 70 % ethanol. Mix the content by gentle shaking.
4. Apply 700 μL to the minispin columns and centrifuge at 12,000 × *g* and room temperature for 20 s.
5. Discard the filtrate, apply the remaining solution to the minispin column and centrifuge at 12,000 × *g* and room temperature for 20 s.
6. Discard the filtrate, apply 350 μL of RW1 buffer to the minispin column and centrifuge at 12,000 × *g* and room temperature for 20 s.
7. Discard the filtrate, apply 10 μL of DNase and 70 μL RDD buffer to the column and incubate for 15 min at room temperature (*see* **Note 3**).
8. Apply 500 μL of RPE wash buffer and centrifuge at 12,000 × *g* and room temperature for 20 s.
9. Discard the filtrate, apply 500 μL of RPE wash buffer and centrifuge at 12,000 × *g* and room temperature for 20 s.
10. Shift the minispin columns to new 2 mL collection tubes and centrifuge at 12,000 × *g* and room temperature for 1 min.
11. Transfer the minispin columns to nuclease-free 1.5 mL tubes, apply 22 μL of nuclease-free water and centrifuge at 12,000 × *g* and room temperature for 1 min.
12. Remove the collection tubes and place them on ice.

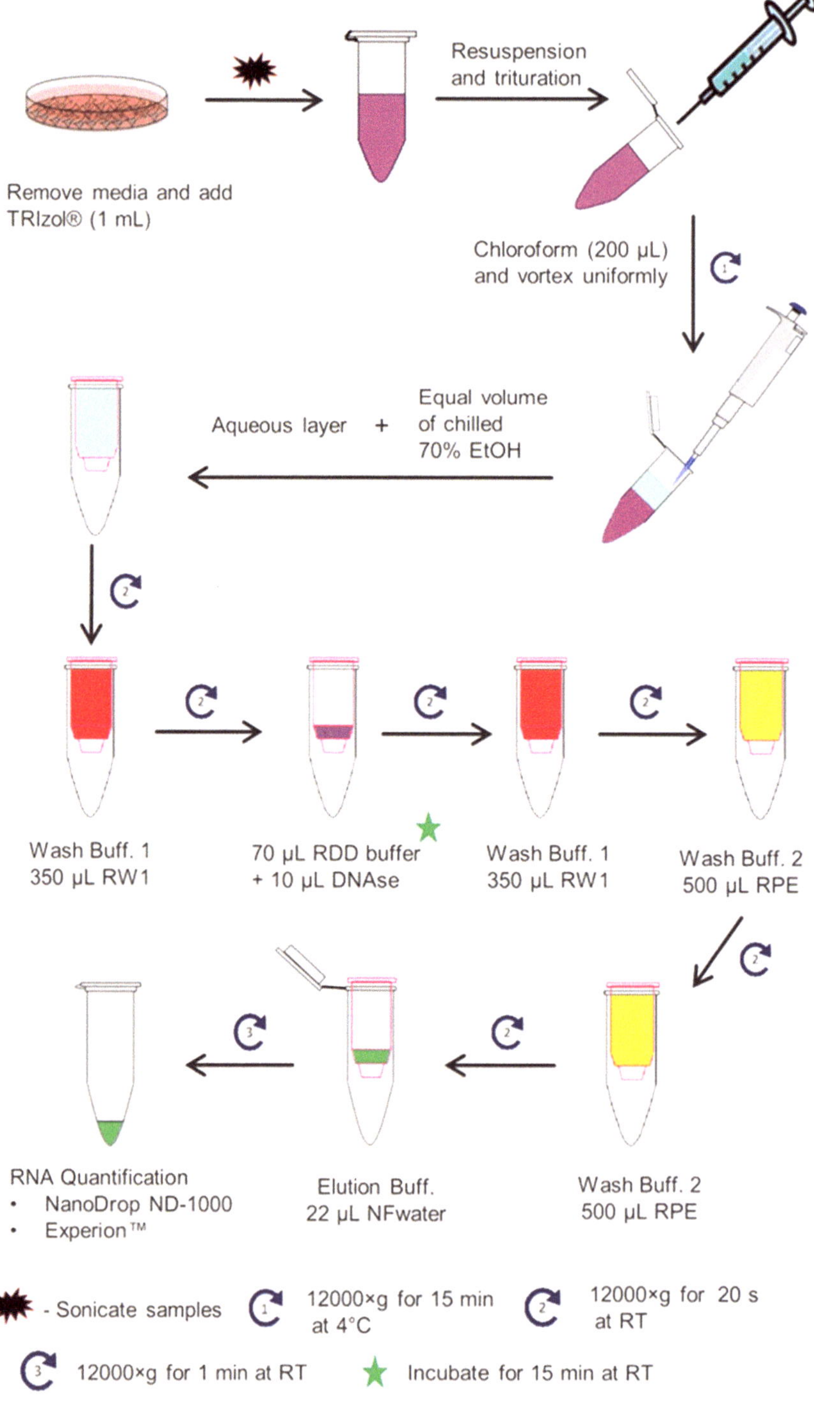

Fig. 3 Schematic representation of the RNA isolation procedure using the RNeasy mini kit

13. Quantify RNA using the Nanodrop ND-1000 device.
14. Dilute the RNA to get a final concentration 33.33 ng/μL (*see* **Note 4**).

3.6 RNA Target Preparation (Fig. 4)

Use the GeneChip 3′ IVT express kit for RNA target preparation (*see* **Note 5**). Carry out all procedures at room temperature unless otherwise specified. Use 100 ng RNA as starting material. The total volume mentioned in the procedure holds for ten samples, including 5 % overage to adjust pipetting errors.

3.6.1 Poly-A RNA Control Preparation

1. Add poly-A control dilution buffer in four nuclease-free tubes (i.e. tube 1: 38 μL, tubes 2 and 3: 98 μL, and tube 4: 36 μL).
2. Add 2 μL of the poly-A control stock to the first tube, mix, spin down and serially transfer 2 μL until tube number 3.
3. Add 4 μL from tube number 3 to tube number 4, mix and spin down (*see* **Note 6**).
4. Take out the ten PCR tubes and transfer 2 μL from tube number 4 to each tube.
5. Add 3 μL of total RNA (i.e. 100 ng) from each sample to the PCR tubes and keep them on ice. These tubes are designated as RNA/poly-A RNA control mixture tubes (*see* **Note** 7).

3.6.2 First-Strand cDNA Synthesis

1. Thaw the first-strand synthesis reagents on ice.
2. Prepare the first-strand master mix by adding 10.5 μL first-strand enzyme mix to 42 μL of first-strand buffer mix in nuclease-free tubes on ice.
3. Add 5 μL of first-strand master mix to the RNA/poly-A RNA control mixture tubes.
4. Gently mix, centrifuge for 5 s and keep on ice.
5. Program the thermal cycler for first-strand cDNA synthesis (i.e. 2 h at 42 °C and indefinite hold at 4 °C).
6. Incubate the PCR tubes on thermal cycler (i.e. 2 h at 42 °C), centrifuge the tubes for 5 s and transfer them on ice. These tubes are designated as the first-strand cDNA samples (*see* **Note 8**).

3.6.3 Second-Strand cDNA Synthesis

1. Thaw the second-strand synthesis reagents on ice.
2. Prepare the second-strand master mix by adding 52.5 μL second-strand buffer mix and 21 μL second-strand enzyme mix to 136.5 μL of nuclease-free water in nuclease-free tubes on ice.
3. Mix gently, centrifuge for 5 s and keep on ice.
4. Add 20 μL of second-strand master mix to the tubes containing first-strand cDNA samples.
5. Mix gently, centrifuge for 5 s and keep on ice.

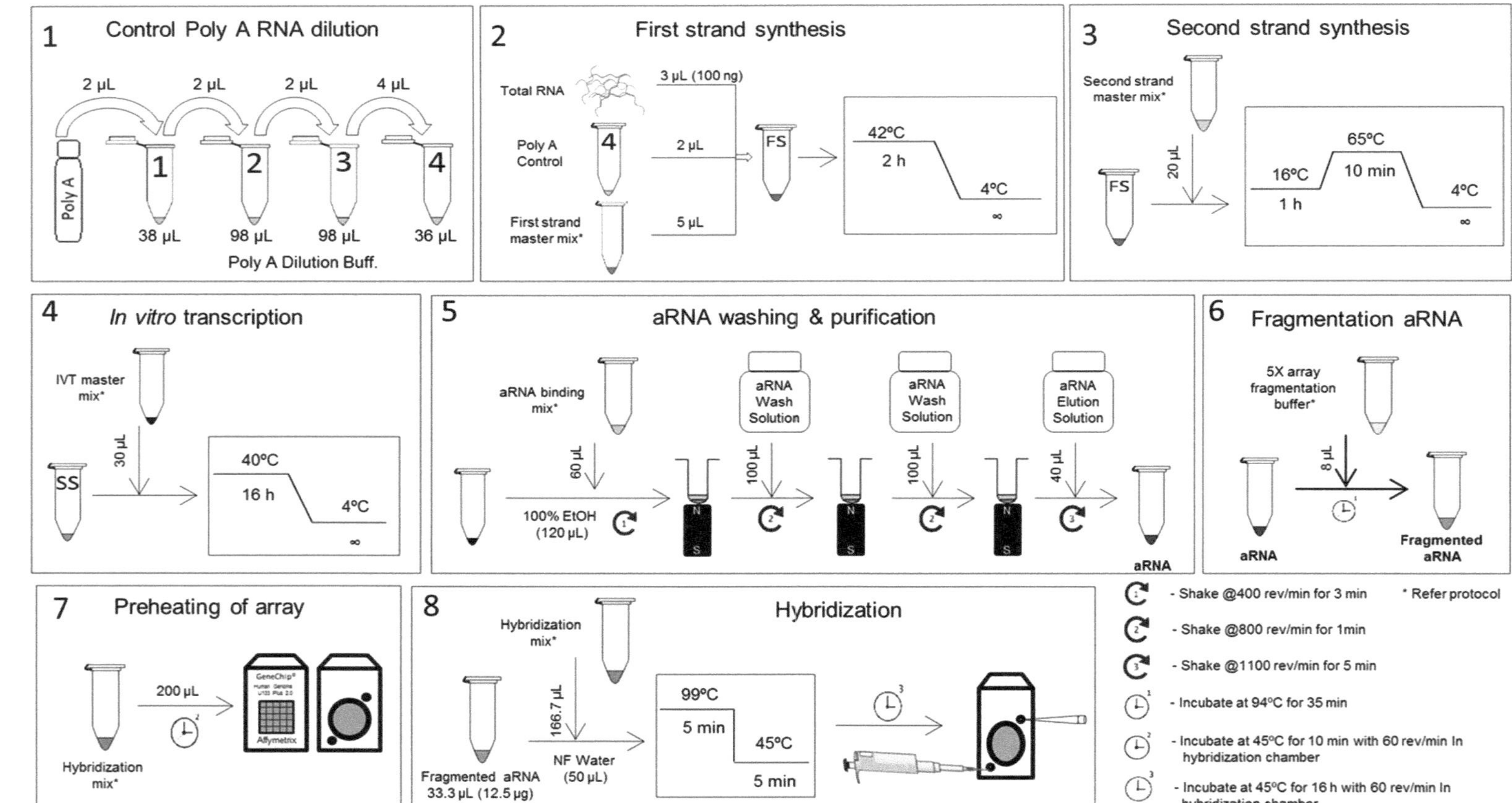

Fig. 4 Schematic representation of target RNA preparation, gene chip hybridization, and scanning for microarray studies using the GeneChip3′ IVT express kit. The *number* indicates the steps to be followed sequentially

6. Program the thermal cycler for second-strand cDNA synthesis (i.e. 1 h at 16 °C, 10 min at 65 °C and indefinite hold at 4 °C).
7. Incubate the PCR tubes in the thermal cycler for second-strand cDNA synthesis, centrifuge the tubes for 5 s and transfer them on ice. These tubes are designated as second-strand cDNA samples (*see* **Note 9**).

3.6.4 Labeled aRNA Synthesis

1. Thaw the IVT enzyme mix on ice, and the IVT biotin label and labeling buffer at room temperature.
2. Prepare the IVT master mix by adding 63 μL IVT enzyme mix and 42 μL IVT biotin label to 210 μL of IVT labeling buffer in nuclease-free tubes.
3. Mix gently, centrifuge for 5 s and keep on ice.
4. Add 30 μL of the IVT master mix to the tubes containing the second-strand cDNA samples.
5. Mix gently, centrifuge for 5 s and keep on ice.
6. Program the thermal cycler for IVT (i.e. 16 h at 40 °C and indefinite hold at 4 °C).
7. Incubate the PCR tubes in the thermal cycler for labeled aRNA synthesis and transfer the tubes on ice. These tubes are designated as labeled aRNA samples (*see* **Note 9**).

3.6.5 aRNA Purification

1. Preheat 550 μL of aRNA elution solution at 60 °C for 10 min.
2. Mix and add 110 μL of aRNA-binding beads to 550 μL aRNA-binding buffer concentrate in a nuclease-free tube at room temperature.
3. Add 60 μL of aRNA-binding mix to labeled aRNA samples.
4. Transfer the samples to the wells of a U-bottom plate (*see* **Note 10**).
5. Mix gently by pipetting up and down.
6. Add 120 μl of absolute ethanol to each sample and mix gently by pipetting up and down.
7. Shake the plate at low velocity on a plate shaker for 3 min.
8. Remove the plate and keep on magnetic stand for 5 min.
9. The aRNA-binding beads are captured in the center as a pellet. Gently aspirate and discard the surrounding supernatant (*see* **Note 11**).
10. Add 100 μL of aRNA wash solution to the samples, gently pipette to mix and shake the plate at medium velocity on a plate shaker for 1 min.
11. Remove the plate and keep on magnetic stand for 5 min.
12. The aRNA-binding beads are captured in the center as a pellet. Gently aspirate and discard the surrounding supernatant.

13. Add 100 μL of aRNA wash solution to the samples, gently pipette to mix and shake the plate at medium velocity on a plate shaker for 1 min.
14. Remove the plate and keep on magnetic stand for 5 min.
15. The aRNA-binding beads are captured in the center as a pellet. Gently aspirate and discard the surrounding supernatant.
16. Shake the plate at high velocity on a plate shaker for 1 min (*see* **Note 12**).
17. Add 50 μL of preheated aRNA elution solution to the samples.
18. Shake the plate at high velocity on a plate shaker for 5 min (*see* **Note 13**).
19. Remove the plate and keep on magnetic stand for 5 min.
20. The aRNA-binding beads are captured in the center as a pellet and purified eluted aRNA in the supernatant. Carefully aspirate and transfer the supernatant to nuclease-free tubes and keep on ice (*see* **Note 14**).
21. Quantify the aRNA using a spectrophotometer.
22. Dilute the aRNA to get a final concentration 15 μg/32 μL in nuclease-free water (*see* **Note 15**).

3.6.6 Fragmentation of Labeled aRNA

1. Add 8 μL of 5× concentrated array fragmentation buffer to 32 μL of aRNA (i.e. 15 μg) sample in a nuclease-free tube.
2. Incubate the samples at 94 °C for 35 min.
3. Transfer the samples on ice (*see* **Note 16**).

3.6.7 Hybridization

1. Incubate the 20× concentrated GeneChip eukaryotic hybridization controls at 65 °C for 5 min.
2. Prepare the hybridization master mix by mixing 42 μL of 3 nM control oligonucleotide, 125 μL of 20× eukaryotic hybridization control, 1.25 mL of 2× concentrated hybridization mix and 250 μL of dimethylsulfoxide in a nuclease-free tube.
3. Add 50 μL of nuclease-free water to ten nuclease-free tubes.
4. Add 166.7 μL of the hybridization master mix to the ten tubes.
5. Add 33.3 μL of fragmented labeled RNA (i.e. 12.5 μg) to the tubes. These samples are designated as hybridization cocktail.
6. Take out the probe array from the refrigerator and equilibrate at room temperature for 30 min.
7. Add 200 μL of the prehybridization mix through one septum of the probe array.
8. Incubate the probe arrays at 45 °C for 10 min with 60 RPM in a hybridization oven.

9. Incubate the hybridization cocktail at 99 °C for 5 min in a heat block.
10. Incubate the hybridization cocktail tubes to 45 °C in a heat block for 5 min.
11. Spin the tubes with hybridization cocktail for 5 min at low velocity.
12. Take out the probe array from the hybridization oven, vent the array with a nuclease-free 200 μL pipette tip from the septum and extract the prehybridization mix from the other septum with a 200 μL pipette tip. Inject 200 μL of hybridization cocktail from the latter septum. Take out the pipette tips from both septa and seal them with tags (*see* **Note 17**).
13. Incubate the probe array into the hybridization oven at 45 °C for 16 h with 60 RPM.

3.7 GeneChip Washing, Staining and Scanning

1. Take out the probe array and remove the tags. Extract the hybridization cocktail from the arrays into the labeled nuclease-free tubes.
2. Immediately proceed to washing and staining using commercially available wash and stain kits with the GeneChip fluidics station 450.
3. Scan the stained array using GeneChip scanner 3000-7G and GeneChip operating software [22].

3.8 Data Analysis

1. Import the .cel files generated from GeneChip operating software into commercially available gene expression analysis software, such as Partek Genomic Suite [23]. Perform a quality control check, summarization and normalization with robust multi array analysis.
2. Obtain differentially expressed gene lists by performing appropriate analysis of variance. Filter the gene lists based on fold change and false discovery rate controlled *p*-value.
3. Perform further analyses, like principal component analysis, gene ontology enrichment or pathway analysis, using the same software or online tools, such as database for annotation, visualization and integrated discovery [24], or onto-tools [25].

4 Notes

1. Proper collagen gelation can be assessed by tilting the plates. The gel should be firm and the surface even and smooth.
2. Handle TRIzol carefully, wear chemical-resistant gloves, safety goggles and perform all steps in a safety cabinet. In general, 5–6 times trituration is sufficient for the cells.

3. Aliquot 10 μL DNase in nuclease-free tubes and store at −20 °C. At the time of addition on the column, thaw DNase on ice and add 70 μL of RDD buffer to each aliquot, mix by pipetting and transfer on the column.
4. Use 5 μL of total RNA for dilution and store the remainder at −80 °C. Confirm the RNA purity and integrity using an Experion™ automated electrophoresis system and the RNA StdSens kit (Bio-Rad, USA). Perform all the procedures according to the manufacturer's instructions. Use Experion™ software to analyze the data. The RNA quality indicator indicates degradation of RNA. A value of 10 indicates intact RNA, while a value of 1 indicates degraded RNA. It also provides RNA concentration expressed in pg/μL, ratio of 28S/18S RNA, RNA quality indicator classification and alert. All details are provided in the instruction manual.
5. Prepare the aRNA wash solution by adding 8 mL absolute ethanol in aRNA wash solution concentrate and indicate the date.
6. Aspirate the supernatant for transfer without disturbing the sediment if any.
7. Poly-A RNA dilution in tube 1 can be used up to 4 weeks at −20 °C.
8. After first-strand cDNA synthesis, immediately proceed to second-strand synthesis and set the thermal cycler for second-strand cDNA synthesis.
9. After second-strand cDNA synthesis or labeled aRNA synthesis, immediately proceed to the next step or store the tubes at −20 °C overnight.
10. Transfer the samples in every alternate well so that they are in direct contact with the magnetic stands probe.
11. The surrounding supernatant contains enzymes, salts, and unincorporated nucleotides. Do not disturb the beads while taking out the surrounding supernatant.
12. This step is performed to remove residual ethanol from the samples.
13. Perform gentle pipetting if the aRNA-binding beads are not dispersed.
14. Purified aRNA can be stored at maximum −20 °C for 1 year.
15. Concentrate by vacuum centrifugation if the aRNA concentration is too low.
16. Store undiluted fragmented aRNA at −20 or −70 °C.
17. Seal the septa in such a way that seals can be easily removed for next steps.

Acknowledgements

We thank Margit Henry, Tamara Rotshteyn, and Susan Rohani for the excellent technical support. This work was supported by grants from the European Union (Project ESNATS) and the German Ministry for Research (BMBF, Project SysDT).

References

1. Dumortier G, Cabaret W, Stamatiadis L et al (2002) Hepatic tolerance of atypical antipsychotic drugs. Encéphale 28:542–551
2. Kaplowitz N (2005) Idiosyncratic drug hepatotoxicity. Nat Rev Drug Discov 4:489–499
3. Need AC, Motulsky AG, Goldstein DB (2005) Priorities and standards in pharmacogenetic research. Nat Genet 37:671–681
4. Hengstler J, Hammad S, Ghallab A et al (2014) *In vitro* systems for hepatotoxicity testing. In: Bal-Price A, Jennings P (eds) *In vitro* toxicology system. Springer, New York, NY, pp 27–44
5. Atienzar FA, Novik EI, Gerets HH et al (2014) Predictivity of dog co-culture model, primary human hepatocytes and HepG2 cells for the detection of hepatotoxic drugs in humans. Toxicol Appl Pharmacol 275:44–61
6. Tolosa L, Pinto S, Donato MT et al (2012) Development of a multiparametric cell-based protocol to screen and classify the hepatotoxicity potential of drugs. Toxicol Sci 127:187–198
7. Soldatow VY, LeCluyse EL, Griffith LG et al (2013) *In vitro* models for liver toxicity testing. Toxicol Res (Camb) 2:23–39
8. Meganathan K, Jagtap S, Wagh V et al (2012) Identification of thalidomide-specific transcriptomics and proteomics signatures during differentiation of human embryonic stem cells. Plos ONE 7:e44228
9. Krug AK, Kolde R, Gaspar JA et al (2013) Human embryonic stem cell-derived test systems for developmental neurotoxicity: a transcriptomics approach. Arch Toxicol 87:123–143
10. Leist M, Ringwald A, Kolde R et al (2013) Test systems of developmental toxicity: state-of-the art and future perspectives. Arch Toxicol 87: 2037–2042
11. Kienhuis AS, van de Poll MCG, Wortelboer H et al (2009) Parallelogram approach using rat-human *in vitro* and rat *in vivo* toxicogenomics predicts acetaminophen-induced hepatotoxicity in humans. Toxicol Sci 107:544–552
12. Liguori MJ, Anderson MG, Bukofzer S et al (2005) Microarray analysis in human hepatocytes suggests a mechanism for hepatotoxicity induced by trovafloxacin. Hepatology 41: 177–186
13. Parmentier C, Truisi GL, Moenks K et al (2013) Transcriptomic hepatotoxicity signature of chlorpromazine after short- and long-term exposure in primary human sandwich cultures. Drug Metab Dispos 41: 1835–1842
14. Gerets HHJ, Tilmant K, Gerin B et al (2012) Characterization of primary human hepatocytes, HepG2 cells, and HepaRG cells at the mRNA level and CYP activity in response to inducers and their predictivity for the detection of human hepatotoxins. Cell Biol Toxicol 28:69–87
15. Heijne WHM, Slitt AL, van Bladeren PJ et al (2004) Bromobenzene-induced hepatotoxicity at the transcriptome level. Toxicol Sci 79: 411–422
16. Hirode M, Omura K, Kiyosawa N et al (2009) Gene expression profiling in rat liver treated with various hepatotoxic-compounds inducing coagulopathy. J Toxicol Sci 34:281–293
17. Kienhuis AS, Bessems JGM, Pennings JLA et al (2011) Application of toxicogenomics in hepatic systems toxicology for risk assessment: acetaminophen as a case study. Toxicol Appl Pharmacol 250:96–107
18. Reilly TP, Bourdi M, Brady JN et al (2001) Expression profiling of acetaminophen liver toxicity in mice using microarray technology. Biochem Biophys Res Commun 282:321–328
19. Heinloth AN, Irwin RD, Boorman GA et al (2004) Gene expression profiling of rat livers reveals indicators of potential adverse effects. Toxicol Sci 80:193–202
20. Suh SK, Jung KK, Jeong YK et al (2006) Gene expression profiling of acetaminophen induced hepatotoxicity in mice. Mol Cell Toxicol 2:236–243

21. Dalma-Weiszhausz DD, Warrington J, Tanimoto EY et al. (2006) The affymetrix GeneChip (R) platform: an overview. DNA microarrays part A: array platforms and wet-bench protocols methods in Enzymology 410, 3–28
22. http://www.affymetrix.com/support/mas/index.affx (consulted July 2014)
23. http://www.partek.com/pgs (consulted July 2014)
24. http://david.abcc.ncifcrf.gov/ (consulted July 2014)
25. http://vortex.cs.wayne.edu/projects.htm (consulted July 2014)

Chapter 17

Global MicroRNA Analysis in Primary Hepatocyte Cultures

Julian Krauskopf, Almudena Espín-Pérez, Jos C. Kleinjans, and Theo M. de Kok

Abstract

MicroRNAs are small non-coding molecules that regulate gene expression and in return affect diverse biological functions, including those involved in toxicity and development of disease. Recent evidence suggests that microRNAs play an important role in liver pathologies, like viral hepatitis, alcoholic liver, hepatocellular carcinoma, or drug-induced liver injury. Furthermore, numerous studies demonstrated the high potential of microRNAs as promising non-invasive biomarkers of liver disease or as relevant targets for therapeutic treatment. This chapter describes a method for global microRNA analysis of primary hepatocytes by high-throughput sequencing. The method comprises the isolation of high-quality total RNA, analysis of microRNA sequencing data, and the validation of the findings by reverse transcriptase quantitative polymerase chain reaction analysis.

Key words microRNAs, Primary hepatocyte, Small RNA sequencing, Total RNA isolation

1 Introduction

MicroRNAs (miRNAs) are small non-coding RNA molecules that regulate gene expression and in return affect diverse biological functions, such as cell differentiation and proliferation, including those in the liver. In most cases, miRNAs bind to the 3′-untranslated region of messenger RNAs (mRNAs), building the RNA-induced silencing complex and resulting in repressed translation of mRNAs to proteins [1]. Each miRNA can target up to hundreds of genes and each gene can be targeted by multiple miRNAs [2]. Currently, more than 2,000 distinct miRNAs have been described and were found to affect all fundamental physiological processes [3]. Several studies indicated the importance of miRNAs in various liver diseases, such as viral hepatitis, alcoholic liver, hepatocellular carcinoma, or drug-induced liver injury [4–6]. The fact that miRNAs are present in all eukaryotic cells and that most of them are conserved between vertebrate species is pointing towards their biological significance [7]. Tissue-specific expression patterns [8]

Mathieu Vinken and Vera Rogiers (eds.), *Protocols in In Vitro Hepatocyte Research*, Methods in Molecular Biology, vol. 1250, DOI 10.1007/978-1-4939-2074-7_17, © Springer Science+Business Media New York 2015

as well as their high stability in different biological fluids, such as urine or blood [9], makes them very promising non-invasive biomarkers for hepatotoxicity or diagnostic disease markers for prediction of liver disease [10, 11]. Currently, miR-122 is emerging as a reliable circulating biomarker for several liver diseases, including hepatitis B [12], hepatitis C [13], hepatocellular carcinoma [14] and non-alcoholic fatty liver disease [15]. Furthermore, miRNAs show great potential as therapeutic targets for the treatment of liver disease. Considering the important role of miRNAs in regulating metabolic homeostasis, miRNAs provide promising therapeutic targets in metabolic pathologies. Recently, a major increase in the incidence of diabetes type 2 and fatty liver disease has raised the interest in deciphering the underlying mechanisms of these conditions. It has been shown that glucose and lipid metabolism are mainly regulated in liver. Xiao and co-workers demonstrated that miR-130a is involved in the regulation of insulin sensitivity and steatosis in liver [16]. Great potential is also seen in the value of miRNAs for the prediction of compound toxicity. Yokoi and group presented an overview of the role of miRNAs in toxicological outcomes of which some derived from in vitro studies [17]. Human hepatocyte cultures are optimal for in vitro screening assays [18]. For instance, miRNAs from the family miR-181, miR-146a, miR-365, miR199b-5p and miR30c-1* are overexpressed in HepG2 cells treated with the carcinogen benzo[*k*]fluoranthene [19]. Furthermore, production of miR-181a, miR-181b, and miR-181d becomes increased in response to exposure to benzo[*a*] anthracene and benzo[*k*]fluoranthene [20].

Numerous studies on miRNAs have been conducted using microarray technology. This technology is relatively cheap and enables high-throughput processing of samples. However, microarray technology holds several drawbacks, like signal saturation, different hybridization properties of probes, cross-hybridization and background signals, all which introduce biases. Since high-throughput sequencing has emerged to present a more comprehensive, sensitive and cost-effective approach, it overcomes these limitations and brings as such the advantages of detecting very low abundant miRNAs, sequence variation and modifications, while also enabling the discovery of novel microRNAs [21]. Due to the limitations of quantitative polymerase chain reaction analysis (qPCR) in terms of capacity, this technique is only of interest for further validation of deregulated microRNAs. Since small RNA sequencing is more and more replacing microarray technology, the protocol described in this chapter will focus on the application of high-throughput sequencing to global miRNA analysis in cultures of primary hepatocytes. However, as the actual sequencing of miRNAs is usually serviced by a sequencing facility, this protocol will focus on the isolation of high-quality total RNA from primary hepatocyte cultures, the analysis of small RNA sequencing data, and the validation of the findings by reverse transcriptase qPCR (RT qPCR).

2 Materials

2.1 Isolation of Total RNA

1. Phosphate-buffered saline (PBS). 137 mM NaCl, 2.7 mM KCl, 10 mM Na_2HPO_4, 1.8 mM KH_2PO_4 in distilled water. Adjust pH to 7.4.
2. 100, 75 and 70 % ethanol.
3. RNeasy Mini Kit (Qiagen, The Netherlands). RNeasy mini spin columns, collection tubes, Buffer RLT, Buffer RW1, Buffer RPE.
4. RNase-free water.
5. 5 M NaCl.
6. Vacuum concentrator.

2.2 RT qPCR Validation

1. 5 ng/μL total RNA.
2. Universal cDNA Synthesis Kit II. 5× concentrated reaction buffer, 10× concentrated enzyme mix, RNA spike-in template (UniSp6).
3. miRCURY LNA microRNA PCR, ExiLENT SYBRGreen master mix (Exiqon, The Netherlands). 2× concentrated ExiLENT SYBRGreen master Mix, control primer set (UniSp6).
4. miRCURY LNA Universal RT miRNA PCR, miRNA primer set (Exiqon, The Netherlands).
5. Nuclease-free water.
6. Thermocycler.
7. 96-well PCR cycler.
8. 96-well plates and optical sealing.
9. 96-well microcentrifuge.

3 Methods

3.1 Isolation of Total RNA

This part describes total RNA isolation from three-dimensional collagen sandwich cultures of primary hepatocytes [22].

1. Add 4 volumes of 100 % ethanol to buffer RPE for a working solution.
2. Remove the cell culture medium for the primary hepatocyte cultures and wash twice with PBS at 4 °C (*see* **Note 1**).
3. Transfer the hepatocytes and the collagen construct into a cooled tube (*see* **Notes 2** and **3**).
4. Spin the tube for 1 min at 6,500 ×*g* and 4 °C.
5. Carefully remove the residual medium and do not disturb the pellet (*see* **Notes 3–5**).

6. Add 600 μL buffer RLT and homogenize by pipetting.
7. Add 1 volume of cold 70 % ethanol and mix well by pipetting.
8. Transfer up to 700 μL of the sample to an RNeasy mini spin column and centrifuge for 15 s at 8,000 × *g* and 4 °C. Discard the eluate.
9. Add 700 μL buffer RW1 to the columns and centrifuge for 15 s at 8,000 × *g* and 4 °C. Discard the eluate.
10. Add 500 μL buffer RPE to the columns and centrifuge for 15 s at 8,000 × *g* and 4 °C. Discard the eluate.
11. Add 500 μL buffer RPE to the columns and centrifuge for 2 min at 8,000 × *g* and 4 °C. Discard the eluate.
12. Place the RNeasy spin columns in new 1.5 mL collection tubes, add 100 mL RNase-free water to the column membrane and elute the RNA for 1 min at 8,000 × *g* and 4 °C.
13. Add 1/10 volume of 5 M NaCl and 3 volumes of 100 % ethanol.
14. Store for 1 h at −80 °C.
15. Thaw at room temperature.
16. Centrifuge the sample for 20 min at 8,000 × *g* and 4 °C and remove the supernatant.
17. Add 1 mL of 75 % ethanol.
18. Centrifuge the sample for 10 min at 8,000 × *g* and 4 °C and remove the supernatant.
19. Dry the pellet in a vacuum concentrator.
20. Resuspend the pellet in 40 μL RNase-free water (*see* **Note 6**).

3.2 Small RNA Sequencing Analysis

This section will focus on data analysis of differentially expressed miRNAs and the discovery of novel miRNAs. However, in the actual sequencing, a library is generated by the TruSeq Small RNA sample preparation kit (Illumina, The Netherlands) based on the isolated total RNA. This step includes the ligation of the sequencing adapter, PCR amplification, and a size selection. In this method, the ligated adapters are hybridized to a solid phase and subsequently the cDNA is amplified and sequenced by incorporating and detecting fluorescently labeled reversible terminator nucleotides [23]. Adding unique index tags to the samples before the sequencing enables several samples to be sequenced in a single multiplexed run. Information about the design of RNA sequencing experiments can be found in the paper of Fang and Cui [24].

3.2.1 Retrieving Software and Sequence Databases

The software miRDeep2 (v2.0.0.5) [25] and the bowtie short read aligner (v1.0.1) [26] can be downloaded and installed on a Linux system, including all prerequisites indicated in the README files. In order to quantify miRNA expression and predict new miRNAs from the obtained sequencing reads, it is essential to download publicly available sequences. The human miRNA sequences can be downloaded in fasta format from the miRNA database miRBase (Release 20; 1,872 precursors and 2,578 mature sequences) [27]. For prediction of novel miRNAs, the human genome assembly GRCh37 can be obtained from the 1000 Genomes Project [28] (*see* **Notes** 7 and **8**).

3.2.2 Sequence Quality Control

The sequencing reads of 100 nucleotides (nt) obtained from the Illumina HiSeq 2000 are automatically stored in a fastq file. This type of file contains the sequence information and the quality score for each base pair. At first, the quality for each file should be assessed using FastQC (v0.10.1) [29] (*see* **Note 9**).

3.2.3 Data Pre-processing and Mapping

The miRDeep2 software consists of several command line scripts. The perl script mapper.pl can be used for pre-processing of the reads (i.e. 3′ adapter trimming, size selection and collapsing of the reads to distinct sequences) followed by mapping the reads to the human genome with no mismatches allowed. The script has to be run with the sequence of the used adapter (*see* **Note 10**), the settings to discard reads shorter than 18 nt and the path to the indexed genome. Subsequently, the script miRdeep2.pl is used to predict novel miRNAs based on default settings. The novel predicted miRNAs should be added to the list of mature and precursor sequences from miRBase to quantify the expression, including the known miRNAs using the script quantifier.pl (*see* **Note 11**).

3.2.4 Detect Differentially Expressed miRNAs

The quantitative expression output can be analyzed using the open source software R (v.2.15.2) [30] and bioconductor [31] (*see* **Note 12**). For screening of differentially expressed miRNAs, the DESeq package (v.1.10.1) [32] should be used. After normalization of the quantitative miRNA counts, the negative binominal test is applied to produce the list of differentially expressed miRNAs (i.e. control versus treated). MiRNAs are considered significantly differentially expressed when showing a *p*-value of less than 0.05, adjusted for multiple testing using Benjamini-Hochberg method as implemented in DESeq (*see* **Note 13**).

3.2.5 Target Gene Prediction and Pathway Analysis

The differentially expressed miRNAs can be analyzed for possible gene targets using the experimentally validated miRNA-target interactions database miRTarBase [33]. These gene targets then can be analyzed for enriched pathways and gene ontology (GO) terms using the gene interaction database ConsensusPathDB [34] for biological interpretation of results.

3.3 RT qPCR Validation

This protocol is according to miRCURY LNA Universal RT miRNA PCR instruction manual v.5.3. To validate novel predicted miRNAs, custom primers can be designed with Exiqon (The Netherlands). While designing the PCR experiment, each reaction should be done at least in duplicate and for each primer, two blank

Table 1
Reverse transcription reaction

Reagent	Volume per reaction (μL)
5× concentrated reaction buffer	2
Nuclease-free water	4.5
Enzyme mix	1
Synthetic RNA spike ins (optional, otherwise nuclease-free water)	0.5
Template total RNA (5 ng/μL)	2
Total volume	10

reactions should be included without any cDNA. Also, a reference miRNA (*see* **Note 14**) should be included for normalization during data analysis.

3.3.1 First-Strand Synthesis

1. Thaw the 5× concentrated reaction buffer and nuclease-free water and place on ice.
2. Resuspend the RNA spike-in template (UniSp6) and leave on ice for 15–20 min (*see* **Note 2**).
3. Dilute each total RNA sample to a concentration of 5 ng/μL using nuclease-free water.
4. Immediately before use, remove the enzyme mix from the freezer, mix and place on ice.
5. Prepare the revere transcription reaction mix as depicted in Table 1.
6. Gently mix the reaction and briefly spin down.
7. Incubate in a thermocycler for 60 min at 42 °C, 5 min at 95 °C and immediately cool down to 4 °C.

3.3.2 qPCR Reaction

1. Thaw cDNA, nuclease-free water and 2× concentrated ExiLENT SYBRGreen master mix on ice for 15–20 min. Minimize exposure to light.
2. Immediately before use, mix the 2× concentrated ExiLENT SYBRGreen master mix by pipetting.
3. Dilute the required amount of 80× concentrated cDNA template in nuclease-free water.
4. Prepare the required amount of primer master mix as shown in Table 2.
5. Add 6 μL to each well and spin the 96-well plate to remove bubbles.

Table 2
qPCR reaction

Reagent	Volume per reaction (μL)
2× concentrated ExiLENT SYBRGreen master mix	5
PCR primer mix	1
Diluted cDNA template	4
Total reaction	10

Table 3
qPCR cycle

Process step	Settings
Polymerase activation/denaturation	95 °C, 10 min
40 amplification cycles	95 °C, 10 s 60 °C, 1 min Ramp rate 1.6 °C/s Optical read
Melting curve analysis	Yes

6. Add cDNA template to each well, mix, spin down and seal the plate with an optical sealing.
7. Perform qPCR amplification according to the program outlined in Table 3.

3.3.3 Data Analysis

Retrieve the Ct values from the PCR instrument and analyze these values by the Livak method [35].

1. The Ct values (*see* **Note 15**) are normalized by subtracting the Ct values of the reference miRNA from the Ct values of the miRNAs of interest:

$$\Delta Ct = Ct(\text{miRNA}) - Ct(\text{reference miRNA})$$

2. The relative expression is calculated as follows:

$$\text{expression}(\text{miRNA}) = 2^{-\Delta Ct}$$

3. The ratio between the two conditions can be calculated according to the following formula:

$$\Delta\Delta Ct = \Delta Ct(\text{miRNA treated}) - \Delta Ct(\text{miRNA control})$$

$$\text{ratio} = 2^{-\Delta\Delta Ct}$$

4 Notes

1. RNA extraction should be performed as quickly as possible and on ice to maintain high-quality total RNA.
2. Use RNase-free tips.
3. Use a slightly truncated pipette tip to avoid the tip to be clogged by floating collagen fibers.
4. Take care to not disturb the pellet, as it is not very well attached.
5. To improve the result, repeat the washing with 1 mL PBS followed by the centrifugation in **step 3**.
6. Measure the RNA concentration using a Nanodrop ND-1000 instrument and asses the RNA quality by means of an Agilent Technologies Bioanalyzer 2100 (Agilent, The Netherlands) or equivalent. It is recommended to use 1 μg of high-quality total RNA with an RNA integrity number above 8 for Illumina sequencing.
7. For data analysis of the sequencing output, it is essential to have a running Linux system and some basic knowledge of Unix and R commands.
8. To make the human genome file available for the alignment with bowtie, the file should be indexed using the bowtie-built program (i.e. bowtie-built human_g1k_v37.fa human_g1k_v37, all whitespaces in the identifier of the fatsa file have to be removed before this step).
9. For small RNA sequencing, the plots for sequence quality and GC content for the first 22 base pairs are of importance.
10. Adapter sequence is delivered in combination with the TruSeq Small RNA sample preparation kit (Illumina, The Netherlands).
11. We strongly recommend to use the option to weight reads that align to multiple sequences in order to get an appropriate distribution of your sequencing reads.
12. Before the import of the expression output to R, all redundant miRNA identifiers have to be removed using for instance a spreadsheet application.
13. A second significance criterion to compensate for bias introduced by very low abundant sequences is excluding miRNAs with less than 10 read counts in 1 of both sample groups. This approach resembles, with small adjustments, previously

published miRNA sequencing studies [36, 37] and is to our experience the most robust approach.

14. The reference miRNA should have little variation in expression across all samples. The miRNA expression data obtained from the sequencing can be screened for miRNAs with low standard deviation between all samples for a reference miRNA.
15. For all duplicate reactions, the average Ct values should be used for further calculations.

References

1. Bartel DP (2009) MicroRNAs: target recognition and regulatory functions. Cell 136: 215–233
2. Bartel DP (2004) MicroRNAs: genomics, biogenesis, mechanism, and function. Cell 116: 281–297
3. Kloosterman WP, Plasterk RHA (2006) The diverse functions of MicroRNAs in animal development and disease. Dev Cell 11:441–450
4. Zhang H, Li QY, Guo ZZ et al (2012) Serum levels of microRNAs can specifically predict liver injury of chronic hepatitis B. World J Gastroenterol 18:5188–5196
5. Zhang Y, Jia Y, Zheng R et al (2010) Plasma microRNA-122 as a biomarker for viral-, alcohol-, and chemical-related hepatic diseases. Clin Chem 56:1830–1838
6. Wang K, Zhang S, Marzolf B et al (2009) Circulating microRNAs, potential biomarkers for drug-induced liver injury. Proc Natl Acad Sci U S A 106:4402–4407
7. van Rooij E (2011) The art of microRNA research. Circ Res 108:219–234
8. Landgraf P, Rusu M, Sheridan R et al (2007) A mammalian microRNA expression atlas based on small RNA library sequencing. Cell 129: 1401–1414
9. Arroyo JD, Chevillet JR, Kroh EM et al (2011) Argonaute2 complexes carry a population of circulating microRNAs independent of vesicles in human plasma. Proc Natl Acad Sci U S A 108:5003–5008
10. Bala S, Marcos M, Szabo G (2009) Emerging role of microRNAs in liver diseases. World J Gastroenterol 15:5633–5640
11. Turchinovich A, Cho WC (2014) The origin, function and diagnostic potential of extracellular microRNA in human body fluids. Front Genet 5:30
12. Chen YN, Shen A, Rider PJ et al (2011) A liver-specific microRNA binds to a highly conserved RNA sequence of hepatitis B virus and negatively regulates viral gene expression and replication. FASEB J 25:4511–4521
13. McDermott AM, Heneghan HM, Miller N et al (2011) The therapeutic potential of microRNAs: disease modulators and drug targets. Pharm Res 28:3016–3029
14. Kutay H, Bai S, Datta J et al (2006) Downregulation of miR-122 in the rodent and human hepatocellular carcinomas. J Cell Biochem 99:671–678
15. Miyaaki H, Ichikawa T, Taura N et al (2014) Significance of serum and hepatic microRNA-122 levels in patients with non-alcoholic fatty liver disease. Liver Int 34(7): e302–e307
16. Xiao F, Yu J, Liu B et al (2014) A novel function of microRNA 130a-3p in hepatic insulin sensitivity and liver steatosis. Diabetes 63(8): 2631–2642
17. Yokoi T, Nakajima M (2013) microRNAs as mediators of drug toxicity. Annu Rev Pharmacol Toxicol 53:377–400
18. Yokoi T, Nakajima M (2011) Toxicological implications of modulation of gene expression by microRNAs. Toxicol Sci 123:1–14
19. Song MK, Song M, Choi HS et al (2012) Benzo[*k*]fluoranthene-induced changes in miRNA-mRNA interactions in human hepatocytes. Toxicol Environ Health Sci 4:143–153
20. Song MK, Park YK, Ryu JC (2013) Polycyclic aromatic hydrocarbon (PAH)-mediated upregulation of hepatic microRNA-181 family promotes cancer cell migration by targeting MAPK phosphatase-5, regulating the activation of p38 MAPK. Toxicol Appl Pharmacol 273:130–139
21. Morin RD, O'Connor MD, Griffith M et al (2008) Application of massively parallel sequencing to microRNA profiling and discovery in human embryonic stem cells. Genome Res 18:610–621
22. Heidebrecht F, Schulz I, Keller M et al (2009) Improved protocols for protein and

RNA isolation from three-dimensional collagen sandwich cultures of primary hepatocytes. Anal Biochem 393:141–144

23. Mardis ER (2008) Next-generation DNA sequencing methods. Annu Rev Genomics Hum Genet 9:387–402
24. Fang Z, Cui X (2011) Design and validation issues in RNA-seq experiments. Brief Bioinform 12:280–287
25. Friedlander MR, Mackowiak SD, Li N et al (2012) miRDeep2 accurately identifies known and hundreds of novel microRNA genes in seven animal clades. Nucleic Acids Res 40:37–52
26. Langmead B, Trapnell C, Pop M et al (2009) Ultrafast and memory-efficient alignment of short DNA sequences to the human genome. Genome Biol 10:R25
27. Griffiths-Jones S (2006) miRBase: the microRNA sequence database. Methods Mol Biol 342:129–138
28. 1000 Genomes Project Consortium, Abecasis GR, Auton A et al (2012) An integrated map of genetic variation from 1092 human genomes. Nature 491:56–65
29. Andrews S (2012) FastQC A quality control tool for high throughput sequence data. http://www.bioinformatics.babraham.ac.uk/projects/fastqc/ (consulted July 2014)
30. The R Core Team (2013) R: a language and environment for statistical computing: reference index. R Foundation for Statistical Computing, Version 3.0.1, 1-3604
31. Gentleman RC, Carey VJ, Bates DM et al (2004) Bioconductor: open software development for computational biology and bioinformatics. Genome Biol 5:R80
32. Anders S, Huber W (2010) Differential expression analysis for sequence count data. Genome Biol 11:R106
33. Hsu SD, Lin FM, Wu WY et al (2011) miRTarBase: a database curates experimentally validated microRNA-target interactions. Nucleic Acids Res 39:D163–D169
34. Kamburov A, Wierling C, Lehrach H et al (2009) ConsensusPathDB-a database for integrating human functional interaction networks. Nucleic Acids Res 37:D623–D628
35. Livak KJ, Schmittgen TD (2001) Analysis of relative gene expression data using real-time quantitative PCR and the 2(T)(-Delta Delta C) method. Methods 25:402–408
36. Kang L, Cui X, Zhang Y et al (2013) Identification of miRNAs associated with sexual maturity in chicken ovary by Illumina small RNA deep sequencing. BMC Genomics 14:352
37. Dhahbi JM, Atamna H, Boffelli D et al (2011) Deep sequencing reveals novel microRNAs and regulation of microRNA expression during cell senescence. PLoS One 6:e20509

Chapter 18

Mass Spectrometry-Based Proteomics for Relative Protein Quantification and Biomarker Identification in Primary Human Hepatocytes

Lisa Dietz and Albert Sickmann

Abstract

Liquid chromatography-tandem mass spectrometry-based proteomics is a highly sensitive and effective tool to identify and quantify potential biomarkers in repeated dose toxicity studies using primary cell culture systems. In this respect, 8-plex isobaric tag for relative and absolute quantification labeling is the method of choice for relative quantification. After cell lysis and tryptic protein digestion, an individual isobaric tag is added to the amine groups of arginine and lysine. Then, up to eight differentially labeled samples are mixed and analyzed together in a mass spectrometry experiment. During peptide fragmentation in the mass spectrometer, the individual tag intensity of each identified peptide could be detected, reflecting the peptide intensities in the eight samples. The identified peptides are matched to their specific protein using specific search engines and finally to eight individual relative protein quantities. The two-dimensional fractionation of complex peptide mixtures minimizes the possibility of co-fragmentation of peptides from different origin in the mass spectrometer, which leads to a higher number of peptide search matches and therefore to better identification and quantification results.

Key words iTRAQ, Relative quantification, Strong anion exchange, Proteomics, Two-dimensional fractionation

1 Introduction

Mass spectrometry (MS)-based relative protein quantification has become an outstanding strategy to investigate complex cellular systems, such as in hepatocyte research. The labeling of proteins or peptides is the key step in relative quantification to enable an equalized sample preparation, mixing the different samples at the earliest possible timepoint (i.e., direct after labeling) to minimize technical bias. The differentially labeled samples are distinguishable by their specific label in the final MS data analysis using specific protein quantification software. Many different labeling strategies exist. In principle, the most important difference is the MS level used for quantification analysis. Enzymatically labeled peptides as performed

Mathieu Vinken and Vera Rogiers (eds.), *Protocols in In Vitro Hepatocyte Research*, Methods in Molecular Biology, vol. 1250, DOI 10.1007/978-1-4939-2074-7_18, © Springer Science+Business Media New York 2015

within the stable isotope labeling with amino acids in cell culture (SILAC) approach [1] as well as label-free quantification approaches are using the precursor integration as a value for relative quantification. Chemical labeling strategies, such as 8-plex isobaric tag for relative and absolute quantification (iTRAQ) [2] or tandem mass tag (TMT) [3], allow quantification on MS2 level using the fragment (i.e., reporter) ion intensity as a quantitative value for relative quantification. The specific benefit of the iTRAQ labeling strategy is that isobaric tags are not visible on MS1 level, but show the sample specific reporter ion on MS2 level. Hence, the MS1 spectra are not raised in complexity, such as in SILAC approaches, and therefore precursor isolation for data-dependent fragmentation is more accurate. Combining 8-plex iTRAQ labeling with further two-dimensional peptide separation strategies as strong anion exchange chromatography improves identification and quantification of hepatocyte proteins. Overall, 3,010 proteins could be quantified in primary hepatocytes in four independent experiments using the method described in this chapter. In comparison, Rowe and group [4] quantified around 754 proteins in four independent experiments in primary hepatocytes using a one-dimensional iTRAQ shotgun approach as quantification method.

In this chapter, a robust 8-plex iTRAQ proteomic workflow, optimized for primary human hepatocytes cultured between collagen sandwich layers, is described. The complete workflow, including hepatocyte lysis, digestion, labeling, desalting steps, quality control as well as stage tip-based strong anion exchange (SAX) fractionation and liquid chromatography-tandem mass spectrometry (LC-MS/MS) analysis, is outlined in detail.

2 Materials

Prepare all solutions using ultrapure water (i.e., prepared by purifying deionized water to attain a sensitivity of 18 MΩ×cm and a total organic carbon content of 1–5 ppb at 25 °C) and analytical grade reagents. Prepare and store all reagents at 4 °C unless indicated otherwise. Diligently follow all waste disposal regulations when disposing waste materials.

2.1 Cell Lysis and DNA Depletion

1. Lysis buffer A (i.e., 150–300 μL/2 million hepatocytes). 2 mM $MgCl_2$ and 1 % sodiumdodecylsulfate (SDS) in 50 mM ammonium bicarbontate (ABC), pH 8.5. Dissolve 39.5 mg ABC in 10 mL ultrapure water, pH 8.5. Add 200 μL 100 mM $MgCl_2$ in 10 mL 50 mM ABC buffer, pH 8.5 (i.e., 100 mM $MgCl_2$ hexahydrate: 20.3 mg/mL H_2O). Add 0.1 g SDS in 10 mL ABC, pH 8.5 (*see* **Note 1**).
2. Lysis buffer B (i.e., 150–300 μL/2 million hepatocytes). 2 mM $MgCl_2$ and 0.1 % rapigest SF (Waters, Germany) in 50 mM ABC, pH 8.5. Dissolve 3.95 mg ABC in 1 mL ultra-

pure water, pH 8.5. Add 20 μL 100 mM $MgCl_2$ in 1 mL 50 mM ABC buffer, pH 8.5 (i.e., 100 mM $MgCl_2$ hexahydrate: 20.3 mg/mL) (*see* **Note 1**). Add 1 mL lysis buffer B to 1 vial of rapigest containing 1 mg rapigest SF resulting in 0.1 % rapigest SF solution.

3. 100 mM $CaCl_2$. Dissolve 111 mg $CaCl_2$ in 10 mL ultrapure water and store at 4 °C.
4. Ultrasonication bath RK 52 (Bandelin Sonorex, Germany).
5. Benzonase (Novagen, Germany).
6. Phosstop tablets (Roche, Germany) (*see* **Note 1**).

2.2 Bicinchoninic Acid Assay

1. Pierce bicinchoninic acid (BCA) protein assay kit (Thermo Scientific, Germany).
2. Pierce 2 mg bovine serum albumin standard ampoule (Thermo Scientific, Germany).
3. Spectrophotometer for wavelength measurement, spectral filter, 570 nm wavelength (Thermo Scientific, Germany).
4. Flat bottom 96-well plate and adhesive foil.
5. 50 mM ABC, pH 8.5.

2.3 Tryptic Digest Using Filter-Associated Sample Preparation (FASP)

1. 100 mM Tris(hydroxymethyl)-aminomethane (Tris)–HCl. Dissolve 605.7 mg Tris–HCl in 50 mL ultrapure water. Adjust to pH 8.5 using 1 N HCl.
2. 2.1 M 1,4-dithiothreitol (DTT). Dissolve 154.3 mg DTT in 1 mL ultrapure water and store in 10 μL aliquots at −20 °C.
3. 0.5 M 2-iodacetamide (IAA). Dissolve 18.5 mg IAA in 200 μL 100 mM Tris–HCl, pH 8.5. Prepare ex tempore and keep in dark.
4. 8 M urea in 100 mM Tris–HCl, pH 8.5. Dissolve 4.8 g urea in 10 mL 100 mM Tris–HCl, pH 8.5.
5. 30 K Omega centrifugal devices (PALL LifeSciences, Germany).
6. SpeedVac concentrator.
7. Trypsin sequencing grade (Promega, Germany). Prepare 1 μg/μL solution with the dilution buffer provided by the manufacturer. Store in 10 μL aliquots at −80 °C. Pre-heat an aliquot 15 min at 30 °C directly before use.
8. 100 mM $CaCl_2$.
9. 50 mM triethyl ammonium bicarbonate (TEAB). Add 1 mL 1 M TEAB solution, pH 8.5 (Sigma-Aldrich, Germany) to 19 mL ultrapure water. Prepare ex tempore.
10. 500 mM TEAB. Add 5 mL 1 M TEAB solution, pH 8.5 (Sigma-Aldrich, Germany) to 5 mL ultrapure water. Prepare ex tempore.

2.4 Digestion Control Using Chromatographic Separation with a Monolithic Column

1. Monolithic column PepSwift, C18, 200 μm, 5 cm (Dionex, Germany).
2. Buffer A. 0.1 % trifluoro acetic acid (TFA) (*see* **Note 2**).
3. Buffer B. 84 % acetonitrile (ACN) (*see* **Note 2**) in 0.1 % TFA.
4. Smartflow UltiMate 3000 (UltiMate 3000 pump, RS Variable wave length detector, flow manager, auto sampler), split 1:100, 1 μL loop, 3 nL flow cell (Dionex, Germany).
5. Sample vials with glass inlet septum for automated sample injection (*see* **Note 3**).

2.5 Determine Peptide Concentration Using NanoDrop

1. NanoDrop (NanoDrop 2000, ThermoScientific, Germany).
2. 0.5 mL low bind Eppendorf tubes.

2.6 Labeling

1. 500 mM TEAB.
2. 8-plex iTRAQ labeling kit (iTRAQ reagent kit P/N 4381663, AB SCiex).
3. Isopropanol (*see* **Note 2**).
4. 0.1 % TFA (*see* **Note 2**).
5. SpeedVac concentrator.

2.7 Desalting Tip Preparation

1. 10 mL plastic syringe.
2. 5 mL pipette tip.
3. Gel loader tip.
4. 200 μL Eppendorf tips.
5. C18 sorbent, 47 mm filter disks (3M Empore, Germany).
6. Reversed phase packing oligo R3 material (Applied Biosystems, Germany).
7. ACN/water 70/30 solution (*see* **Note 2**).
8. Oligo R3 solution. Dilute a spatula tip of oligo R3 material in 0.5 mL ACN/water 70/30. Prepare ex tempore and mix before use (*see* **Note 4**).

2.8 Desalting

1. C18 solid phase extraction (SPE) stage tip columns.
2. Centrifuge adapter (Glygen Corp, USA).
3. 10 % TFA (*see* **Note 2**).
4. 0.1 % TFA (*see* **Note 2**).
5. 100 % ACN (*see* **Note 2**).
6. ACN/0.1 % TFA 60/40 (*see* **Note 2**).
7. SpeedVac concentrator.

2.9 SAX SPE Stage Tip Column Preparation

1. Gel loader tip.
2. 200 µL Eppendorf tips.
3. Strong anion exchange material. SAX-SR, 47 mm, P/N 14-386-6, No 2252, 3M Empore (Fisher Scientific, Germany).
4. Centrifuge adapter (Glygen Corp, USA).

2.10 SAX SPE Stage Tip Fractionation

1. SAX-SPE stage tip columns.
2. Britton Robinson (BR) buffers with pH 11, 8, 6, 5, 4 and 3 [7]. 20 mM acetic acid, 20 mM phosphoric acid, 20 mM boric acid. Add 114.4 µL acetic acid, 104.8 µL and 123.6 mg boric acid to 100 mL ultrapure water.
3. Adjust the pH of the BR buffers with 1 M NaOH.
4. Add 146.1 mg NaCl to 10 mL BR buffer solution, pH 3, to gain a final concentration of 0.25 M NaCl.
5. 10× concentrated BR buffer, pH 11.
6. 100 % methanol (*see* **Note 2**).
7. 1 M NaOH. Dissolve 39.9 mg NaOH in 1 mL ultrapure water.
8. 10 % TFA (*see* **Note 2**).

2.11 Final Sample Preparation Prior to MS Analysis

1. C18 SPE stage tip columns.
2. Centrifuge adapter (Glygen Corp, USA).
3. 10 % TFA (*see* **Note 2**).
4. 0.1 % TFA (*see* **Note 2**).
5. 100 % ACN (*see* **Note 2**).
6. ACN/0.1 % TFA 60/40 (*see* **Note 2**).
7. SpeedVac concentrator.
8. Sample vials with glass inlet and septum for automated sample injection (*see* **Note 4**).

2.12 LC-MS Analysis

1. Trapping solvent. 0.1 % TFA (*see* **Note 2**).
2. Solvent A. 0.1 % formic acid (FA) (*see* **Note 2**).
3. Solvent B. 84 % ACN/0.1 % FA (*see* **Note 2**).
4. 10 % NH_4OH.
5. C18 trapping column. acclaim PepMap 100 nano trap, ID 100 µm, length 2 cm, particle size 3 µm, pore size 100 Å (Dionex Thermo-Scientific, Germany).
6. C18 separating column. acclaim PepMap RSLC, ID 75 µm, length 50 cm, C18 particle size 2 µm, pore size 100 Å (Dionex Thermo-Scientific, Germany).

7. NanoHPLC system. UltiMate 3000 rapid separation liquid chromatography (RSLC) System (Dionex Thermo-Scientific, Germany).
8. MS system. qExactive mass spectrometer (Thermo Scientific, USA).

3 Methods

3.1 Cell Lysis and Depletion of DNA

1. Add 150–300 μL lysis buffer A or B to 1–2 million harvested hepatocytes (*see* **Note 1**).
2. Sonicate the samples six times for 15 s and cool on ice between each sonication step.
3. Repeat sonication until the solution is clear or add additional lysis buffer up to maximum 300 μL.
4. Add 25 U (i.e., 1 μL) benzonase/mg DNA.
5. Incubate the samples for 30 min at 37 °C.
6. Centrifuge samples at 14,000×*g* and room temperature for 15 min to pellet the cell debris.
7. Transfer the supernatant to a new 0.5 mL low-bind Eppendorf tube for further analysis.
8. Store and deliver lysate samples at least at −80 °C.
9. Discard pellet.

3.2 BCA Assay

1. Add 2 and 5 μL of each lyzed sample in a new low-bind Eppendorf tube to 98 and 95 μL 50 mM ABC buffer, pH 8.5, and mix gaining a sample dilution of 1–20 and 1–50, respectively.
2. Prepare a standard dilution series of bovine serum albumin with 0 (i.e., blank), 5, 25, 50, 125, 250 μg/mL bovine serum albumin using the albumin standard ampoule.
3. Dilute 100 μL 2 mg/ml bovine serum albumin stock solution in 700 μL 50 mM ABC solution (i.e., diluent 1).
4. Add 400 μL of dilution 1–400 μL 50 mM ABC (i.e., diluent 2) and mix.
5. Add 300 μL of diluent 2–450 μL 50 mM ABC (i.e., diluent 3) and mix.
6. Add 400 μL of diluent 3–400 μL 50 mM ABC (i.e., diluent 4) and mix.
7. Add 100 μL of diluent 4–400 μL 50 mM ABC (i.e., diluent 5) and mix.
8. 50 mM ABC is used as a blank (i.e., diluent 6).
9. Pipette 25 μL of each standard and sample concentration in triplicates in a flat bottom 96-well plate (*see* **Note 5**).

10. Add 200 μL working solution per well. The working solution is prepared ex tempore by mixing 50 parts of BCA Reagent A with 1 part of BCA Reagent B of the Pierce BCA assay kit.
11. After covering the 96-well plate with adhesive foil, incubate the plate at 60 °C for 30 min.
12. Cool the plate to room temperature, remove the adhesive foil and measure the absorbance using a spectrophotometer set to an optimum wavelength of 562 nm according to the manufacturer's protocol.
13. Subtract the average 562 nm absorbance measurement of the blank standard replicates from the 562 nm measurements of all other individual standard and unknown sample replicates.
14. Prepare a standard curve by plotting the average blank-corrected 562 nm measurement for each BSA standard versus its concentration expressed in μg/mL. Use the standard curve to determine the protein concentration of each unknown sample (*see* **Note 6**).

3.3 Carbamidomethylation and FASP

The FASP protocol is performed with minor changes as described by Wisniewski and group [5].

1. Reduce 100 μg of each sample by adding 1 M DTT to a final concentration of 10 mM DTT and mix.
2. Incubate for 30 min at 60 °C on a thermomixer.
3. Add 8 M urea buffer to each sample to gain a final sample volume of 200 μL, mix and spin.
4. Flush the 30 kDa cut-off filter (i.e., FASP filter) two times with 100 μL ultrapure water.
5. Discard the flow through.
6. Transfer the samples to the FASP filter and centrifuge for 10 min at 14,000 × *g* and room temperature.
7. Add 96 μL 8 M urea buffer and 4 μL 0.5 M IAA to each filter device, mix gently and incubate the samples for 20 min at room temperature in the dark.
8. Spin 15 min at 14,000 × *g* and room temperature.
9. Add 100 μL urea buffer.
10. Spin 15 min at 14,000 × *g* and room temperature.
11. Repeat **steps 9** and **10** twice.
12. Discard the flow through.
13. Add 100 μL 50 mM TEAB.
14. Spin 15 min at 14,000 × *g* and room temperature.
15. Repeat **steps 13** and **14** twice.
16. Discard the flow through.

17. Prepare the digestion solution by mixing 902 μL 50 mM TEAB, pH 8.5, with 9.4 μL 100 mM $CaCl_2$ to a final concentration of 1 mM $CaCl_2$.
18. Add 76 μL digestion solution to each sample and boil for 2 min in a thermomixer at 80 °C.
19. Add 2 μL 1 μg/μL trypsin sequencing grade solution to the samples gaining a trypsin to protein proportion of 1:50.
20. Gently mix and incubate the samples in an oven at 37 °C overnight.
21. Transfer the filter in a low-bind Eppendorf tube and centrifuge at 14,000 × *g* and room temperature for 15 min.
22. Add 50 μL 50 mM TEAB and spin 15 min at 14,000 × *g* and temperature.
23. Add 1 μL of each digested sample to 14.3 μL 0.1 % TFA in glass vials for digestion control analysis on a monolithic separation system (Fig. 1).
24. Transfer each sample to a fresh 0.5 mL low-bind Eppendorf tube, freeze at −80 °C and dry the samples in a speedVac concentrator.
25. Store the dried samples at −80 °C until use.

3.4 Digestion Quality Control

To control if the protein digestion protocol works properly, a monolithic reverse phase separation system is used. Alternatively, a one-dimensional gel could be used if no monolithic system is present. A monolithic column is able to separate peptides, polypeptides, and proteins in a single step. Due to their reduced hydrophobicity, peptides elute first, polypeptides with missed cleavage sites elute in between, and proteins elute at the end of the gradient [6] (Fig. 1).

1. Use a linear gradient from 5 to 50 % solvent B over 34 min ending with a final washing step of 5 min. The last 8 min are used to equilibrate the column for the next run.
2. Perform the separation with a flow rate of 2.2 μL/min at 60 °C.
3. Peptide/protein separation: 0–3 min: 5 % B; 3–6 min: linear increase from 5 to 10 % B; 6–34 min: linear increase from 10 to 50 % B; 34–37 min: linear increase from 50 to 95 % B; 37–41 min: 95 % B; 41–42 min: linear decrease from 95 to 5 % B; 42–50 min: 95 % B.

3.5 Determination of Peptide Concentration Using NanoDrop

Determination of the final peptide concentration per sample is obligatory to reduce bias resulting from sample preparation to enable equal sample proportions after mixing. Here, a wavelength-based peptide concentration measurement is performed using the NanoDrop spectrophotometer instrument, which is capable to

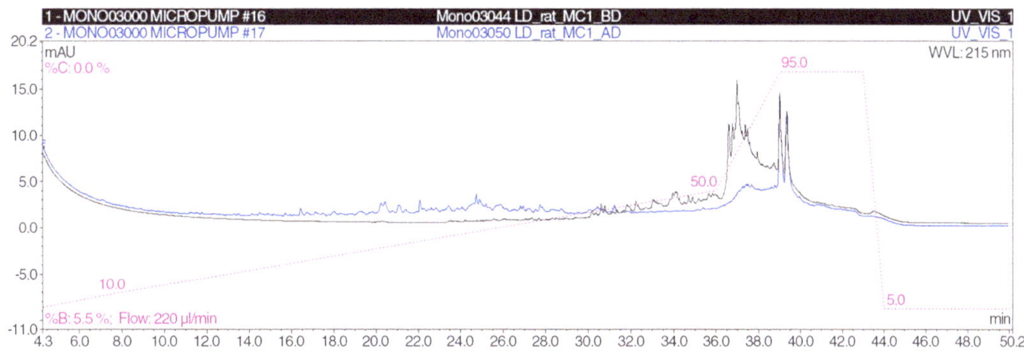

Fig. 1 Digest control via overlay of two monolithic UV chromatograms. The UV chromatogram of 1 μg primary hepatocyte lysate before digest is shown in *black*. The UV chromatogram of one primary hepatocyte lysate after tryptic digest is shown in *blue*. The binary gradient is shown with *dashed line*

analyze 0.5–2 μL sample volumes. This minimizes sample consumption and is therefore optimal for peptide concentration control during sample preparation.

1. Add 30 μL 500 mM TEAB to the dried sample.
2. Dilute 1 μL sample in 19 μL ultrapure water, mix and spin. Water is used as a blank. Store the remaining 29 μL at −80 °C until proceeding with the next preparation step.
3. Add 2 μL ultrapure water on the Nanodrop measurement point and use as a blank.
4. Add 2 μL of each sample dilution to the measurement point.
5. Analysis is performed automatically.

3.6 iTRAQ 8-plex Labeling and Sample Mix

Perform the 8-plex iTRAQ labeling and mixing according to the manufacturer's instructions.

1. Add 50 μL isopropanol to each label, mix and spin.
2. Transfer the complete label to the sample and incubate for 2 h at room temperature.
3. Freeze the labeled samples at −80 °C and dry them using a speedVac concentrator.
4. Dissolve the samples in adequate volumes of 0.1 % TFA according to the peptide concentration measured with the NanoDrop device to gain 1 μg/μL concentration.
5. Mix 5 μg/sample (i.e., 5 μL) resulting in one 8-plex mix-sample containing 40 μg peptides (i.e., 8 times 5 μg) in total and store mixed samples at −80 °C until use.

3.7 Desalting

1. Cut the very end of a 5 mL pipette tip and press it on the tip of the 10 mL plastic syringe to construct a tight adapter between syringe and 200 μL tip.
2. Cut the very end of a 200 μL pipette tip, discard the very end and use the rest of the tip to cut out three layers per sample from the C18 filter material with layer size of approximately 0.5 mm diameter.
3. Put the cut out layers in a new 200 μL pipette tip and push layer by layer down with a gel loader tip to the end of the 200 μL pipette tip and try to avoid air between the layers.
4. Drop 10 μL of oligo R3 material on top of the layers in the pipette tip and press the solution through the filter layers using the 10 mL syringe with the adapter.
5. Store the C18 SPE stage tip columns in a closed tipbox at room temperature until use.
6. Check the pH of the mixed iTRAQ samples. The pH should be lowered to pH 3 using 10 % TFA.
7. Place the filter tip in a 2 mL Eppendorf tube using the centrifuge adapter.
8. Equilibrate the C18 SPE stage tip column by adding twice 100 μL ACN on top of the tip column and centrifuge for 1 min at 200 × g and room temperature.
9. Discard the flow through.
10. Wash the filter by adding twice 0.1 % TFA and centrifuge for 5 min at 300 × g and room temperature.
11. Discard the flow through.
12. Replace 2 mL Eppendorf tubes with 1.5 mL low-bind Eppendorf tubes.
13. Add the complete amount of sample (i.e., maximum 100 μL) and centrifuge for 5 min at 300 × g and room temperature.
14. Add the flow through again to the C18 SPE stage tip column and centrifuge for 5 min at 300 × g and room temperature.
15. Discard the flow through.
16. Wash the C18 SPE stage tip column by adding twice 0.1 % TFA and centrifuge for 5 min at 300 × g and room temperature.
17. Discard the flow through.
18. Elute the sample from the C18 SPE stage tip column by adding twice 100 μL ACN/0.1 % TFA 60/40.
19. Freeze the samples at −80 °C and dry them frozen in the speedVac concentrator until dryness.
20. Store dried samples at −80 °C until usage.

3.8 SAX SPE Stage Tip Fractionation

Perform the SAX stage tip fractionation according to the method described by Wisniewski and group [7].

1. Cut the very end of a 200 μL pipette tip, discard the very end and use the rest of the tip to cut out two layers per sample from the SAX filter material with layer size of approximately 0.5 mm diameter.
2. Put the cut out layers in a new 200 μL pipette tip and push layer by layer down with a gel loader tip to the end of the 200 μL pipette tip (*see* **Note** 7).
3. Store the SAX SPE stage tip columns in a closed tipbox at room temperature until use.
4. Dilute the samples in BR buffer, pH 11.
5. Check the pH of the desalted iTRAQ samples. The pH should be raised to pH 11 by using 10× concentrated BR buffer, pH 11.
6. Place the SAX SPE stage tip column in a 2 mL Eppendorf tube using the centrifuge adapter.
7. Equilibrate the SAX column twice with 100 % methanol and centrifuge at 6,000 × *g* and room temperature for 3 min. Discard flow through.
8. Equilibrate the SAX column twice with 1 M NaOH and centrifuge at 6,000 × *g* and room temperature for 3 min. Discard flow through.
9. Equilibrate the SAX column twice with BR buffer, pH 11, and centrifuge at 6,000 × *g* and room temperature for 3 min. Discard flow through.
10. Replace 2 mL Eppendorf tubes with 1.5 mL low-bind Eppendorf tubes.
11. Add the complete amount of sample (i.e., maximum 100 μL) to the SAX column and centrifuge at 6,000 × *g* and room temperature for 5 min.
12. Reduce the pH of the flow through sample to pH 3 using 10 % TFA for further desalting and store at −80 °C until use.
13. Place the SAX SPE stage tip column in a new 1.5 mL low-bind Eppendorf tube and add 100 μL BR buffer, pH 8, to the SAX column and centrifuge at 6,000 × *g* and room temperature for 5 min.
14. Reduce the pH of the flow through sample to pH 3 by using 10 % TFA for further desalting and store at −80 °C until use.
15. Place the SAX SPE stage tip column in a new 1.5 mL low-bind Eppendorf tube, add 100 μL BR buffer, pH 6, to the SAX column and centrifuge at 6,000 × *g* and room temperature for 5 min.

16. Reduce the pH of the flow through sample to pH 3 by using 10 % TFA for further desalting and store at -80 °C until use.
17. Place the SAX SPE stage tip column in a new 1.5 mL low-bind Eppendorf tube, add 100 μL BR buffer, pH 5, to the SAX column and centrifuge at 6,000 × *g* and room temperature for 5 min.
18. Reduce the pH of the flow through sample to pH 3 by using 10 % TFA for further desalting and store at −80 °C until use.
19. Place the SAX SPE stage tip column in a new 1.5 mL low-bind Eppendorf tube, add 100 μL BR buffer, pH 4, to the SAX column and centrifuge at 6,000 × *g* and room temperature for 5 min.
20. Reduce the pH of the flow through sample to pH 3 by using 10 % TFA for further desalting and store at −80 °C until use.
21. Place the SAX SPE stage tip column in a new 1.5 mL low-bind Eppendorf tube, add 100 μL BR buffer, pH 3, to the SAX column and centrifuge at 6,000 × *g* and room temperature for 5 min and store the samples at −80 °C until use.

3.9 Final Sample Preparation Prior to MS Analysis

1. Desalt samples using C18 SPE stage tip columns as described and dry the samples using the speedVac concentrator.
2. Dissolve samples in 20 μL 0.1 % TFA.
3. Add 1–14 μL of 0.1 % TFA in a MS sample vial.
4. Measure each fraction on a monolithic column system, 200 μm, gradient 5 to 50 %, 50 min as described to calculate the correct sample proportion for the LC-MS/MS system using 10 mAU peak intensity as a maximum.

3.10 MS Analysis

The SAX fractions are analyzed on a qExactive instrument coupled to an RSLCnano using the following conditions and MS methods.

1. Automatically inject 15 μL sample and load onto a C18 trap column with a flow rate of 20 μL/min in 0.1 % TFA. Peptide separation is performed on a 75 μm ID RP column using a binary gradient ranging from 3 to 42 % of solvent B at a flow rate of 250 nL/min in 177 min. MS full scans are acquired over 145 min in the range of 300 to 1,500 m/z at a resolution of 70,000 using the polysiloxane m/z 371.101236 as lock mass and an AGC target of $3e^6$ ions with a maximal ion time of 120 ms. Data-dependent MS2 scans are acquired with 17,500 resolution and with an AGC target of $1e^6$ ions accepting a 10 % under fill ratio. Maximal ion time is set to 250 ms. Dynamic exclusion is set to 12 s. The 15 most intensive signals (i.e., isolation window 2 m/z) are subjected to HCD-MS/MS analysis. HCD spectra are acquired with normalized collision energy of 32 % (*see* **Note 8**).

2. Additionally, use of 5–10 % NH_4OH in front of the ESI source is recommended for better ionization efficiency (*see* **Note 9**).
3. For washing and reconstitution of the HPLC column, the gradient rises to 95 % of solvent B for 8 min and finally equilibrates for 13 min at 5 % of solvent B.
4. To reduce memory effects and improve column lifetime, a 45-min wash-blank program after each sample is highly recommended.

3.11 Data analysis using ProteomDiscoverer 1.3

Adequate software is necessary to analyze the raw files delivered from the qExactive. Analysis results can be exported to an Excel file.

1. Analyze the different SAX fractions together in one mud pit analysis to gain as much information as possible.
2. Spectrum selector settings: precursor selection: use MS1 precursor, spectrum properties filter: lower RT limit: 0; upper RT limit: 0; first scan: 0; last scan: 0; charge state: 2–5; minimum precursor mass: 300; maximum precursor mass: 5,000; total intensity threshold: 0; minimum peak count: 10.
3. Scan event filter settings: mass analyzer: IS ITMS, FTMS; MS order: is MS2; activation typ: any; scan type: full.
4. Peak filter settings: S/N threshold (FT only): 3.
5. Replacement of unrecognized properties: unrecognized charge replacement: automatic; unrecognized mass analyzer replacement: FTMS; unrecognized MS order replacement: MS2; unrecognized activation type replacement: HCD; unrecognized polarity replacement: positive.
6. Mascot: protein database: uniprot_human_fw; enzyme: trypsin; Instrument: EsiTrap; use decoy data base; precursor mass tolerance: 10 ppm; fragmentation mass tolerance: 0.5 Da; use average precursor mass: false; dynamic modification: oxidation, iTRAQ 8plex N-term, K, Y; fixed modification: carbamidomethylation.
7. Peptide validator: search against decoy database: true; target FDR (strict): 0.01; target FDR (relaxed): 0.05; reporter ions quantifier; quantification method: iTRAQ 8-plex (custom).
8. Peak integration: integration window tolerance: 20 ppm; integration method: most confident centroid.
9. Quantification method: experimental bias: normalize on protein median (minimum protein count: 20); normalize for each ratio pair singularly, because samples are not dependent from each other and may have total opposite bias. Show the raw quan values. Replace missing quan values with minimum intensity. Apply quan value corrections.
10. Consolidate the single ratio lists in excel to one big file and put two cut-off levels to the lists:

    relaxed: $0.66 > \text{ratio} > 1.5$; strict: $0.5 > \text{ratio} > 2$.

4 Notes

1. If FASP and iTRAQ labeling is planned, use lysis buffer A (i.e., 1 % SDS in 2 mM $MgCl_2$/50 mM ABC buffer, pH 8.5) for cell lysis. If label-free analysis is planned, use lysis buffer B (i.e., 0.1 % rapigest SF lysis buffer in 2 mM $MgCl_2$/50 mM ABC buffer, pH 8.5) for cell lysis. If label-free is planned for overview results and FASP is planned for further experiments, use first lysis buffer B for preliminary results. Afterwards, add SDS to a final concentration of 1 %. If phosphor analyses are planned, add 15–30 μL 10× concentrated phosstop stock solution (i.e., one tablet in 1 mL ultrapure water, storable for 1 month at 4–8 °C) to 150–300 μL lysis buffer.
2. Use only organic solvents of ULC/MS grade.
3. Sample vials should be silanized to reduce peptide binding to the glass vial.
4. The material is used as a suspension since it will not dissolve.
5. 75 μL (i.e., three times 25 μL) are needed from each dilution per BCA assay. The unused dilution series can be stored at −20 °C for at least 3 months.
6. The coefficient of variation should be lower than 10 %.
7. Do not press too hard and try to avoid air holes between the layers.
8. Collision energy needs to be optimized for each mass spectrometer separately.
9. Add 5–10 % NH_4OH in a 2 mL tube, close it and make a little hole in the lid. Place the tube close to the emitter until iTRAQ samples are finished. Remove the NH_4OH during standard sample runs without iTRAQ labeling.

Acknowledgements

The financial support by the European Commission's FP7 Health Program and Cosmetics Europe are gratefully acknowledged (SEURAT-1 project DETECTIVE No 266838).

References

1. Ong SE, Blagoev B, Kratchmarova I et al (2002) Stable isotope labeling by amino acids in cell culture, SILAC, as a simple and accurate approach to expression proteomics. Mol Cell Proteomics 1:376–386
2. Ross PL, Huang YN, Marchese JN et al (2004) Multiplexed protein quantitation in *Saccharomyces cerevisiae* using amine-reactive isobaric tagging reagents. Mol Cell Proteomics 3:1154–1169

3. Thompson A, Schafer J, Kuhn K et al (2003) Tandem mass tags: a novel quantification strategy for comparative analysis of complex protein mixtures by MS/MS. Anal Chem 75:1895–1904
4. Rowe C, Goldring CE, Kitteringham NR et al (2010) Network analysis of primary hepatocyte dedifferentiation using a shotgun proteomics approach. J Proteome Res 9:2658–2668
5. Wisniewski JR, Zougman A, Nagaraj N et al (2009) Universal sample preparation method for proteome analysis. Nat Methods 6:359–362
6. Premstaller A, Oberacher H, Walcher W et al (2001) High-performance liquid chromatography-electrospray ionization mass spectrometry using monolithic capillary columns for proteomic studies. Anal Chem 73:2390–2396
7. Wisniewski JR, Zougman A, Mann M (2009) Combination of FASP and StageTip-based fractionation allows in-depth analysis of the hippocampal membrane proteome. J Proteome Res 8:5674–5678

Chapter 19

Targeted Metabolomics for Homocysteine-Related Metabolites in Primary Hepatocytes

Irena Selicharová and Marek Kořínek

Abstract

Liquid chromatography-tandem mass spectrometry has become the most convenient method to identify and quantify low molecular weight metabolites from various sources. Metabolomics studies of hepatocytes hold promise for the identification of the mechanisms of toxicant-related disease processes. In this chapter, we present a rapid and sensitive liquid chromatography-tandem mass spectrometry method for the quantification of intracellular concentrations of nine homocysteine-based metabolites, namely homocysteine, methionine, cysteine, dimethylglycine, cystathionine, *S*-adenosylmethionine, *S*-adenosylhomocysteine, choline, and betaine. The method is specifically designed for the analysis of cultured primary hepatocytes.

Key words Cell culture, Homocysteine, Intracellular metabolites, LC-MS/MS

1 Introduction

Fatty liver disease, whether alcoholic [1] or non-alcoholic (i.e., associated with obesity, type II diabetes, and hypertension) [2, 3], and cardiovascular diseases account for serious health problems in developed countries. Underlying these pathologies are mutually interrelated metabolic and endocrine disorders, which may be conditioned by genetic predispositions. Metabolomics studies of hepatocytes hold promise for identifying the mechanisms of disease processes.

Methionine (Met) metabolism is known to be disrupted in toxicant-related liver diseases [4]. Homocysteine (Hcy) is an intermediate in Met metabolism and lies at a central position between Met production *via* re-methylation and its removal *via* trans-sulfuration to produce cysteine (Cys). Moreover, hyperhomocysteinemia (i.e., a plasma Hcy concentration of more than 15 μM) is a strong independent cardiovascular disease predictor [5]. Precise, reproducible and highly sensitive methods for measuring the concentrations of compounds associated with Hcy metabolism are desired.

Mathieu Vinken and Vera Rogiers (eds.), *Protocols in In Vitro Hepatocyte Research*, Methods in Molecular Biology, vol. 1250, DOI 10.1007/978-1-4939-2074-7_19, © Springer Science+Business Media New York 2015

Liquid chromatography-tandem mass spectrometry (LC-MS/MS) has become the most convenient method to accomplish the task of rapid, simultaneous, sensitive, and selective determination of metabolites from various sources [6–9]. However, few analyses of metabolites have been performed in cultured cells [10]. Cell cultures require special precautions during manipulation and data interpretation [11]. Several aspects, such as metabolite stability, possible enzymatic or spontaneous conversion of compounds and biological matrix effects, should be taken into account when optimizing metabolite extraction protocols. Special attention should be paid to changes in incubation media [12].

We present here a rapid and sensitive LC-MS/MS method for the quantification of intracellular concentrations of nine Hcy-based metabolites, including Hcy, Met, Cys, dimethylglycine (DMG), cystathionine (Cysta), *S*-adenosylmethionine (SAM), *S*-adenosylhomocysteine (SAH), choline (Cho), and betaine (Bet). We designed this method specifically for the analysis of cells in culture and application to primary hepatocyte cultures [12, 13]. This method is useful for assessment of concentrations in the micromolar range with detection limits in the nanomolar range and can be adjusted to specific needs. In particular, hepatocytes can be exposed to any compound of interest in an appropriate concentration for a desired period of time. Other validated LC-MS/MS methods [14] for determination of different groups of metabolites can be applied similarly, but the protocol might require adjustments due to different requirements for cell lysis, solubility of specific compounds, column, and mass spectrometer specifications.

2 Materials

Prepare all solutions using ultrapure water (i.e., 18.2 MΩ) and analytical grade reagents. Store chemicals and solutions in a refrigerator or freezer according to the manufacturer's instructions.

2.1 Cell Culture

Cells and cultivation media should be handled in sterile conditions (i.e., in a laminar flow cabinet). Autoclave the equipment and filter solutions of test compounds through a 0.45 μm syringe filter.

1. Freshly prepared hepatocyte monolayers seeded in 6-well plates at a plating density of 1.8×10^6 cells/well (*see* **Note 1**).
2. William's E medium with GlutaMAX supplemented with 100 IU/mL penicillin, 100 μg/mL streptomycin, 4 μg/mL bovine insulin, and 50 μM hydrocortisone hemisuccinate. Aliquots of the medium are supplemented with test compounds to reach desired final concentrations (*see* **Note 2**).

3. 0.9 % NaCl solution containing 2 μL/mL protease inhibitor cocktail.
4. Liquid nitrogen.
5. Thermostatic bath.
6. Incubator (37 °C ± 1 °C, 90 % ± 5 % humidity, 5 % ± 1 % CO_2).
7. Laminar flow cabinet.

2.2 Cell Lysis

1. 90 % methanol with 0.03 % trifluoroacetic acid (TFA). To prepare 10 mL, mix 9 mL methanol with 1 mL ultrapure water and add 3 μL TFA.
2. 0.1 M dithiothreitol (DTT) in 80 % methanol. To prepare 1 mL, dissolve 1.54 mg DTT in 200 μL ultrapure water and mix with 800 μL methanol.
3. 0.1 % formic acid in water. To prepare 100 mL, add 100 μL HPLC-grade formic acid to 100 mL ultrapure water.
4. Centrifuge.
5. Sonicator.
6. Vacuum concentrator.

2.3 LC-MS/MS

Operating a mass spectrometer might require experienced staff. Otherwise, follow the manufacturer's instructions.

1. Reversed-phase Aquasil C18 column, 150 × 2.1 mm ID (Thermo Scientific, United States of America) (*see* **Note 3**).
2. Triple Quadrupole Quattro Premier XE mass spectrometer (Waters Micromass, United States of America) with electrospray ionization (ESI) coupled with an Agilent 1200 LC system (Agilent, United States of America) and CTC autosampler (CTC Analytics AG, Switzerland).
3. Mobile phase A. 0.1 % formic acid. To prepare 1 L, add 1 mL HPLC-grade formic acid to 1 L ultrapure water.
4. Mobile phase B. 0.15 % formic acid in a solution consisting of 90 % acetonitrile and 10 % deionized water. To prepare 1 L, mix 900 mL acetonitrile with 98.5 mL water and add 1.5 mL formic acid.
5. Stock solutions of individual analytes in a methanol:water 1:1 mixture (*see* **Note 4**). Cysta must be dissolved in 0.1 M aqueous HCl. Weigh the indicated amount of the compounds (Table 1) and dissolve in the indicated volume of methanol:water or HCl. Use volumetric flasks.
6. Standard addition mixture. Add the indicated volumes of each stock solution (Table 1) to a 100 mL volumetric flask together with 800 μL of 0.1 M DTT in 80 % methanol. Fill the flask with 0.1 % formic acid in water. The resulting solution will contain

Table 1
Preparation of stock solutions of compounds used as standards for LC-MS/MS quantification of the compounds in cellular extracts

Compound	Weight of compound (mg)	Weight of base (mg)	Volume (mL)	Concentration (μg/mL)	Volume for standard addition mixture (μL)
DMG·HCl	3.39	2.5	5.0	500	5
L-Cho	12.50	12.5	5.0	2,500	250
dCho·HCl	2.65	2.0	5.0	400	400
L-Bet	12.50	12.5	5.0	2,500	80
L-Cys·HCl·H_2O	36.20	25.0	5.0	5,000	120
L-Hcy	6.25	6.25	25.0	250	80
L-Met	5.00	5.0	5.0	1,000	150
Cysta (90 %)	2.50	2.5	5.0	450	5
SAM·HCl	2.72	2.5	5.0	500	250
SAH	2.50	2.5	5.0	500	150

200 ng/mL Hcy, 1,500 ng/mL Met, 6,000 ng/mL Cys, 1,250 ng/mL SAM, 750 ng/mL SAH, 2,000 ng/mL Bet, 6,250 ng/mL Cho, 25 ng/mL DMG, 22.5 ng/mL Cysta, and 12 μg/mL DTT in 0.1 % aqueous formic acid (*see* **Note 5**).

7. Blank sample. Mixed control cell lysate that is repeatedly used during calibration of the instrument.
8. Quality control sample. Mixed control cell lysate with standard addition that is repeatedly used during calibration of the instrument.

3 Methods

Use a single well of a 6-well plate for the preparation of each sample. Replicate samples should originate from different wells. Distribute the replicate samples on different plates. For example, if you perform two treatment types, use two wells of one plate as controls, two wells for the first treatment (i.e., 2 mM Hcy) and the remaining two wells for the second treatment (i.e., 0.1 mM Hcy). Prepare a second plate in the same manner to obtain four replicate samples overall. This layout helps control for technical variability introduced during handling of the plates and wells.

3.1 Cell Culture

Cultivate the hepatocytes at 37 °C in a humidified atmosphere of 5 % CO_2. Treat cells with Hcy for 48 h or use toxicants of interest. Medium should be exchanged every 24 h. *Prior* to use, heat the medium to 37 °C in a thermostated bath. To exchange the medium, the well is tilted, a pipette tip is pressed to the wall of the well, and fluid is slowly aspirated to avoid disturbing the cells (*see* **Note 6**). Fresh medium is added by slowly pipetting down the side of the wall.

3.2 Sample Preparation

1. Remove the cultivation medium and add 1 mL fresh medium, including the test compounds (i.e., control, 2 mM Hcy or 0.1 mM Hcy) (*see* **Note 7**).
2. Scrape the cells, diffuse them in the medium and transfer to test tubes. Wash the wells with an additional 0.5 mL of medium and combine the samples.
3. Pellet the cells by centrifugation at 140 × *g* and room temperature for 5 min. Remove the supernatants.
4. Diffuse the cells in 0.5 mL ice-cold 0.9 % NaCl containing 2 μL/mL of protease inhibitor cocktail.
5. Pellet the cells by centrifugation at 500 × *g* and 4 °C for 2 min. Remove the supernatant and immerse the tube in liquid nitrogen. Store the cell pellets at −70 °C until further use.

3.3 Cell Lysis and Metabolite Extraction

1. Disperse the cell pellet in 1.5 mL 90 % methanol with 0.03 % TFA and 20 μL 0.1 M DTT in 80 % methanol.
2. Incubate at ambient temperature for 10 min. Alternate agitation with sonication in pulse mode for 1 min. Repeat twice.
3. Pellet the cell debris and denatured proteins by centrifugation at 16,000 × *g* and 4 °C for 13 min. Collect the supernatants.
4. Dry the supernatants in a vacuum concentrator.
5. Reconstitute the extract by adding 200 μL of 0.1 % aqueous formic acid to each cell lysate, mix for 10 min and centrifuge at 16,000 × *g* and 4 °C for 3 min.

3.4 Metabolite Determination

1. Split each supernatant into two aliquots (i.e., 50 μL and 45 μL).
2. Add 5 μL of the standard addition mixture to the 45 μL aliquot.
3. Run the LC-MS/MS. Equilibrate and test the instrument by analyzing at least two blanks and three quality control samples [14]. Analyze both the spiked and non-spiked sample aliquots. Run at least three replicate samples for each treatment. We use a positive ESI technique in the selected multi-reaction monitoring mode (Table 2). Typical ion chromatograms of the metabolites in primary human hepatocytes are shown in Fig. 1. The capillary voltage is set to 0.7 kV. The cone and desolvation gas flows are 50 L/h and 800 L/h, respectively. The source and desolvation temperatures are 120 °C and 400 °C, respectively. The flow rate is adjusted to 0.2 mL/min. The method

Table 2
Compound-specific settings

Metabolite	Retention time (min)	Precursor ion (*m/z*)	Product ion (*m/z*)	Cone voltage (eV)	Collision energy (eV)
Cysta	1.8	223.0	134.1	20	14
Cys	2.0	122.0	76.0	15	12
DMG	2.4	104.0	58.2	18	12
Hcy	2.6	136.0	90.0	17	13
SAM	2.9	399.1	250.0	22	18
Cho	2.9	104.0	60.0	37	17
dCho	2.9	113.0	69.0	37	17
bet	3.6	118.1	59.2	31	18
Met	3.7	150.0	104.0	18	11
SAH	6.1	385.1	134.0, 136.0	22	21

Dwell time and span were set to 0.3 and 0.5, respectively

starts with a constant flow of 100 % mobile phase A for the first 0.5 min, followed by a linear gradient to 40 % mobile phase B in 6 min. The column is then flushed with 90 % mobile phase B for 1.5 min and re-equilibrated with 100 % mobile phase A for an additional 7.5 min. The total analysis time is 15.5 min. The sample temperature is maintained at 10 °C in the autosampler and the injection volume is 2.5 μL for each run.

4. Calculate the concentrations of the metabolites from the difference between the spiked and non-spiked samples using the formula here below. An example of such calculation is shown in Fig. 2 (*see* **Note 8**).

$$x_i = \frac{V_s \times x_s \times A}{(A_i - 0.9 \times A) \times V_i}$$

x_i = concentration of metabolite (ng/mL)
A_i = average peak area of spiked sample
A = average peak area of non-spiked sample
V_i = sample volume (i.e., 50 μL)
V_s = volume of standard addition (i.e., 5 μL)
x_s = concentration of the metabolite in the standard addition (ng/mL)

5. Calculate the significance of the changes induced by the treatments relative to the control sample using an appropriate statistical method (e.g. 2-tailed *t*-test for independent samples).

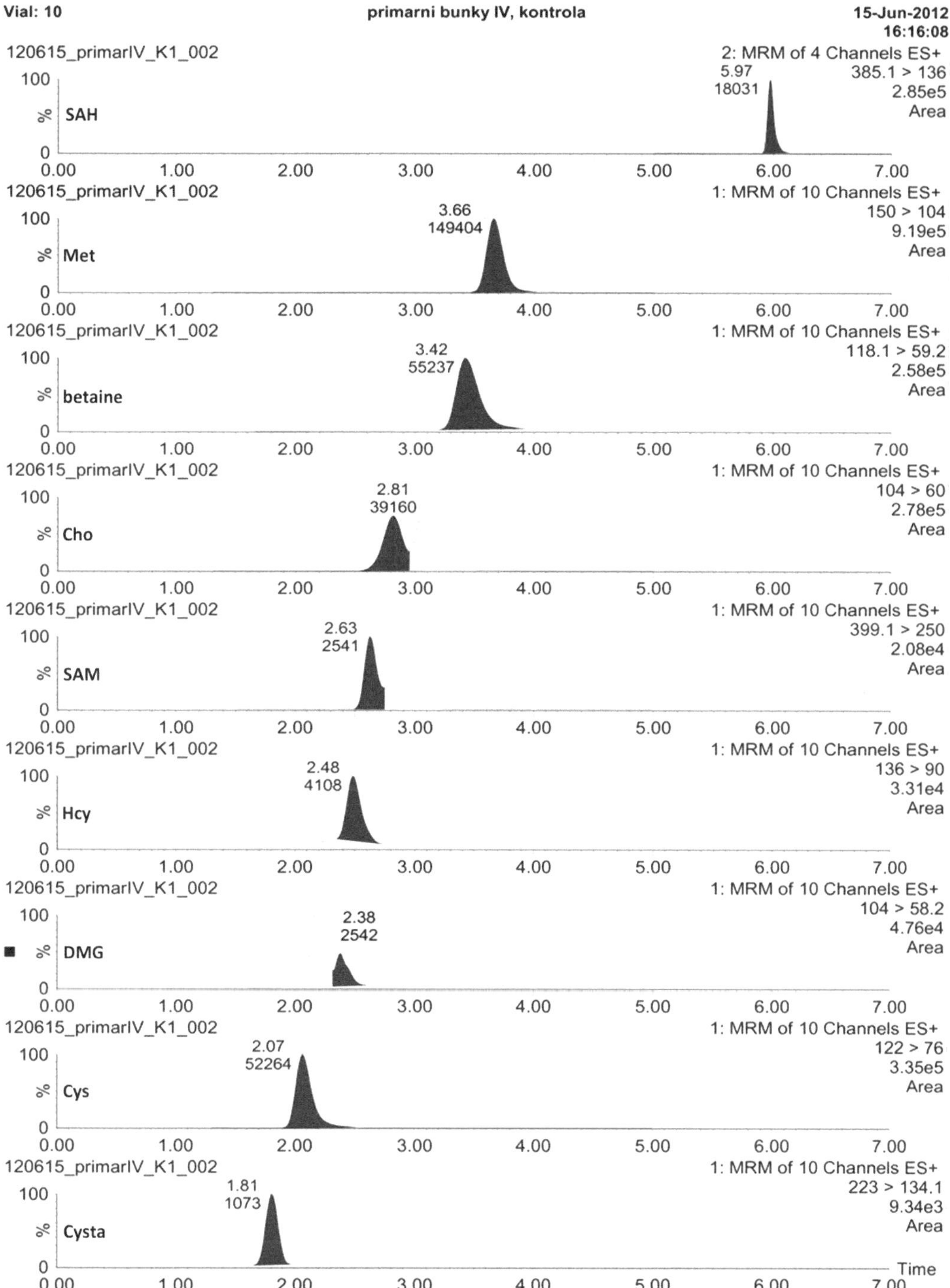

Fig. 1 Representative ion chromatograms. Extracts from primary hepatocytes were analyzed using LC-MS/MS. Original outputs from the Triple Quadrupole Quattro Premier XE mass spectrometer (Waters Micromass, United States of America), including the date of analysis and informal sample name, are shown. Each trace represents one metabolite (Table 2). The retention time of the metabolite and area of the integrated signal are shown next to the peaks (e.g. 5.97 and 18,031 for SAH). Monitored *m*/*z* transitions and the value of the highest reachable signal on the chromatogram (i.e., absolute scale) are listed on the right (e.g. 385.1 more than 136 and 2.85e5 for SAH). Time expressed in min is shown on the *x*-axis and the height of the multiple reaction monitoring transition signal (%) is shown on the *y*-axis (*Bet* betaine, *Cho* choline, *Cys* cysteine, *Cysta* cystathionine, *dCho* deuterium labeled choline, *DMG* dimethylglycine, *Hcy* homocyteine, *Met* methionine, *SAH* *S*-adenosylhomocysteine, *SAM* *S*-adenosylmethionine)

$$x_i = \frac{V_s \times x_s \times A}{(A_i - 0.9 \times A) \times V_i}$$

$V_i = 50\ \mu l$

$V_s = 5\ \mu l$

$x_s = 750$ ng/mL

control $A_1 = 121047$

$A_{1i} = 185892$

2mM Hcy $A_2 = 652861$

$A_{2i} = 704718$

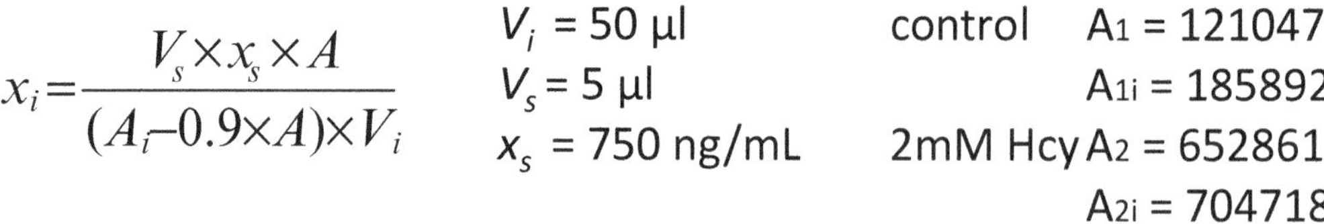

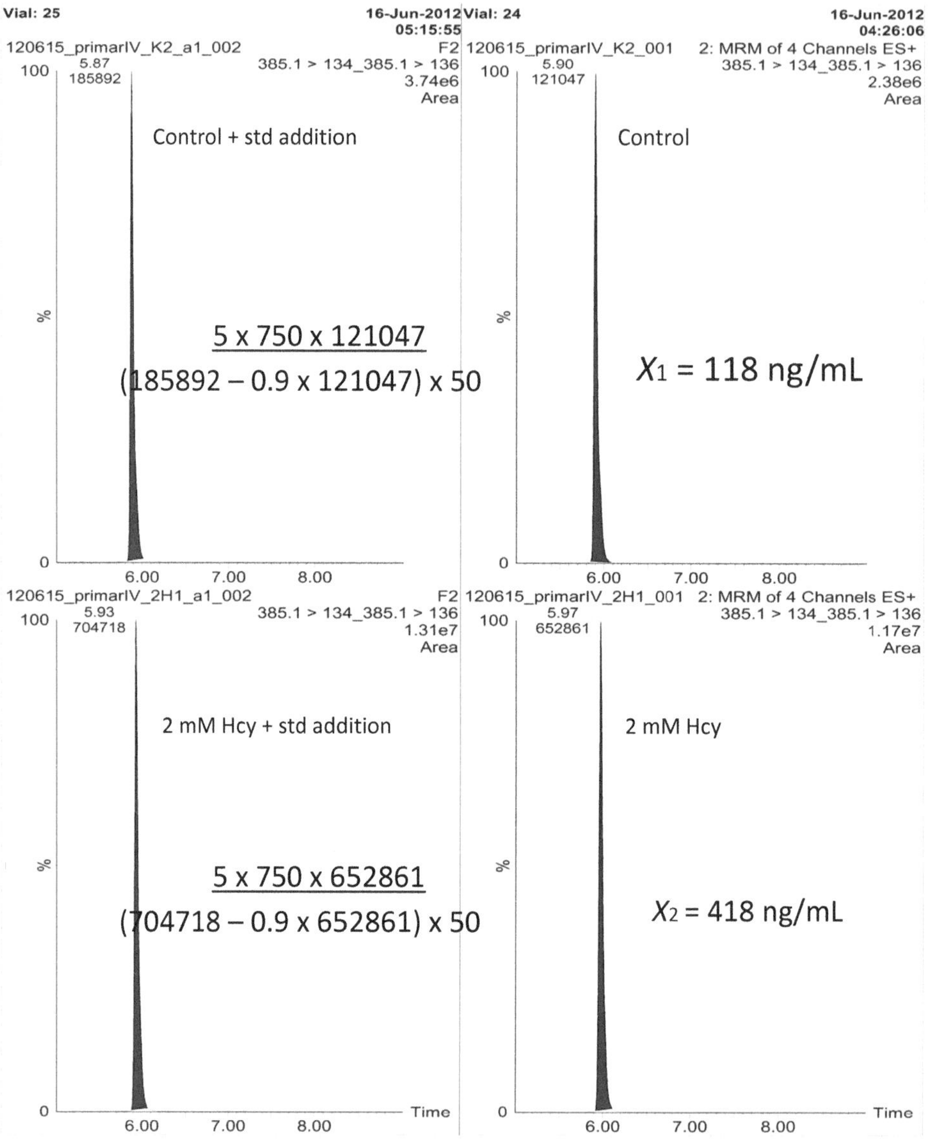

Fig. 2 Calculation of *S*-adenosylhomocysteine (SAH) concentration from LC-MS/MS data. Representative ion chromatograms of SAH in a control sample with and without standard addition and in a sample treated for 48 h with 2 mM homocysteine (Hcy) are shown. Calculations of SAH concentrations in the respective samples are overlaid on the chromatograms, which are shown as original outputs from the Triple Quadrupole Quattro Premier XE mass spectrometer (Waters Micromass, United States of America), including the date of analysis and informal sample name. The retention time of SAH and area of the integrated signal are shown next to the peaks. Monitored *m*/*z* transitions and the value of the highest reachable signal on the chromatogram (i.e., absolute scale) are listed on the *right*. Time expressed in min is shown on the *x*-axis and the height of the MRM transition signal (%) is shown on the *y*-axis (*Bet* betaine, *Cho* choline, *Cys* cysteine, *Cysta* cystathionine, *dCho* deuterium labeled choline, *DMG* dimethylglycine, *Hcy* homocyteine, *Met* methionine, *SAH* *S*-adenosylhomocysteine, *SAM* *S*-adenosylmethionine)

4 Notes

1. We use commercially available human hepatocytes purchased from Biopredic (France). The cells were isolated from liver by a standard 2-step collagenase method and delivered within 2 days after seeding. Upon arrival, the medium is changed and the cells are left to recover for 2 h in a humidified incubator at 37 °C and 5 % CO_2.
2. Prepare stock solution of the toxicant or test compounds in an appropriate concentration so that it can be diluted with a substantial excess of cultivation medium to reach a working concentration.

   We prepare a 200 mM stock solution of Hcy (i.e., 10 mg Hcy dissolved in 370 μL ultrapure water with 0.1 % TFA). The stock solution is diluted 1/100 (i.e., 10 μL per 1 mL) in cultivation medium to reach a final Hcy concentration of 2 mM and 1/2,000 (i.e., 0.5 μL per 1 mL) to reach a final concentration of 0.1 mM. The final volumes of the solutions depend on the number of replicate samples intended for the specific experimental design. Use 2 mL medium per well in the 6-well plate. Prepare approximately 5 % extra volume to allow for pipetting error.
3. An Aquasil C18 column is chosen because of its compatibility with 100 % aqueous mobile phases and its sufficient retention of polar metabolites.
4. The cells are lyzed in 90 % aqueous methanol containing 0.03 % TFA to stabilize SAM and 1.3 mM DTT to prevent thiol oxidation. We use a 90 % methanol concentration because protein precipitation is still efficient, while partition of highly hydrophilic metabolites into the protein fraction does not seriously affect metabolite recoveries. Use of radiolabeled standards is common in recently developed LC-MS/MS-based methods for measurement of low molecular weight compounds [8, 7, 10]. This helps to correct for recoveries. However, we have several reasons to use the method of standard addition, although it is in general less precise. Our method enables simultaneous measurement of nine metabolites. Adding radiolabeled standard for each metabolite would not only increase the costs, but also raise the number of multiple reaction monitoring transitions to 18, which would cause a loss of precision. Moreover, an unpredictable matrix effect is the major limitation of this method and can be different for labeled standards.
5. Concentrations of the compounds in the standard addition are estimated during validation of the method with human hepatoma HepG2 cells [12]. The concentrations are adjusted so that a 100 % increase in the peak area of a specific compound

is achieved. However, basal concentrations of some metabolites (i.e., SAM and Cysta) dramatically differ between HepG2 cells and primary human hepatocytes [13]. We changed the standard addition concentrations accordingly. When implementing a method that was validated using a different cell type or switching to a different cell lysis and extraction method, be aware of possible changes in concentrations of the measured compounds or changes in the matrix effect.

6. Hepatocytes can partially detach from the support during cultivation. Nevertheless, they tend to stick together and form a film. The medium can be successfully aspirated without drawing in the cells. If you swirl the cells, centrifuge the medium at $140 \times g$ and room temperature for 5 min, discard the supernatant, disperse the pelleted cells in fresh medium and transfer back to the well.

7. The test compound continues its action on the cells during scraping and centrifugation (i.e., **steps 2–5**). Thus, you must incorporate this time into your schedule and keep it constant for all samples. This procedure is not applicable for short treatments (i.e., less than 30 min). Provided that the freshly prepared hepatocytes are well attached to the support, the procedure can be modified for short treatments. In this case, the cells are not scraped. The medium is aspirated to finish the stimulation, cells are washed in the wells with ice-cold 0.9 % NaCl, the fluid is immediately aspirated and liquid nitrogen is delivered directly to the wells. Cell lysis is then performed in the wells.

8. Calculated concentrations in ng/mL can be expressed as pmol/10^6 cells. Such values are more suitable for interpreting the data in a physiological context. To perform the computation, divide the value in ng/mL by 5 (i.e., the cell extracts were reconstituted in 200 μL). Then, divide this value by the number of millions of cells in the sample, *in casu* 1.8, and by the relative molecular weight of the particular metabolite. For example, the concentrations calculated in Fig. 2 can be expressed as 34 pmol/10^6 cells SAH in the control sample and 120 pmol/10^6 cells SAH in the sample treated with 2 mM Hcy.

Acknowledgements

This work was supported by the Research Project of the Academy of Sciences of the Czech Republic (RVO: 61388963) and by grant from the Grant Agency of the Czech Republic (P207/10/1277).

References

1. Newton BW, Russell WK, Russell DH et al (2009) Liver proteome analysis in a rodent model of alcoholic steatosis. J Proteome Res 8:1663–1671
2. Thomas A, Stevens AP, Klein MS et al (2012) Early changes in the liver-soluble proteome from mice fed a nonalcoholic steatohepatitis inducing diet. Proteomics 12:1437–1451
3. Qureshi K, Abrams GA (2007) Metabolic liver disease of obesity and role of adipose tissue in the pathogenesis of nonalcoholic fatty liver disease. World J Gastroenterol 13: 3540–3553
4. Conde-Vancells J, Rodriguez-Suarez E, Embade N et al (2008) Characterization and comprehensive proteome profiling of exosomes secreted by hepatocytes. J Proteome Res 7:5157–5166
5. Blom HJ, Smulders Y (2011) Overview of homocysteine and folate metabolism with special references to cardiovascular disease and neural tube defects. J Inherit Metab Dis 34:75–81
6. Godat E, Madalinski G, Muller L et al (2010) Mass spectrometry-based methods for the determination of sulfur and related metabolite concentrations in cell extracts. In: Cadenas E., Packer L. (eds). Methods Enzymol 473:41–76
7. Rafii M, Elango R, House JD et al (2009) Measurement of homocysteine and related metabolites in human plasma and urine by liquid chromatography electrospray tandem mass spectrometry. J Chromatogr B 877:3282–3291
8. Krijt J, Duta A, Kozich V (2009) Determination of *S*-adenosylmethionine and *S*-adenosylhomocysteine by LC-MS/MS and evaluation of their stability in mice tissues. J Chromatogr B 877:2061–2066
9. Liang XP, Liang QL, Xia JF et al (2009) Simultaneous determination of sixteen metabolites related to neural tube defects in maternal serum by liquid chromatography coupling with electrospray tandem mass spectrometry. Talanta 78:1246–1252
10. Huang Y, Lu Z-Y, Brown KS et al (2007) Quantification of intracellular homocysteine by stable isotope dilution liquid chromatography/tandem mass spectrometry. Biomed Chromatogr 21:107–112
11. Leon Z, Garcia-Canaveras JC, Donato MT et al (2013) Mammalian cell metabolomics: experimental design and sample preparation. Electrophoresis 34:2762–2775
12. Korinek M, Sistek V, Mladkova J et al (2013) Quantification of homocysteine-related metabolites and the role of betaine-homocysteine *S*-methyltransferase in HepG2 cells. Biomed Chromatogr 27:111–121
13. Selicharova I, Korinek M, Demianova Z et al (2013) Effects of hyperhomocysteinemia and betaine-homocysteine *S*-methyltransferase inhibition on hepatocyte metabolites and the proteome. Biochim Biophys Acta 1834:1596–1606
14. Braggio S, Barnaby RJ, Grossi P et al (1996) A strategy for validation of bioanalytical methods. J Pharm Biomed Anal 14:375–388

Chapter 20

Measurement of Cytochrome P450 Enzyme Induction and Inhibition in Human Hepatoma Cells

Robim M. Rodrigues, Joery De Kock, Tatyana Y. Doktorova, Vera Rogiers, and Tamara Vanhaecke

Abstract

Cytochrome P450 enzymes are a diverse group of catalytic enzymes in the liver that are mainly responsible for the biotransformation of organic substances. Cytochrome P450 activity as well as both its induction and inhibition are key factors in drug biotransformation and can be involved in deactivation, activation, detoxification and toxification processes. Thus, the modulation of cytochrome P450 activity is an important parameter when evaluating the potential toxicity of chemical compounds using an in vitro system. The cytochrome P450 3A subfamily proteins are among the most important drug-metabolizing enzymes in human liver and are responsible for about half of all cytochrome P450-dependent drug oxidations. In vitro, these enzymes are active not only in primary human hepatocyte cultures, but also in differentiated human hepatoma HepaRG cells. The present protocol describes the culture of cryopreserved differentiated HepaRG cells and the evaluation of its cytochrome P450 activity upon exposure to a chemical compound using a commercially available luminogenic cytochrome P450 assay. This in vitro model can be used to monitor the induction and inhibition of cytochrome P450 3A following exposure to a particular test compound.

Key words Biotransformation, P450-Glo, HepaRG, CYP induction and inhibition

1 Introduction

Hepatocytes are the most important cells in the body involved in the detoxification of xenobiotics transported by the blood. This biotransformation process allows the liver to convert hydrophobic compounds to more hydrophilic molecules, which can be easier excreted. Nevertheless, biotransformation may also result in the formation of highly reactive intermediates, leading to toxicity. The liver enzymes responsible for biotransformation are typically divided in two groups, namely phase I and phase II biotransformation enzymes [1, 2]. Phase I reactions include hydrolysis, reduction, oxidation, hydration and dethioacetylation, whereas phase II reactions bind covalently endogeneously hydrophilic substances to a reactive chemical group, often a

Mathieu Vinken and Vera Rogiers (eds.), *Protocols in In Vitro Hepatocyte Research*, Methods in Molecular Biology, vol. 1250, DOI 10.1007/978-1-4939-2074-7_20, © Springer Science+Business Media New York 2015

product of the phase I reactions. The most prominent phase I enzymes are the cytochrome P450-dependent enzymes (CYP) and the flavin-containing monooxygenases. However, other phase I enzymes, including alcohol and aldehyde dehydrogenase, aldehyde oxidase, monoamine oxidase, cyclooxygenase and xanthine oxidase, also play an important role in oxidative reactions [3–5]. CYPs catalyze the majority of known oxidative drug-metabolizing reactions. They are therefore important in the bioavailability and clearance rate of drugs. Altered CYP activities may be at the basis of adverse drug–drug interactions [6, 7]. For example, if a drug inhibits the CYPs that metabolize a co-administered drug, the clearance of the latter could become slower and lead to toxic accumulation. On the contrary, if a drug induces the expression of the CYPs that metabolize a co-administered drug, the clearance of the latter could be accelerated and its efficacy reduced. In vitro cell-based methods to test for CYP induction and inhibition are interesting tools, as they are less time-consuming and less expensive than animal-based studies [8]. In this respect, human primary hepatocytes are the gold standard. However, due to their limited availability and short lifespan, alternative cellular models are increasingly being used [9]. Immortalized human liver cell lines, including HepaRG cells, have become popular in vitro models. HepaRG cells are readily available and possess sufficient metabolic capacity to be used as surrogates in liver biotransformation studies [10, 11]. They display CYP1A2, CYP2C9, and CYP3A4 activities [12]. In man, CYP3A is the most important drug-metabolizing CYP family, as it is responsible for about half of all CYP-dependent drug oxidations. The presented protocol describes the use of cryopreserved HepaRG cells as a human in vitro system to evaluate the effects of drugs on CYP3A activity. It relies on the use of a luminogenic CYP3A substrate, namely luciferin-6′-pentafluorobenzyl ether (luciferin-PFBE) [13]. Upon biotransformation by CYP3A, luciferin-PFBE is converted into luciferin, which generates light when combined with a luciferin detection reagent. The normalization of the data relies on the cell number and cell viability and is evaluated by another bioluminescence reaction in which the levels of adenosine-5′-triphosphate (ATP), the basic energy source of living cells, is measured. The same protocol can also be applied to evaluate CYP1A2 and CYP2C9 activities by using an alternative set of substrates.

2 Materials

2.1 General Equipment

1. Luminescence plate reader (Thermofisher Scientific, Belgium).
2. White opaque 96-well plates.

2.2 HepaRG Cell Culture

1. Cryopreserved differentiated HepaRG cells (Biopredic International, France).
2. Basal hepatic cell medium (Biopredic International, France).
3. Additives for thaw, seed and general purpose HepaRG medium (Biopredic International, France).
4. Additives for HepaRG induction medium (Biopredic International, France).
5. Collagen-coated 96-well plates (Biopredic International, France).
6. Incubator (37 ± 1 °C, 90 ± 5 % humidity, 5 ± 1 % CO_2).

2.3 P450-Glo assay

2.3.1 Luciferin Detection Reagent (Promega, Belgium)

1. Reconstitution buffer (i.e. reagent 1).
2. Luciferin detection reagent (i.e. reagent 2). Transfer the contents of one bottle of reconstitution buffer to the bottle containing the lyophilized luciferin detection reagent. Mix by swirling or inverting several times to obtain a homogeneous solution. Store at room temperature until ready to use (*see* **Note 1**).
3. Luciferin-PFBE substrate. Thaw the substrate solutions and keep on ice protected from light. Store unused substrate at or below −20 °C and protected from light.

2.3.2 CellTiter Reagent Mixture (Promega, Belgium)

1. CellTiter-Glo buffer. Thaw and equilibrate to room temperature *prior* to use. For convenience, the CellTiter-Glo buffer may be thawed and stored at room temperature for up to 48 h *prior* to use.
2. CellTiter-Glo substrate. Equilibrate to room temperature and transfer the entire liquid volume of the CellTiter-Glo buffer bottle to the CellTiter-Glo substrate vial. Mix by gently mixing, swirling or inverting the contents to obtain a homogeneous solution.

2.4 Phosphate-Buffered Saline (PBS) (See Note 2)

1. Mix 200 mg KCl, 200 mg KH_2PO_4, 2.8 g NaCl, 3.1 g $Na_2HPO_4{\cdot}12H_2O$ and 1,000 mL ultrapure water.
2. Adjust the pH to 7.4.
3. Sterilize by passing through a 0.22 μm filter.
4. Store at 4 °C for up to 6 months.

3 Methods

3.1 HepaRG Cell Seeding

1. Prepare general purpose HepaRG medium by mixing the necessary additives to the basal hepatic cell medium.
2. Pre-warm the HepaRG medium in a waterbath at 37 °C.
3. Add 9 mL per HepaRG cryovial of the HepaRG medium to a sterile 50 mL centrifuge tube.
4. Remove the cryovial from the liquid nitrogen vessel.

5. Twist the cap a quarter turn, but do not open the cryovial completely, to release the internal pressure, and then close it again.
6. Quickly transfer the cryovial to the waterbath at 37 °C without submerging completely.
7. While holding the tip of the cryovial, gently shake the vial for 1–2 min. Small ice crystals should remain when the vial is removed from the waterbath (*see* **Note 3**).
8. Transfer the HepaRG cell suspension into the tube containing 9 mL of the pre-warmed HepaRG medium resulting in a 1:10 ratio of cell suspension to total volume.
9. Rinse the cryovial once with approximately 1 mL of the HepaRG medium and add the resulting suspension to the 50 mL centrifuge tube.
10. Centrifuge the differentiated HepaRG cell suspension for 2 min at 357 × *g* and room temperature.
11. Aspirate the supernatant and resuspend the differentiated HepaRG cell pellet with 5 mL HepaRG medium.
12. Determine the cell seeding densities for each cell recipient using the cell densities provided by the supplier for each independent vial. Do not count the cells (*see* **Note 4**).
13. Seed the cells in 96-multiwell plates at 72,000 cells/well with an automated pipette (i.e. 100 μL per well) (*see* **Note 5**).
14. Once seeded, place the plate in a humidified incubator at 37 °C and 5 % CO_2.
15. Refresh cell culture medium 24 h after seeding and then every 2–3 days.
16. Cryopreserved differentiated HepaRG cells are cultured for 7 days before use for toxicity and metabolism studies.

3.2 CYP3A Assay

3.2.1 Measurement of CYP3A-Activity in 96-Well Plates (See Note 6)

1. Prepare HepaRG induction medium by mixing the necessary additives to the basal hepatic cell medium.
2. Expose the cells to the test compounds solubilized in HepaRG induction medium for 24–72 h (*see* **Note 7**).
3. Ensure that you have equivalent numbers of controls for each test compound.
4. If desired, change the medium and control wells with test compounds once daily for the duration of the treatment time (*see* **Note 8**).
5. Wash the cells with PBS at 37 °C.
6. Replace the culture medium with 100 μL of general purpose HepaRG medium containing 50 μM luciferin-PFBE substrate at 37 °C.
7. Incubate the culture plate at 37 °C and 5 % CO_2 for 4 h.

8. Transfer 50 μL of the medium from each sample well to a 96-well opaque white luminometer assay plate at room temperature. Keep the culture plate for determination of the cell number.
9. Prepare a no-cell control by adding 50 μL fresh general purpose HepaRG medium containing luciferin-PFBE substrate to a number of wells of the 96-well opaque white luminometer assay plate.
10. Add 50 μL of luciferin detection reagent to initiate the luminescent reaction.
11. Shake the assay plate gently on an orbital shaker.
12. Incubate the assay plate at room temperature for 30 min.
13. Read luminescence with integration 1 s.

3.2.2 Measurement of Cell Number in 96-Well Plates

1. Equilibrate the culture plate at room temperature for 30 min.
2. Remove all remaining medium of the wells.
3. Add 50 μL CellTiter-Glo reagent to each sample well.
4. Mix the content for 2 min on an orbital shaker to induce cell lysis.
5. Transfer 50 μL supernatant from each sample well to the 96-well opaque white luminometer assay plate.
6. Prepare a no-cell control by adding 25 μL fresh maintenance medium and 25 μL CellTiter-Glo reagent to a number of wells of the 96-well opaque white luminometer assay plate.
7. Incubate the plate for 10 min at room temperature.
8. Read luminescence with integration 1 s.

3.3 Processing of the Results (See Note 9)

1. Calculate net signals for CYP3A and CellTiter assay by subtracting background luminescence values (i.e. no-cell control) from sample values.
2. Calculate normalized CYP3A signals by dividing CYP3A net signals by the respective CellTiter-Glo net signals.
3. Calculate fold changes by dividing normalized CYP3A net values of the exposed samples by normalized untreated CYP3A net values of the control samples.

4 Notes

1. The reconstituted Luciferin detection reagent can be stored at room temperature for 24 h or at 4 °C for 1 week without loss of activity. For long-term storage, storage at −20 °C is recommended. Be sure to mix the thawed luciferin detection reagent well before use.
2. Commercially available, sterile, cell culture-grade PBS may also be used.

3. Do not repeatedly remove the vial from the waterbath during thawing process, as this can result in low cell viability.
4. Thawed differentiated HepaRG cells consist mainly of small cell clusters and not of single cells. This makes counting difficult. Disruption of the cell clusters to obtain a precise cell count may damage the cells and can result in lower cell viability. We suggest using the cell densities provided by the supplier for each vial. Typically, each vial contains 5 or 8 million living cells. Alternatively, a small volume of the cell suspension can be enzymatically disrupted into a single cell suspension which can be easier counted. This procedure, however, is aggressive and results in a cell suspension with a lower viability.
5. Alternatively, cells can be seeded in 24-well plates.
6. This protocol can also be applied to evaluate CYP1A2 and CYP2C9 activities. If so, kits V8751 and V8791 from Promega (Belgium) should be used.
7. If applicable, before starting the experiments, subcytotoxic concentrations of the compounds to be tested should be determined for the duration of the exposure time.
8. Phenobarbital can be used as a prototypical CYP3A4 inducer. Exposure to a concentration of 500 μM for 24 h shows a threefold increase of CYP3A4 activity. Ketoconazole at a concentration of 0.2 μM will inhibit the CYP3A4 activity.
9. From a statistical point of view, it is strongly recommended to perform the experiments on cells of at least three different experiments. Results can be processed and evaluated by 1-way analysis of variance (i.e. repeated measures) followed by post hoc Bonferroni tests.

Acknowledgements

This work has received funding from grants of the Fund for Scientific Research in Flanders (FWO-Vlaanderen, GO16312N), the European Community's Seventh Framework Programme (FP7/2007–2013) under grant agreement no. 266838 (DETECTIVE) and no. 266777 (HEMIBIO) and from the Brussels research fund INNOVIRIS (Brustem).

References

1. Pelkonen O (2002) Human CYPs: *in vivo* and clinical aspects. Drug Metab Rev 34:37–46
2. Wrighton S, Stevens J (1992) The human hepatic cytochromes P450 involved in drug metabolism. Crit Rev Toxicol 22:1–21
3. Gibson GG, Skett P (2001) Introduction to drug metabolism. Nelson and Thornes, Cheltenham, UK
4. Rettie AE, Fisher MB (1999) Transformation enzymes: oxidative; non P450. In: Woolf TF

(ed) Handbook of drug metabolism. Marcel Dekker, New York, NY, pp 131–145

5. Benedetti MS (2001) Biotransformation of xenobiotics by amine oxidases. Fundam Clin Pharmacol 15:75–84
6. Zlokarnik G, Grootenhuis P, Watson J (2005) High-throughput P450 screens in early drug discovery. Drug Discov Today 10:1443–1450
7. Weinkers L, Heath G (2005) Predicting *in vivo* drug interactions from *in vitro* drug discovery data. Nat Rev Drug Discov 4:825–833
8. Vermier M, Annaert P, Mamidi R et al (2005) Cell-based models to study hepatic drug metabolism and enzyme induction in humans. Expert Opin Drug Metab Toxicol 1:75–90
9. Guguen-Guillouzo C, Guillouzo A (2010) General review on *in vitro* hepatocyte models and their applications. Methods Mol Biol 640:1–40
10. Guillouzo A, Corlu A, Aninat C et al (2007) The human hepatoma HepaRG cells: a highly differentiated model for studies of liver metabolism and toxicity of xenobiotics. Chem Biol Interact 168:66–73
11. Aninat C, Piton A, Glaise D et al (2006) Expression of cytochromes P450, conjugating enzymes and nuclear receptors in human hepatoma HepaRG cells. Drug Metab Dispos 34:75–83
12. Rodrigues RM, Bouhifd M, Bories G et al (2013) Assessment of an automated *in vitro* basal cytotoxicity test system based on metabolically-competent cells. Toxicol in Vitro 27:760–767
13. Cali JJ, Ma D, Sobol M et al (2006) Luminogenic cytochrome P450 assays. Expert Opin Drug Metab Toxicol 2:629–645

Chapter 21

Analysis of Sinusoidal Drug Uptake Transporter Activities in Primary Human Hepatocytes

Marc Le Vée, Elodie Jouan, Claire Denizot, Yannick Parmentier, and Olivier Fardel

Abstract

Hepatic drug transporters play an important role in pharmacokinetics and drug–drug interactions. Among these membrane transporters, the sodium taurocholate cotransporting polypeptide (NTCP/SLC10A1), the organic anion transporting polypeptides (OATPs) 1B1 (SLCO1B1), 1B3 (SLCO1B3) and 2B1 (SLCO2B1), the organic anion transporter 2 (OAT2/SLC22A7) and the organic cation transporter 1 (OCT1/SLC22A1) are likely major ones, notably mediating sinusoidal uptake of various drugs or endogenous compounds, like bile acids, from blood into hepatocytes. Studying putative interactions of drugs, including those in development processes, with these transporters is an important issue. For this purpose, cultured human hepatocytes, that exhibit functional expression of NTCP, OATPs, OAT2 and OCT1, are considered as a relevant in vitro cellular model. This chapter describes a method allowing to accurately analyze NTCP, OATP, OAT2 and OCT1 transport activities in primary human hepatocyte cultures, which can be applied to the determination of potential interactions of drugs with these hepatic uptake transporters.

Key words Membrane transporters, Uptake, Drugs, Primary human hepatocytes, Pharmacokinetics

1 Introduction

Hepatic drug transporters are now well recognized as key actors involved in drug disposition and drug–drug interactions [1, 2]. These membrane transporters belong either to the solute carrier (SLC) transporter family or to the adenosine triphosphate-binding cassette (ABC) transporter family [3]. They are expressed at the sinusoidal pole of hepatocytes, where they mediate the uptake of drugs from blood into hepatocytes (i.e. the so-called phase 0 of the hepatic detoxifying system) or at the canalicular pole of hepatocytes, where they secrete drugs or drug metabolites into the bile (i.e. the so-called phase 3a of the hepatic detoxifying system) [4]. Additionally, some sinusoidal ABC transporters can secrete drug metabolites back into the blood (i.e. the so-called phase 3b for a secondary renal elimination) [5].

Mathieu Vinken and Vera Rogiers (eds.), *Protocols in In Vitro Hepatocyte Research*, Methods in Molecular Biology, vol. 1250, DOI 10.1007/978-1-4939-2074-7_21, © Springer Science+Business Media New York 2015

Among hepatic SLC uptake drug transporters, the sodium taurocholate cotransporting polypeptide (NTCP/SLC10A1), the organic anion transporting polypeptides (OATPs) 1B1 (SLCO1B1), 1B3 (SLCO1B3) and 2B1 (SLCO2B1), the organic anion transporter 2 (OAT2/SLC22A7) and the organic cation transporter 1 (OCT1/SLC22A1) are major ones. Indeed, NTCP mediates sodium-dependent influx of bile acids from blood into hepatocytes [6] and, in this way, plays a major role in the enterohepatic circulation of bile salts [7]. NTCP also handles some drugs, like pitavastatin and rosuvastatin [8]. Various compounds, such as cyclosporine A, inhibit NTCP activity, which may perturb bile salt transport and may cause cholestasis [9]. OATPs transport many drugs, including statins, antibiotics, the antihistaminic drug fexofenadine and the antihypertensive agent bosentan, and are considered as key determinants of hepatic clearance of such compounds [10]. Known genetic polymorphisms of these OATP transporters have therefore clinically relevant functional consequences for pharmacokinetics and pharmacodynamics [11, 12]. In addition, hepatic OATPs have been involved in various clinical drug–drug interactions, especially for statins, whose OATPs-mediated liver uptake can be inhibited by several co-administered drugs, like cyclosporine A or gemfibrozil [13], leading to increased blood statin concentrations and therefore putatively contributing to subsequent toxicity, such as rhabdomyolysis [14]. OAT2 is involved in the transportation of various marketed drugs, including salicylates, the antibiotic erythromycin, the antiviral agent zidovudin and the loop diuretic bumetanide [15]. OAT2 can be inhibited by a wide range of compounds, especially cephalosporins, probenecid and non-steroidal anti-inflammatory drugs. However, no clinical drug–drug interaction involving hepatic OAT2 has yet been reported [16]. OCT1 handles several cationic drugs, like the oral antidiabetic drug metformine and the antihistaminic compound cimetidine [17]. As for OATPs, genetic polymorphisms of OCT1 as well as inhibition of OCT1 activity by various compounds are thought to impair pharmacokinetics and therapeutic efficacy of drug substrates [18].

Due to the relevance of transporters in drug disposition and drug–drug interactions, regulatory drug agencies, like the US Food and Drug Agency and the European Medicine Agency, have recently established guidelines for the characterization of interactions of new molecular entities (NMEs) with some transporters, including hepatic OATP1B1 and OATP1B3, during drug development processes in pharmaceutical companies [19, 20]. OCT1 may also deserve interest, at least according to the European Medicine Agency. NTCP and OAT2, although not formally cited in regulatory agency guidelines, should ideally also be considered, given the major role played by NTCP in enterohepatic circulation of bile salts and because of the fact that OAT2 dictates transportation of various marketed drugs.

Cellular in vitro models are useful for studying putative interactions of drugs, including NMEs, with the hepatic human uptake transporters described above [21, 22]. Such models include transporter-transfected cells, cultured human hepatoma cells, and cultured human hepatocytes [23–25]. Primary human hepatocytes that express drug transporters, including NTCP, OATPs, OAT2 and OCT1, in a more physiological manner than transporter-transfected cells or human hepatoma cells, represent a first choice in vitro system for studying transporter activities [26]. In the present chapter, we describe an established method for measuring transport activity of NTCP, OATPs, OAT2 and OCT1 in primary human hepatocytes [26], using reference substrates, such as taurocholic acid (TC) for NTCP [6], estrone-3-sulfate (ES) for OATPs [27], *para*-aminohippuric acid (PAH) for OAT2 [15] and tetraethylammonium (TEA) for OCT1 [18], as well as reference inhibitors, namely probenecid for OATPs [28] and OAT2 [29], and verapamil for OCT1 [18]. Applications of this method to the determination of potential inhibitory effects of NMEs toward these uptake transporters and potential contribution of the transporters to NME uptake into hepatocytes are proposed.

2 Materials

2.1 Primary Human Hepatocyte Cultures

1. Suspension of freshly isolated human hepatocytes (*see* **Note 1**).
2. Hepatocyte seeding medium. William's medium E, supplemented with 5 μg/mL bovine pancreatic insulin, 200 mM glutamine, 10 UI/mL penicillin, 10 μg/mL streptomycin and 10 % fetal calf serum. Prepare in a laminar air flow cabinet and store for maximum 7 days at 4 °C. *Prior* to use, the hepatocyte seeding medium should be placed for 30 min in a thermostated bath at 37 °C.
3. Incubator (37 °C ± 1 °C, 90 % ± 5 % humidity, 5 % ± 1 % CO_2).
4. Plastic 24-well cell culture plates (*see* **Note 2**).
5. Phase-contrast microscope (Olympus, France).

2.2 Uptake Transport Assays

1. Radiolabeled substrates (Perkin Elmer, France). [1-^{14}C]-TEA bromide (specific activity 1–5 mCi/mmol), [6,7-^{3}H(N)]-ES ammonium salt (specific activity: 40–60 Ci/mmol), [Glycyl-2-^{3}H]-PAH (specific activity 1–5 Ci/mmol) and [^{3}H(G)]-TC (specific activity at least 3 Ci/mmol).
2. Transport assay medium. 5.3 mM KCl, 1.1 mM KH_2PO_4, 0.8 mM $MgSO_4$, 1.8 mM $CaCl_2$, 11 mM d-glucose, 10 mM 4-(2-hydroxyethyl)-1-piperazineethanesulfonic acid and 136 mM *N*-methyl glucamine (i.e. sodium-free buffer) or 136 mM NaCl (i.e. sodium-containing standard buffer) in

deionized water adjusted to pH 7.4 [26] and stored at 4 °C for maximum 2 weeks before use.

3. Phosphate-buffered saline (PBS) 137 mM NaCl, 2.7 mM KCl, 10 mM Na_2HPO_4, 1.8 mM KH_2PO_4 in deionized water, adjusted to pH 7.4 and stored for maximum 6 months at 4 °C.
4. 0.01 N NaOH solution in deionized water.
5. 50 mM verapamil in dimethylsulfoxide. This stock solution can be stored at −20 °C for maximum 12 months before use.
6. 1 M probenecid in 2 N NaOH solution in deionized water. This stock solution can be stored at −20 °C for maximum 12 months before use.
7. Scintillation counting vials (Sarstedt, France).
8. Liquid scintillation cocktail Ultima-Gold (Perkin Elmer, France).
9. Liquid scintillation counter LS-6500 (Beckman-Coulter, France).

2.3 Protein Determination

1. Bovine serum albumin. Prepare a solution of 1 mg/mL bovine serum albumin in deionized water and store it at −20 °C for maximum 12 months before use.
2. Bradford reagent (Sigma-Aldrich, France).
3. Non-sterile polystyrene 96-well flat bottom plates (Greiner Bio-One, France).
4. Spectrostar Nano absorbance microplate reader (BMG Labtech, France).

3 Methods

3.1 Establishment of a Monolayer Culture of Human Hepatocytes

1. Prepare a suspension of freshly isolated human hepatocytes at a final concentration of 500,000 cells/mL in hepatocyte seeding medium.
2. Plate human hepatocytes in 24-well cell culture plates at a final concentration of 250,000 cells per well, corresponding to 0.5 mL/well of human hepatocyte suspension (*see* **Note 3**).
3. Place the cell cultures in an incubator at 37 °C and 5 % CO_2 for 24 h.
4. Remove the 24-well plates from the incubator and carefully examine the culture wells with a phase-contrast microscope to verify that human hepatocytes have adhered to the plastic culture surface and form a monolayer culture (*see* **Note 4**).

3.2 Uptake Drug Transport Assays

1. Prepare the radiolabeled substrate incubation solutions (*see* **Note 5**) for the NTCP activity assay (i.e. 43.4 nM $[^3H(G)]$-TC in sodium-containing and sodium-free transport assay buffers), the OATP activity assay (i.e. 3.4 nM

[6,7-^{3}H(N)]-ES in sodium-free transport assay buffer (*see* **Note 6**) whether or not supplemented with the OATP inhibitor probenecid used at 2 mM), the OAT2 activity (i.e. 200 nM [Glycyl-2-^{3}H]-PAH in sodium-containing transport assay buffer whether or not supplemented with the OAT2 inhibitor probenecid used at 10 mM) and the OCT1 activity assay (i.e. 40 μM [1-^{14}C]-TEA in sodium-containing transport assay buffer whether or not supplemented with the OCT1 inhibitor verapamil used at 50 μM). To determine the potential effect of an NME on transporter activities, prepare additional radiolabeled substrate solutions supplemented with the NME at various concentrations (*see* **Note 7**). To investigate the potential contribution of influx transporters to NME uptake in hepatocytes, prepare radiolabeled NME solutions (*see* **Notes 7** and **8**) in sodium-containing transport assay buffer whether or not supplemented with 2 mM probenecid (*see* **Note 9**) or 50 μM verapamil and in sodium-free transport assay buffer.

2. Place the 24-well primary hepatocyte plates on ice.
3. Remove the culture medium from the well and replace by ice-cold PBS (i.e. 0.5 mL/well).
4. Remove the PBS and add the different solutions of radiolabeled substrates described above (i.e. 0.5 mL radiolabeled substrate solution/well) (*see* **Note 10**) or, for wells for protein determination and background scintillation counting, add substrate-free sodium-containing transport buffer (i.e. 0.5 mL/well).
5. Incubate the 24-well plates for 10 min at 37 °C in a thermostated bath (*see* **Note 11**).
6. Place the 24 well plates on ice.
7. Remove the radiolabeled substrate solution and substrate-free solution from each well.
8. Add 0.5 mL ice-cold PBS to each well.
9. Remove the PBS and add again 0.5 mL ice-cold PBS to each well.
10. Remove the PBS, add 0.15 mL 0.01 N NaOH to each well and keep the 24-well plates at least for 5 min at room temperature (*see* **Note 12**).
11. Homogenize the cellular lysates through vigorous pipetting.
12. Collect cell lysates and add to scintillation counting vials, except those from wells dedicated to protein determination.
13. Add 2 mL liquid scintillation cocktail to each counting vial.
14. Proceed to radioactivity count of the vials using the scintillation counter. Results are expressed as count per min (cpm).

15. Subtract the background cpm value mean obtained from control wells not incubated with radiolabeled substrates from each cpm value of the wells incubated with radiolabeled substrates. This gives radiolabeled substrate-related cpm for each well.

3.3 Protein Determination

1. Prepare a series of bovine albumin protein standards diluted in 1 mL Bradford reagent to final concentrations of 0 (i.e. blank with Bradford reagent only), 2, 5, 10, 15 and 20 μg/mL using the 1 mg/mL bovine albumin stock solution.
2. Collect 15 μL of each cell lysate from wells dedicated to protein determination and add this lysate volume to 1 mL of Bradford reagent in plastic tubes. Mix the samples.
3. Deposit 200 μL of each Bradford/bovine albumin standard and Bradford/cell lysate mix samples in 96-well microplates.
4. Read the spectrophotometric absorbance at 595 nm using an absorbance microplate reader.
5. Plot the absorbance of the standards versus their concentration. Compute the extinction coefficient and calculate the concentrations of the cell lysate samples (*see* **Note 13**). Deduce the amount of protein/well and express as mg/well.

3.4 Calculations

1. Convert radiolabeled substrate-related cpm values for each well to disintegration per min (dpm) values according to the following equation:
   substrate dpm/well = substrate cpm/well/efficiency factor (*see* **Note 14**).
2. Transform substrate dpm values per well to Ci per well according to the following equation:
   substrate Ci/well = substrate dpm/well/2.22×10^{12}.
3. Transform substrate Ci per well to mol per well according to the following equation:
   substrate mol/well = substrate Ci/well/specific activity (*see* **Note 15**).
4. Convert values of substrate mol per well to mol per mg protein according to the following equation:
   substrate mol/mg protein = substrate mol/well/amount of protein/well.
5. NTCP activity corresponds with TC accumulation in the presence of sodium (TC^{+Na}) *minus* TC accumulation in the absence of sodium (TC^{-Na}) [26] (Fig. 1) (*see* **Note 16**).
6. Determination of the TC accumulation in the presence of 1 concentration of an NME with sodium ($TC^{+Na/+NME}$) or without sodium ($TC^{-Na/+NME}$) allows to calculate the potential inhibitory effect of this NME concentration toward NTCP activity using the following equation:

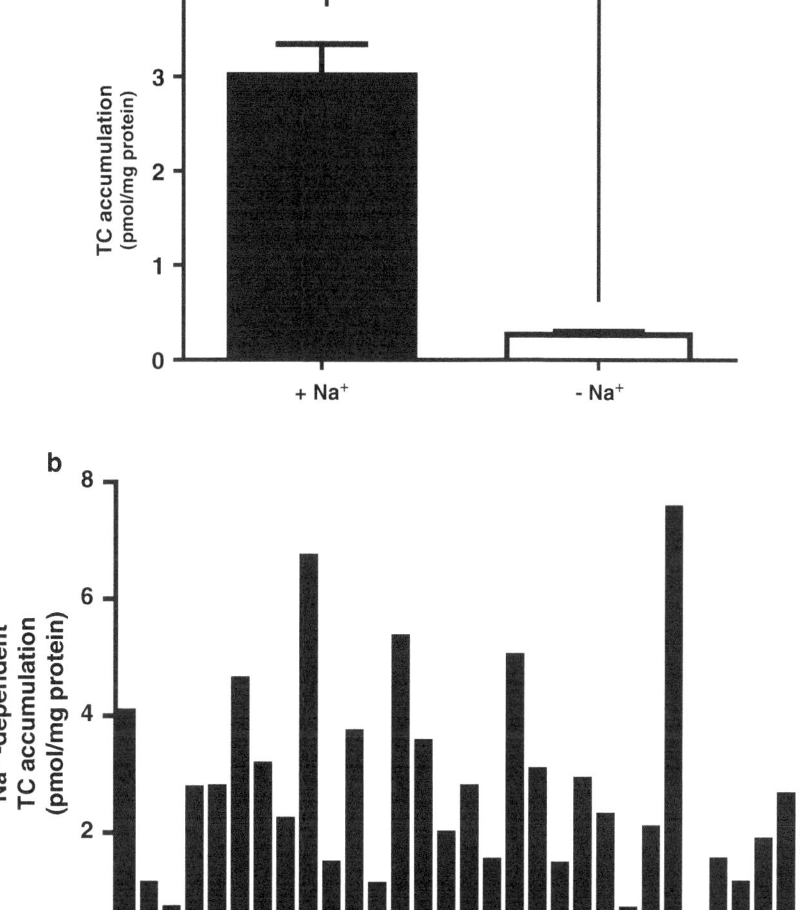

Fig. 1 NTCP-mediated uptake of TC in primary human hepatocytes. Primary human hepatocytes were incubated with 43.4 nM [^{3}H(G)]-TC at 37 °C for 10 min in the presence or absence of sodium. Intracellular accumulation of TC was next determined by scintillation counting. Data are expressed as histograms corresponding to mean ± standard error of TC accumulation in 30 independent human hepatocyte populations (**a**) or to sodium-dependent TC accumulation value, reflecting NTCP activity, in each of the 30 independent hepatocyte populations (**b**) (*$p < 0.0001$; paired Student's *t*-test)

NTCP inhibition $\% = 100 - [(TC^{+Na/+NME} - TC^{-Na/+NME}) \times 100/(TC^{+Na} - TC^{-Na})]$.

7. Determination of the NME uptake in the presence (NME^{+Na}) or absence (NME^{-Na}) of sodium permits to evaluate the putative contribution of NTCP to NME uptake using the following equation:

   NTCP contribution $\% = 100 \times (NME^{+Na} - NME^{-Na})/NME^{+Na}$ (*see* **Notes 17** and **18**).

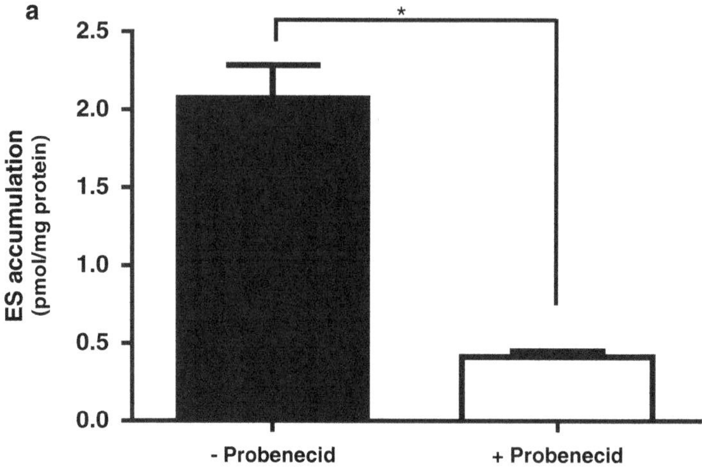

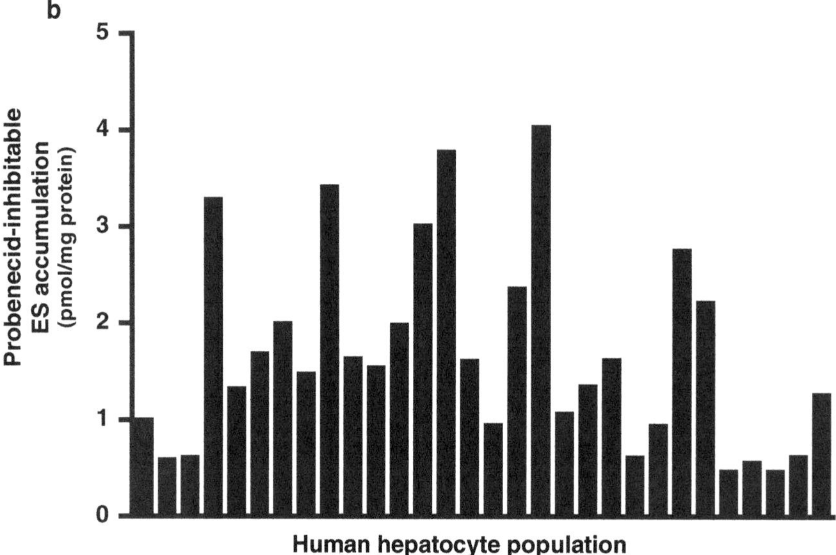

Fig. 2 OATPs-mediated uptake of ES in primary human hepatocytes. Primary human hepatocytes were incubated with 3.4 nM [6,7-^{3}H(N)]-ES at 37 °C for 10 min in the presence or absence of 2 mM probenecid. Intracellular accumulation of ES was next determined by scintillation counting. Data are expressed as histograms corresponding to mean ± standard error of ES accumulation in 30 independent human hepatocyte populations (**a**) or to probenecid-inhibitable ES accumulation value, reflecting OATP activity, in each of the 30 independent hepatocyte populations (**b**) (*$p < 0.0001$; paired Student's *t*-test)

8. OATP activity corresponds with ES accumulation in the absence of probenecid (ES) *minus* ES accumulation in the presence of probenecid ($ES^{+Probenecid}$) [26] (Fig. 2) (*see* **Note 16**).
9. Determination of the ES accumulation in the presence of 1 concentration of an NME (ES^{+NME}) allows to calculate the potential inhibitory effect of this NME concentration towards OATP activity using the following equation:

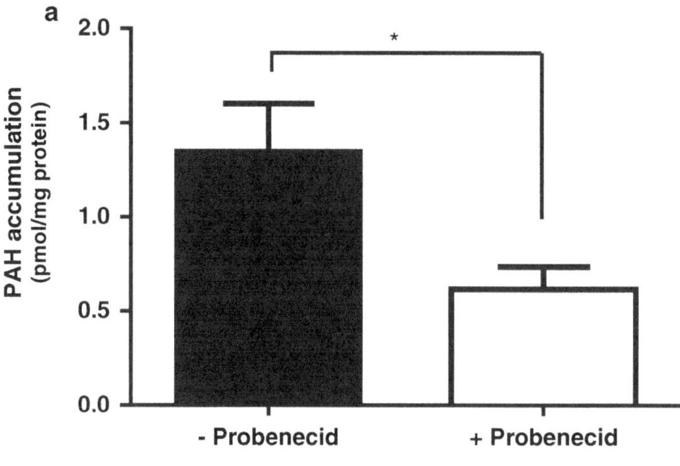

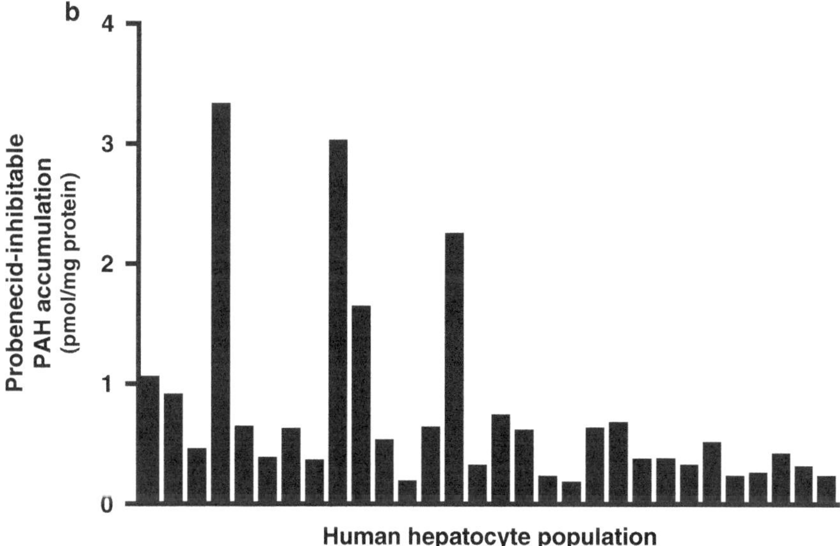

Fig. 3 OAT2-mediated uptake of PAH in primary human hepatocytes. Primary human hepatocytes were incubated with 200 nM [Glycyl-2-^{3}H]-PAH at 37 °C for 10 min in the presence or absence of 10 mM probenecid. Intracellular accumulation of PAH was next determined by scintillation counting. Data are expressed as histograms corresponding to mean ± standard error of PAH accumulation in 30 independent human hepatocyte populations (**a**) or to probenecid-inhibitable PAH accumulation value, reflecting OAT2 activity, in each of the 30 independent hepatocyte populations (**b**) (*$p < 0.0001$; paired Student's *t*-test)

OATP inhibition $\% = (ES - ES^{+NME}) \times 100 / (ES - ES^{+Probenecid})$.

10. OAT2 activity corresponds with PAH accumulation in the absence of probenecid (PAH) *minus* PAH accumulation in the presence of probenecid ($PAH^{+Probenecid}$) [26] (Fig. 3) (*see* **Note 16**).

11. Determination of PAH accumulation in the presence of 1 concentration of an NME (PAH^{+NME}) allows to calculate the

potential inhibitory effect of this NME concentration towards OAT2 activity using the following equation:

OAT2 inhibition $\% = (\text{PAH} - \text{PAH}^{+\text{NME}}) \times 100 / (\text{PAH} - \text{PAH}^{+\text{Probenecid}})$.

12. Determination of NME uptake in the absence (NME) or presence of probenecid ($\text{NME}^{+\text{Probenecid}}$) permits to investigate the possible contribution of OATPs or OAT2 to NME uptake using the following equation:

    OATP/OAT2 contribution $\% = 100 \times (\text{NME} - \text{NME}^{+\text{Probenecid}})/\text{NME}$ (*see* **Notes 17** and **19**).
13. OCT1 activity corresponds with TEA accumulation in the absence of verapamil (TEA) *minus* TEA accumulation in the presence of verapamil ($\text{TEA}^{+\text{Verapamil}}$) (Fig. 4) [26] (*see* **Note 16**).
14. Determination of TEA accumulation in the presence of 1 concentration of an NME ($\text{TEA}^{+\text{NME}}$) allows to calculate the potential inhibitory effect of this NME concentration towards OCT1 activity using the following equation:

    OCT1 inhibition $\% = (\text{TEA} - \text{TEA}^{+\text{NME}}) \times 100 / (\text{TEA} - \text{TEA}^{+\text{verapamil}})$.
15. Determination of NME uptake in the absence (NME) or presence of verapamil ($\text{NME}^{+\text{Verapamil}}$) permits to investigate the possible contribution of OCT1 to NME uptake using the following equation:

    OCT1 contribution $\% = 100 \times (\text{NME} - \text{NME}^{+\text{Verapamil}})/\text{NME}$ (*see* **Note 17**).
16. NME concentration inhibiting 50 % of NTCP, OATP, OAT2 or OCT1 transporter activity (i.e. IC_{50}) can be determined through nonlinear regression analysis of the curve percentage of transporter inhibition versus log([NME]) using a 4-parameter logistic equation (*see* **Note 20**).

4 Notes

1. Human hepatocytes are isolated from liver tissue or fragments obtained from patients undergoing resection for removal of liver tumors or from resected tissue from whole livers obtained from multi-organ donors, unused or rejected for transplantation, using a 2-step collagenase perfusion technique [30–34]. Ethical approvals and signed consent forms must be obtained *prior* to processing of any tissues, and the appropriate rules and regulations for human tissue processing, cell handling and storage must be followed. Freshly isolated human hepatocyte suspensions can be purchased from some commercial distributors. This also holds for plateable human cryopreserved hepatocytes or ready-to-use 24-well plated human hepatocytes,

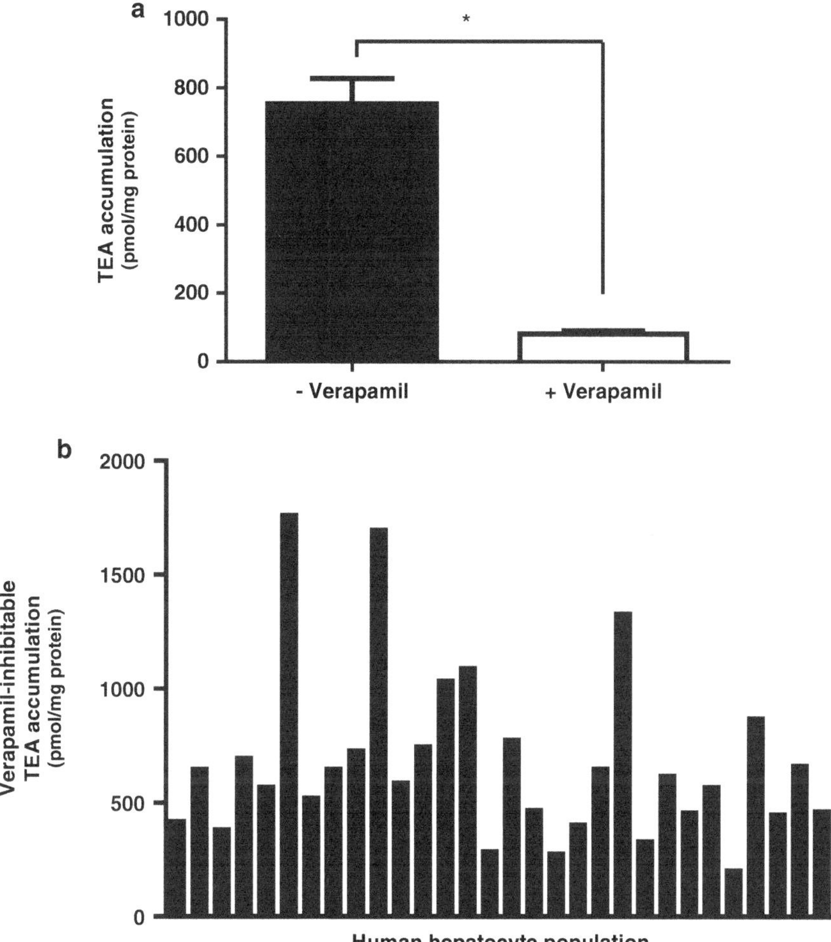

Fig. 4 OCT1-mediated uptake of TEA in primary human hepatocytes. Primary human hepatocytes were incubated with 40 μM [1-^{14}C]-TEA at 37 °C for 10 min in the presence or absence of 50 μM verapamil. Intracellular accumulation of TEA was next determined by scintillation counting. Data are expressed as histograms corresponding to mean ± standard error of TEA accumulation in 30 independent human hepatocyte populations (**a**) or to verapamil-inhibitable TEA accumulation value, reflecting OCT1 activity, in each of the 30 independent hepatocyte populations (**b**) (*$p < 0.0001$; paired Student's *t*-test)

which can be used as an alternative to freshly isolated human hepatocyte suspensions.

2. Collagen-coated 24-well plates (BD Biosciences, France) may be used instead of conventional plastic 24-well plates in order to improve attachment of hepatocytes.

3. The total number of plated wells depends on the numbers of required replicates. Additional wells for protein determination have also to be taken into account.

4. Cultures exhibiting no or only a low proportion of attached hepatocytes (i.e. less than 50 %) after the 24 h period attachment

cannot be used for uptake transport assays. Because sinusoidal transporter expression and activities are rather well-preserved during primary cultures of human hepatocytes [35], transport assays can also be performed using human hepatocytes kept in primary conventional culture up to 5 days after cell seeding [26].

5. All experiments with radiolabeled substrates have to be performed in agreement with safety and legal rules related to the use of radioactivity.
6. ES is a common substrate for OATP1B1, OATP1B3, and OATP2B1 [27]. Because NTCP can also transport ES in a sodium-dependent manner [36], the use of a sodium-free transport buffer is recommended for evaluating OATPs-mediated ES uptake.
7. The concentrations of NMEs may be chosen from pre-existing pharmacokinetics data if available.
8. The use of unlabeled NMEs instead of its radiolabeled form is possible, but requires an adequate analytical method, such as high-performance liquid chromatography or combined liquid chromatography/mass spectrometry, for measuring intracellular NME accumulation in primary human hepatocytes.
9. Rifamycin SV in concentration of 100 μM can also be used as a reference OATP inhibitor [37] in NME uptake experiments.
10. These steps have to be performed quickly in order to minimize potential passive diffusion of substrates across cell membranes.
11. A shorter incubation time may be selected for NME uptake experiments in order to remain in the initial linear phase of influx, which has to be previously characterized through determining intracellular NME uptake in hepatocytes over a 2.5–30 min range of incubation times.
12. An additional step of neutralization through adding 1.5 μL/well of 1 N HCl can be performed.
13. This step may be automatically performed through dedicated software of the microplate reader.
14. Efficiency factors depend on the nature of isotope and on the type of scintillation counter.
15. Use the specific activity value given by the radiolabeled substrate supplier.
16. Activities of NTCP (i.e. sodium-dependent TC accumulation) (Fig. 1), OATPs (i.e. probenecid-inhibitable ES accumulation) (Fig. 2), OAT2 (i.e. probenecid-inhibitable PAH accumulation) (Fig. 3) and OCT1 (i.e. verapamil-inhibitable TEA accumulation) (Fig. 4) have been clearly detected in

each of 30 independent primary human hepatocyte populations. Nevertheless, levels of sodium-dependent TC accumulation, probenecid-inhibitable ES accumulation, probenecid-inhibitable PAH accumulation and verapamil-inhibitable TEA accumulation exhibit some differences according to the hepatocyte populations, which may reflect interindividual variations in transporter expression or activity in primary human hepatocytes, as classically reported for drug detoxifying proteins [38, 39]. Interestingly, NTCP activity was found to correlate with that of OATPs (Fig. 5), suggesting that these transporters share common regulation factors in primary human hepatocytes. Similarly, OCT1 activity is correlated to that of OAT2 (Fig. 5). By contrast, levels of OCT1 activity in cultured human hepatocytes are not correlated to those of NTCP activity or to those of OATP activity, and OAT2 activity is not correlated to that of NTCP or OATPs (Fig. 5).

17. Involvement of a SLC transporter in NME uptake has to be complementary investigated by adequate fitting of the curve initial uptake velocity versus NME concentration in the medium, permitting to discriminate the non-saturable and transporter-unrelated passive part of NME uptake from its saturable and transporter-mediated active part, characterized by the Michaelis–Menten kinetics parameters V_{max} and K_m [40]. The balance of active versus passive uptake of drugs can be additionally evaluated through determination of the temperature-dependent fraction of drug influx, knowing that active transport is temperature-dependent. However, caution has to be taken when interpreting data from temperature-related transport experiments, because permeability across plasma membrane through passive diffusion, generally thought to be temperature-independent, may in fact be influenced by temperature-mediated changes in physico-chemical properties of the plasma membrane [40]. Finally, implication of a specific transporter in NME uptake can be fully confirmed through comparing NME accumulation in a cell clone overexpressing this transporter and in the corresponding control cell clone.

18. Complementary NME uptake experiments in the presence or absence of NTCP inhibitors, like cyclosporine A [9], may additionally be useful to fully establish a putative sodium-dependent NTCP-mediated transport of the NME into primary hepatocytes and to discard any sodium-dependent, but NTCP-unrelated, transport.

19. Because probenecid-mediated inhibition of NME uptake may be caused either by probenecid-mediated inhibition of OATPs or by probenecid-mediated inhibition of OAT2, complementary experiments using additional inhibitors of transporters,

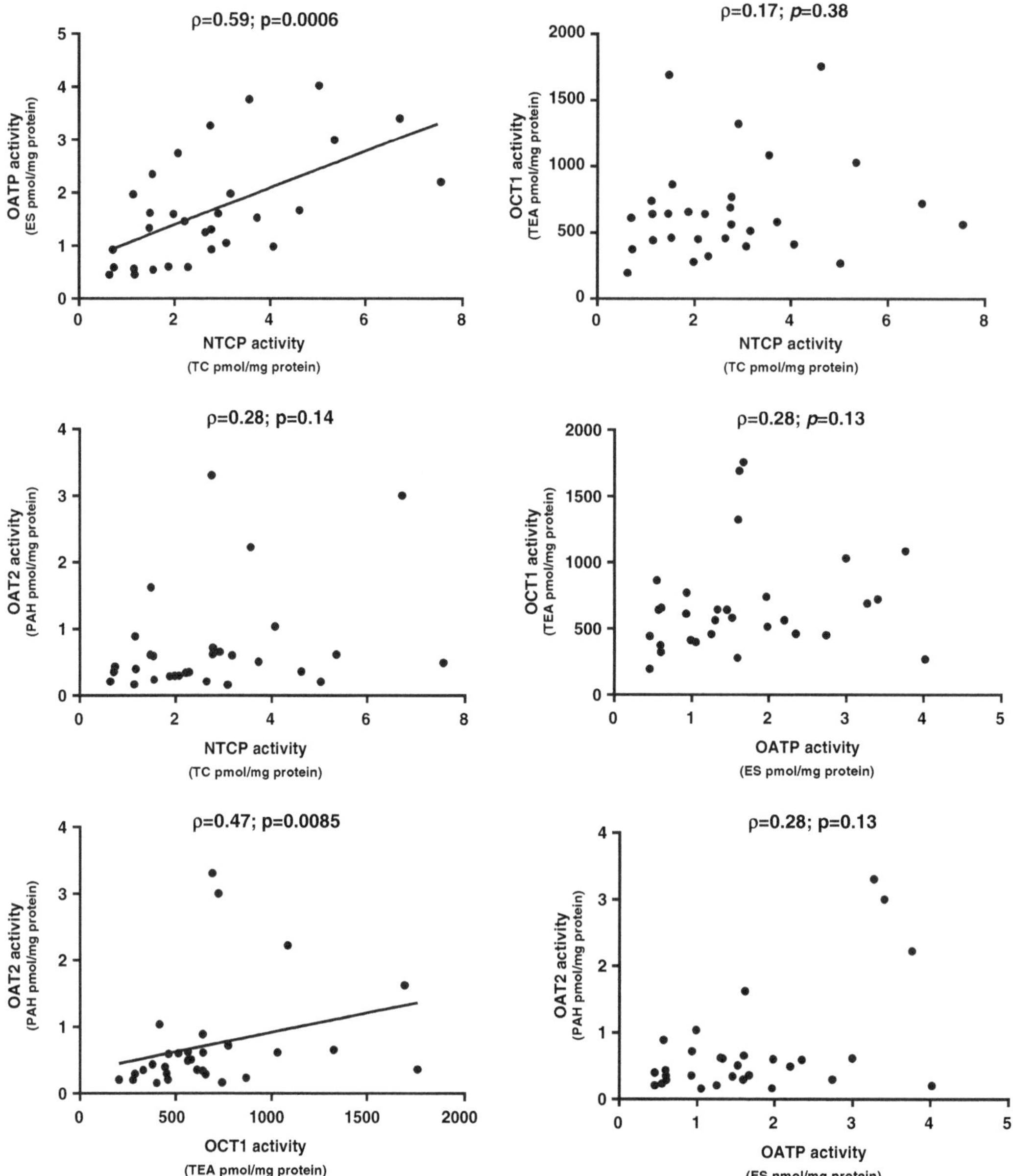

Fig. 5 Correlation analysis of uptake transporter activities in primary human hepatocytes. Correlation analysis between NTCP activity (i.e. sodium-dependent TC accumulation), OATP activity (i.e. probenecid-inhibitable ES accumulation), OAT2 activity (i.e. probenecid-inhibitable PAH accumulation), and OCT1 activity (i.e. verapamil-inhibitable TEA accumulation) in primary human hepatocytes from independent donors ($n = 30$) was performed using the non-parametric Spearman's rank correlation method. Spearman's rank coefficients (ρ) and p values are provided on the top of each correlation graph, knowing that correlation was considered to be statistically significant when $p < 0.05$

such as rifamycin SV inhibiting OATPs or using transfected cell lines, are specifically needed to fully establish and characterize putative OATP- or OAT2-mediated uptake of NMEs in primary human hepatocytes.

20. This nonlinear regression analysis can be performed through dedicated functions of software, such as SigmaPlot (Systat software, United States of America) or Prism (GraphPad software, United States of America).

Acknowledgements

We would like to thank the Biological Resource Center (CHU Rennes) for providing us with human hepatocyte suspensions.

References

1. Giacomini KM, Huang SM, Tweedie DJ et al (2010) Membrane transporters in drug development. Nat Rev Drug Discov 9:215–236
2. Muller F, Fromm MF (2011) Transporter-mediated drug-drug interactions. Pharmacogenomics 12:1017–1037
3. Klaassen CD, Aleksunes LM (2010) Xenobiotic, bile acid, and cholesterol transporters: function and regulation. Pharmacol Rev 62:1–96
4. Funk C (2008) The role of hepatic transporters in drug elimination. Expert Opin Drug Metab Toxicol 4:363–379
5. Pfeifer ND, Hardwick RN, Brouwer KL (2014) Role of hepatic efflux transporters in regulating systemic and hepatocyte exposure to xenobiotics. Annu Rev Pharmacol Toxicol 54:509–535
6. Hagenbuch B, Meier PJ (1994) Molecular cloning, chromosomal localization, and functional characterization of a human liver Na^+/bile acid cotransporter. J Clin Invest 93:1326–1331
7. Kullak-Ublick GA, Stieger B, Meier PJ (2004) Enterohepatic bile salt transporters in normal physiology and liver disease. Gastroenterology 126:322–342
8. Ho RH, Tirona RG, Leake BF et al (2006) Drug and bile acid transporters in rosuvastatin hepatic uptake: function, expression, and pharmacogenetics. Gastroenterology 130:1793–1806
9. Mita S, Suzuki H, Akita H et al (2006) Inhibition of bile acid transport across Na^+/taurocholate cotransporting polypeptide (SLC10A1) and bile salt export pump (ABCB 11)-coexpressing LLC-PK1 cells by cholestasis-inducing drugs. Drug Metab Dispos 34:1575–1581
10. Fenner KS, Jones HM, Ullah M et al (2012) The evolution of the OATP hepatic uptake transport protein family in DMPK sciences: from obscure liver transporters to key determinants of hepatobiliary clearance. Xenobiotica 42:28–45
11. Hagenbuch B, Gui C (2008) Xenobiotic transporters of the human organic anion transporting polypeptides (OATP) family. Xenobiotica 38:778–801
12. Niemi M, Pasanen MK, Neuvonen PJ (2011) Organic anion transporting polypeptide 1B1: a genetically polymorphic transporter of major importance for hepatic drug uptake. Pharmacol Rev 63:157–181
13. Karlgren M, Vildhede A, Norinder U et al (2012) Classification of inhibitors of hepatic organic anion transporting polypeptides (OATPs): influence of protein expression on drug-drug interactions. J Med Chem 55:4740–4763
14. Neuvonen PJ, Niemi M, Backman JT (2006) Drug interactions with lipid-lowering drugs: mechanisms and clinical relevance. Clin Pharmacol Ther 80:565–581
15. Burckhardt G (2012) Drug transport by organic anion transporters (OATs). Pharmacol Ther 136:106–130
16. Burckhardt G, Burckhardt BC (2011) *In vitro* and *in vivo* evidence of the importance of organic anion transporters (OATs) in drug therapy. Handb Exp Pharmacol 201:29–104
17. Koepsell H, Lips K, Volk C (2007) Polyspecific organic cation transporters: structure, function, physiological roles, and biopharmaceutical implications. Pharm Res 24:1227–1251
18. Nies AT, Koepsell H, Damme K et al (2011) Organic cation transporters (OCTs, MATEs), *in vitro* and *in vivo* evidence for the importance in drug therapy. Handb Exp Pharmacol 201:105–167
19. Prueksaritanont T, Chu X, Gibson C et al (2013) Drug-drug interaction studies: regulatory guidance and an industry perspective. AAPS J 15:629–645

20. Giacomini KM, Huang SM (2013) Transporters in drug development and clinical pharmacology. Clin Pharmacol Ther 94:3–9
21. Kindla J, Fromm MF, Konig J (2009) *In vitro* evidence for the role of OATP and OCT uptake transporters in drug-drug interactions. Expert Opin Drug Metab Toxicol 5:489–500
22. Brouwer KL, Keppler D, Hoffmaster KA et al (2013) *In vitro* methods to support transporter evaluation in drug discovery and development. Clin Pharmacol Ther 94:95–112
23. Bi YA, Kimoto E, Sevidal S et al (2012) *In vitro* evaluation of hepatic transporter-mediated clinical drug-drug interactions: hepatocyte model optimization and retrospective investigation. Drug Metab Dispos 40:1085–1092
24. Ramboer E, Vanhaecke T, Rogiers V et al (2013) Primary hepatocyte cultures as prominent *in vitro* tools to study hepatic drug transporters. Drug Metab Rev 45:196–217
25. Le Vee M, Noel G, Jouan E et al (2013) Polarized expression of drug transporters in differentiated human hepatoma HepaRG cells. Toxicol In Vitro 27:1979–1986
26. Jigorel E, Le Vee M, Boursier-Neyret C et al (2005) Functional expression of sinusoidal drug transporters in primary human and rat hepatocytes. Drug Metab Dispos 33:1418–1422
27. Konig J (2011) Uptake transporters of the human OATP family: molecular characteristics, substrates, their role in drug-drug interactions, and functional consequences of polymorphisms. Handb Exp Pharmacol 201:1–28
28. Tahara H, Kusuhara H, Maeda K et al (2006) Inhibition of oat3-mediated renal uptake as a mechanism for drug-drug interaction between fexofenadine and probenecid. Drug Metab Dispos 34:743–747
29. Enomoto A, Takeda M, Shimoda M et al (2002) Interaction of human organic anion transporters 2 and 4 with organic anion transport inhibitors. J Pharmacol Exp Ther 301:797–802
30. Strom SC, Jirtle RL, Jones RS et al (1982) Isolation, culture, and transplantation of human hepatocytes. J Natl Cancer Inst 68:771–778
31. Guguen-Guillouzo C, Campion JP, Brissot P et al (1982) High yield preparation of isolated human adult hepatocytes by enzymatic perfusion of the liver. Cell Biol Int Rep 6:625–628
32. David P, Viollon C, Alexandre E et al (1998) Metabolic capacities in cultured human hepatocytes obtained by a new isolating procedure from non-wedge small liver biopsies. Hum Exp Toxicol 17:544–553
33. Kim HM, Han SB, Hyun BH et al (1995) Functional human hepatocytes: isolation from small liver biopsy samples and primary cultivation with liver-specific functions. J Toxicol Sci 20:565–578
34. Lecluyse EL, Alexandre E (2010) Isolation and culture of primary hepatocytes from resected human liver tissue. Methods Mol Biol 640:57–82
35. Schaefer O, Ohtsuki S, Kawakami H et al (2012) Absolute quantification and differential expression of drug transporters, cytochrome P450 enzymes, and UDP-glucuronosyltransferases in cultured primary human hepatocytes. Drug Metab Dispos 40:93–103
36. Schroeder A, Eckhardt U, Stieger B et al (1998) Substrate specificity of the rat liver Na(+)-bile salt cotransporter in Xenopus laevis oocytes and in CHO cells. Am J Physiol 274:G370–G375
37. Vavricka SR, Van Montfoort J, Ha HR et al (2002) Interactions of rifamycin SV and rifampicin with organic anion uptake systems of human liver. Hepatology 36:164–172
38. Gomez-Lechon MJ, Donato MT, Castell JV et al (2004) Human hepatocytes in primary culture: the choice to investigate drug metabolism in man. Curr Drug Metab 5:443–462
39. Goyak KM, Johnson MC, Strom SC et al (2008) Expression profiling of interindividual variability following xenobiotic exposures in primary human hepatocyte cultures. Toxicol Appl Pharmacol 231:216–224
40. Sugano K, Kansy M, Artursson P et al (2010) Coexistence of passive and carrier-mediated processes in drug transport. Nat Rev Drug Discov 9:597–614

Chapter 22

Measurement of Albumin Secretion as Functionality Test in Primary Hepatocyte Cultures

Karolien Buyl, Joery De Kock, Jennifer Bolleyn, Vera Rogiers, and Tamara Vanhaecke

Abstract

One of the most important functions of hepatocytes is the synthesis of serum proteins, more specifically of serum albumin. Albumin secretion in serum is essential, since it maintains the oncotic pressure in the body. Measurement of albumin secretion is used as a liver function test to indicate potential liver injury and liver pathology. In this chapter, a protocol for the measurement of albumin secretion in the supernatant of cultured rat hepatocytes is described. The procedure relies on an enzyme-linked immunosorbent assay allowing rat albumin to be quantitatively measured.

Key words Albumin, Primary hepatocyte, Liver functionality, Enzyme-linked immunosorbent assay

1 Introduction

Since the liver is a main target of drug-induced liver injury, a lot of attention is paid to the development of in vitro liver models to preclinically screen out potential toxicants [1]. Most preferably, these in vitro liver models are of human origin [2]. However, human hepatocytes are very scarce, since they are optimally used for liver transplantation [3]. Consequently, the availability of human hepatocytes for the development of in vitro models is limited [4] and, as a result, most in vitro models are based on rodent hepatocytes [5]. Hepatocytes comprise approximately 80 % of the total liver mass and fulfill various vital functions in the organism, such as glycogen storage, cholesterol synthesis and transport, urea metabolism, drug detoxification and lipid and serum protein biosynthesis [6]. The most abundant serum protein is albumin, which maintains the oncotic pressure in the body and is also responsible for the transport of hydrophobic molecules [7].

The determination of albumin concentration in serum is used to evaluate the synthetic function of the liver and consequently to

Mathieu Vinken and Vera Rogiers (eds.), *Protocols in In Vitro Hepatocyte Research*, Methods in Molecular Biology, vol. 1250, DOI 10.1007/978-1-4939-2074-7_22, © Springer Science+Business Media New York 2015

detect acute and chronic liver injury [8]. In this chapter, a protocol is presented to evaluate hepatocyte function in vitro. It is based upon a quantitative sandwich enzyme immunoassay technique and measures albumin in the culture medium of primary hepatocytes (Fig. 1). Standards and samples are sandwiched between an immobilized polyclonal antibody and a biotinylated polyclonal antibody specific for rat albumin. The latter is recognized by a streptavidin-peroxidase conjugate and after adding a peroxidase enzyme substrate, a blue color is developed. The blue product, on its turn, is converted towards a yellow derivative of which the absorbance can be measured at 450 nm.

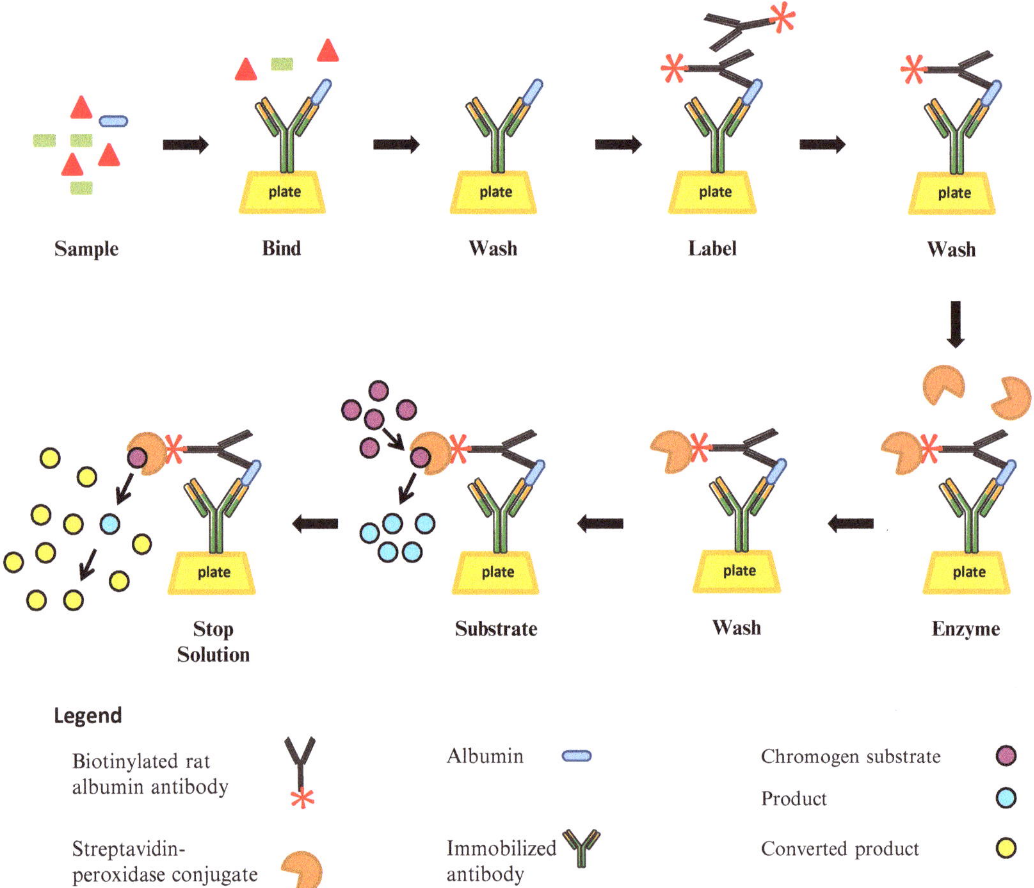

Fig. 1 Albumin enzyme-linked immunosorbent assay. This assay employs a quantitative sandwich enzyme immunoassay technique in which standards and samples are sandwiched by an immobilized polyclonal antibody and a biotinylated polyclonal antibody specific for rat albumin. The latter is recognized by a streptavidin-peroxidase conjugate. After adding a peroxidase enzyme substrate, in this case the chromogen substrate, a blue color is developed. After adding a stop solution, the blue color will change to yellow and the absorbance can be measured at 450 nm (the antibody cartoons are from Servier Medical Art)

2 Materials

2.1 Establishment of Primary Hepatocytes in Culture and Sampling

1. Hepatocyte seeding medium. William's E medium containing 7 ng/mL glucagon, 292 mg/mL l-glutamine, 7.33 IE/mL sodium benzylpenicillin, 50 μg/mL kanamycin monosulfate, 10 μg/mL sodium ampicillin, 50 μg/mL streptomycin sulfate, and 10 % fetal bovine serum. Prepare in a laminar air flow cabinet and store for maximum 7 days at 4 °C. Place, *prior* to use, the hepatocyte seeding medium for 30 min in a thermostated bath at 37 °C.
2. Hepatocyte culture medium. Serum-free hepatocyte seeding medium supplemented with 25 μg/mL hydrocortisone sodium hemisuccinate and 0.5 μg/mL insulin. Prepare in a laminar air flow cabinet and store for maximum 7 days at 4 °C. Place, *prior* to use, the hepatocyte culture medium for 30 min in a thermostated bath at 37 °C (*see* **Note 1**).
3. Laminar air flow cabinet.
4. Incubator (37 ± 1 °C, 90 ± 5 % humidity, 5 ± 1 % CO_2).
5. Thermostated bath at 37 °C.
6. Plastic culture dishes with diameter 3.5 and 10 cm.

2.2 Measurement of Albumin Secretion

1. AssayMax rat albumin enzyme-linked immunosorbent assay Kit (Assaypro, USA).
2. Multiplate reader (Thermofisher Scientific, Belgium).

3 Methods

3.1 Establishment of a Monolayer Culture of Primary Hepatocytes and Sampling

1. Use freshly isolated primary rat hepatocytes [9] (*see* **Note 2**).
2. Evenly plate the hepatocytes on plastic culture dishes at a density of 0.56×10^5 cells/cm^2 in hepatocyte seeding medium. Place the cell cultures in an incubator at 37 °C and 5 % CO_2 for 4 h.
3. Remove the hepatocyte seeding medium and replace by identical volumes of hepatocyte culture medium. Place the cell cultures in an incubator at 37 °C and 5 % CO_2 for 24 h.
4. Collect cell culture media at a number of chosen timepoints (*see* **Note 3**) and centrifuge at 3,000 × *g* and room temperature for 10 min to remove debris.
5. Collect supernatants.

3.2 Measurement of Albumin Secretion

1. Prepare all reagents, working standards and samples as follows:
   - MIX diluent. Dilute the MIX diluent concentrate 1/10 with reagent-grade water. If crystals are formed in the concentrate, mix gently until the crystals are completely dissolved. Store for up to 1 month at 2–8 °C.

- 200 ng/mL rat albumin stock solution. Solubilize the rat albumin standard with MIX diluent. Stir the standard gently during 10 min *prior* to making the other dilutions. Store the remaining solution at −20 °C (*see* **Note 4**).
- Biotinylated rat albumin antibody. Spin the antibody briefly before dilution. Then, dilute the biotinylated rat albumin antibody 1/100 with MIX diluent. Store the remaining solution at −20 °C.
- Wash buffer. Dilute the wash buffer concentrate 1/20 with reagent-grade water.
- Streptavidin-peroxidase conjugate. Spin the conjugate briefly before dilution. Then, dilute the concentrated streptavidin-peroxidase conjugate 1/100 with MIX diluent. Store the remaining solution at −20 °C.

2. Bring all reagents to room temperature before use. Perform the assay at room temperature (i.e. 20–30 °C) (*see* **Note 5**).
3. Prepare a standard curve with known amounts of albumin taken from the rat albumin stock solution of 200 ng/mL using MIX diluent (Table 1).
4. Remove excess microplate strips from the plate frame and return immediately to the foil pouch with desiccant inside. Reseal the pouch securely to minimize exposure to water vapor and store in a vacuum desiccator.
5. Add 50 μL of either standard or sample per well, cover wells and incubate at room temperature for 1 h. Start the timer after the last sample addition (*see* **Note 6**).
6. Wash five times with 200 μL wash buffer. Invert the plate and decant the contents. At each step, tap the plate 4–5 times on absorbent paper to completely remove all liquid (*see* **Note 7**).

Table 1
Preparation of the albumin standard curve. The albumin standard curve was prepared by diluting the rat albumin stock solution (i.e. 200 ng/mL) using MIX diluent

Standard number	Concentration albumin (ng/mL)	Dilution
P1	200.000	Standard (200 ng/mL)
P2	50.000	1 part P1 and 3 parts MIX diluent
P3	12.500	1 part P2 and 3 parts MIX diluent
P4	3.125	1 part P3 and 3 parts MIX diluent
P5	0.781	1 part P4 and 3 parts MIX diluent
P6	0.000	MIX diluent

7. Add 50 μL biotinylated rat albumin antibody to each well and incubate at room temperature for 30 min.
8. Wash five times with 200 μL wash buffer.
9. Add 50 μL streptavidin-peroxidase conjugate to each well and incubate at room temperature for 30 min.
10. Wash five times with 200 μL wash buffer.
11. Add 50 μL chromogen substrate per well and incubate at room temperature for about 10 min or until the optimal blue color density develops. Gently tap the plate to ensure homogeneous mixing and break any bubble in the wells with pipette tips.
12. Add 50 μL stop solution to each well. The color will change from blue to yellow.
13. Read the absorbance immediately using a microplate reader at a wavelength of 450 nm (*see* **Note 8**).

3.3 Data Analysis

1. Calculate the mean value of the triplicate readings for each standard and sample.
2. To generate a standard curve, plot the graph using the standard concentrations on the *x*-axis and the corresponding mean absorbance measured at 450 nm on the *y*-axis. Determine by regression analysis the best fit curve using log–log curve fit. From this equation, the concentration of the unknown sample can be determined (*see* **Note 9**).

4 Notes

1. The sterility of the cell culture media can be checked by adding 1 mL medium to 25 mL sterile thioglycollate medium. This is placed for 2 days in the incubator at 37 °C and 5 % CO_2.
2. Cell viability following hepatocyte isolation should be checked using trypan blue exclusion, as the isolation procedure may cause considerable harm to the cells. Cell viability should be at least 85 %.
3. After collecting cell culture media, the samples can be snap-frozen in liquid nitrogen and stored at −80 °C. Protein determination can be performed later in time.
4. The rat albumin stock solution contains crystals. Make sure that these are completely dissolved before making further dilutions.
5. In this assay, temperature is a critical parameter, since it strongly influences enzyme activity. Thus, it is important to ensure a constant room temperature (i.e. 20–30 °C) during incubation of the 96-well plate.
6. When the standard or sample is added to the wells, make sure that the bottom of the wells is not touched in order not to

disturb the immobilized polyclonal antibody. Pipetting should be done at the side of the wells.

7. Air bubbles formed during the different pipetting steps should be removed as they can interfere with the final measurement.
8. About 10 min after the reaction is stopped, black particles may be formed at high concentrations. These will reduce the readings.
9. From a statistical point of view, it is strongly recommended to perform the experiments with hepatocytes from three different rats. Results can be processed and evaluated by 1-way analysis of variance (i.e. repeated measures) followed by post hoc Bonferroni tests.

Acknowledgements

This work was financially supported by the grants of the European Union (FP7) and Cosmetics Europe (projects DETECTIVE and HeMiBio), the Fund for Scientific Research Flanders (FWO-Vlaanderen) and the Research Council (OZR) of the Vrije Universiteit Brussel.

References

1. Baranczewski P, Stanczak A, Sundberg K et al (2006) Introduction to *in vitro* estimation of metabolic stability and drug interactions of new chemical entities in drug discovery and development. Pharmacol Rep 58:453–472
2. Mandenius CF, Andersson TB, Alves PM et al (2011) Toward preclinical predictive drug testing for metabolism and hepatotoxicity by using *in vitro* models derived from human embryonic stem cells and human cell lines: a report on the vitrocellomics EU-project. Altern Lab Anim 39:147–171
3. Szabo M, Veres Z, Baranyai Z et al (2013) Comparison of human hepatoma HepaRG cells with human and rat hepatocytes in uptake transport assays in order to predict a risk of drug induced hepatotoxicity. Plos ONE 8:1–8
4. Vanhaecke T, Rogiers V (2006) Hepatocyte cultures in drug metabolism and toxicological research and testing. Methods Mol Biol 320:209–227
5. Beigel J, Fella K, Kramer PJ et al (2008) Genomics and proteomics analysis of cultured primary rat hepatocytes. Toxicol In Vitro 22:171–181
6. Si-Tayeb K, Lemaigre FP, Duncan SA (2010) Organogenesis and development of the liver. Dev Cell 16:175–189
7. Gekle M (2005) Renal tubule albumin transport. Annu Rev Physiol 67:573–594
8. Woreta TA, Algahtani SA (2014) Evaluation of abnormal liver tests. Med Clin North Am 98:1–16
9. Papeleu P, Vanhaecke T, Henkens T et al (2006) Isolation of rat hepatocytes. Methods Mol Biol 320:229–237

Chapter 23

Measurement of Blood Coagulation Factor Synthesis in Cultures of Human Hepatocytes

Stefan Heinz and Joris Braspenning

Abstract

An important function of the liver is the synthesis and secretion of blood coagulation factors. Within the liver, hepatocytes are involved in the synthesis of most blood coagulation factors, such as fibrinogen, prothrombin, factor V, VII, IX, X, XI, XII, as well as protein C and S, and antithrombin, whereas liver sinusoidal endothelial cells produce factor VIII and von Willebrand factor. Here, we describe methods for the detection and quantification of most blood coagulation factors in hepatocytes in vitro. Hepatocyte cultures indeed provide a valuable tool to study blood coagulation factors. In addition, the generation and expansion of hepatocytes or hepatocyte-like cells may be used in future for cell-based therapies of liver diseases, including blood coagulation factor deficiencies.

Key words Human hepatocytes, Blood coagulation factors

1 Introduction

Coagulation is a complex process that is driven by interactions between damaged endothelium, platelets, and blood-derived coagulation factors. After initial adhesion and activation to collagen of platelets from injured tissue, secondary hemostasis proceeds in three phases that enable the controlled and local production of fibrin for clot formation [1, 2]. During the initiation phase, factor VIIa binds to tissue factor from the subendothelium, which activates small amounts of factor IX and X. Factor Xa binds factor Va (i.e. the prothrombinase complex), which in turn triggers thrombin (i.e. factor IIa). In the amplification phase, the initially generated thrombin activates platelets, factor V, VIII, V and XI on the platelet membrane. In the propagation phase, which also mainly takes place on the platelet membrane, factor IXa binds factor VIIIa (i.e. the tenase complex) to produce factor Xa that in turn binds factor Va (i.e. the prothrombinase complex), yielding more thrombin in a positive feedback loop. Thrombin subsequently cleaves fibrinogen to fibrin, which forms strands that strengthen the platelet plug.

Mathieu Vinken and Vera Rogiers (eds.), *Protocols in In Vitro Hepatocyte Research*, Methods in Molecular Biology, vol. 1250, DOI 10.1007/978-1-4939-2074-7_23, © Springer Science+Business Media New York 2015

Next to the endothelium and blood cells, the liver plays a major role in the synthesis and secretion of blood coagulation factors. This is evidenced by the fact that hepatic injury and dysfunction, such as liver cirrhosis, severely affect hemostasis [3]. In addition, liver transplantations have been reported to correct congenital bleeding disorders [4, 5]. Direct evidence for the hepatic synthesis of blood coagulation factors has been obtained from studies on perfused liver [6, 7] and primary hepatocytes in culture [8–10]. The organ-specific and cell type-specific origin of factor VIII has been controversial, but there is accumulating evidence that liver sinusoidal endothelial cells and extrahepatic endothelial cells are responsible for factor VIII production [11–14]. Another important blood coagulation factor secreted by endothelial cells is von Willebrand factor that stabilizes factor VIII in the blood and which captures platelets [15].

In this chapter, experimental procedures are described to study blood coagulation factor synthesis and secretion by human hepatocytes in culture. These techniques can also be used to evaluate propagated hepatocytes or hepatocyte-like cells that may be used in future for cell-based therapies of liver diseases, including blood coagulation factor deficiencies.

2 Materials

2.1 Plating and Culture of Primary Hepatocytes

1. Cryopreserved plateable human primary hepatocytes frozen after isolation at 6×10^6 cells/vial or upcyte® hepatocytes (Medicyte, Germany) at the same cell number.
2. Collagen-coated plates.
3. 0.02 M acetic acid solution. Add 1 mL glacial acetic acid to 870 mL ultrapure water.
4. Collagen type I (Sigma-Aldrich, Germany).
5. Fetal bovine serum, heat-inactivated (Pan-Biotech, Germany).
6. Hepatocyte culture medium. William's E medium-based culture medium, termed upcyte® hepatocyte high performance medium, designed for the culture and endpoint measurement of upcyte® and primary hepatocytes (Medicyte, Germany).
7. Percoll solution (Sigma-Aldrich, Germany).
8. 0.1 % trypan blue solution (Life Technologies, Germany).
9. Neubauer hemocytometer (neoLab, Germany).
10. Incubator (37 ± 1 °C, 90 ± 5 % humidity, 5 ± 1 % CO_2).
11. Laminar air flow cabinet.
12. Thermostated bath (37 °C).

2.2 Quantification of Blood Coagulation Factors in Supernatants

2.2.1 1-Stage Clotting Assay

1. 25 mM $CaCl_2$ solution. Dissolve 2.775 g $CaCl_2$ in 1 L ultrapure water.
2. Immuno-depleted coagulation factor deficient plasma (Dade Behring, Germany).
3. Activated partial thromboplastin time reagent (Dade Behring, Germany). Reconstitute as described by the supplier.
4. Reference plasma (Haemochrom Diagnostica, Germany).
5. MC1 fibrometer (Merlin, Germany).

2.2.2 Chromogenic Assay

1. Coatest SP4 chromogenic assay (Chromogenix, Italy).
2. Reference plasma (Haemochrom Diagnostica, Germany).
3. Spectrophotometer (Tecan, Switzerland).

2.2.3 Enzyme-Linked Immunosorbent (ELISA) Assay for the Quantification of Blood Coagulation Factors

1. 96-well immulon microplates (Thermo Scientific, United States of America).
2. ELISA reader (Tecan, Switzerland).
3. Coating buffer. Dissolve 1.5 g Na_2CO_3 and 2.93 g $NaHCO_3$ in 1 L distilled water. Adjust to pH 9.6.
4. Phosphate-buffered saline (PBS). 137 mM NaCl, 2.7 mM KCl, 10 mM $Na_2HPO_4{\cdot}2H_2O$, 1.8 mM KH_2PO_4 in deionized water. Adjust to pH 7.4.
5. Blocking buffer. Add 1 g bovine serum albumin to 100 mL PBS.
6. Wash buffer. Add 50 μL Tween 20 to 100 mL PBS.
7. 3,3′,5,5′-Tetramethylbenzidine (Sigma-Aldrich, Germany).
8. 2 M H_2SO_4 solution. Mix 10 mL 96 % H_2SO_4 with ultrapure water up to 1 L.
9. Para-nitrophenylphosphate (Sigma-Aldrich, Germany).
10. Affinity-purified rabbit anti-human factor V polyclonal antibody [16].
11. Factor VII ELISA kit (Diagnostica Stago, France).
12. Factor 8C-EIA factor VIII:C kit (Affinity Biologicals, Canada) with 1 IU of ReFacto, corresponding to 100 ng factor VIII protein/mL, as control.
13. Monoclonal human factor IX antibody, clone HIX-1 (Sigma-Aldrich, Germany).
14. Peroxidase-conjugated polyclonal goat anti-human factor IX (Affinity Biologicals, United States of America).
15. E-80FIB sandwich ELISA kit (Biotrend, Germany).

3 Methods

3.1 Plating and Culture of Primary Hepatocytes

1. For coating of cell culture plates, dilute collagen type I with 0.02 M acetic acid solution to a final concentration of 50 μg/mL. Add 0.1 mL/cm^2 of the diluted collagen type I solution to the cell culture dishes and incubate for 1 h at room temperature. Wash the plate twice with PBS and use directly or dry to air before storing at 4 °C (*see* **Note 1**).
2. Thaw the hepatocytes by completely immersing the vial in a waterbath at 37 °C, taking care not to agitate the cells during the thawing process. There should be no ice remaining in the vial.
3. Add the contents of the vial to 50 mL pre-warmed (i.e. 37 °C) hepatocyte culture medium supplemented with 10 % fetal bovine serum. Rinse the vial twice with 1 mL pre-warmed (i.e. 37 °C) hepatocyte culture medium containing 10 % fetal bovine serum and add to the same tube. No fetal bovine serum should be used from this step onwards.
4. Centrifuge the tube for 5 min at 60 × *g* and room temperature and resuspend the pellet in 4 mL hepatocyte culture medium.
5. Add the 4 mL cell suspension to the Percoll solution in hepatocyte culture medium to give a final concentration of 29 %. Wash the cells from the tube with this solution once to minimize cell loss.
6. Centrifuge the tube for 20 min at 60 × *g* and room temperature.
7. Discard the supernatant and add 15 mL hepatocyte culture medium. Centrifuge for 5 min at 90 × *g* and room temperature. Discard the supernatant and add 4 mL hepatocyte culture medium. Resuspend the pellet gently by swirling the tube.
8. Add 10 μL cell suspension to 90 μL 0.1 % trypan blue solution and mix gently. Add the cells to the chamber of a Neubauer hemocytometer and count all cells in the 4 squares. Calculate the viable cell number.
9. Seed 150,000 cells/cm^2 for confluent monolayers in collagen-coated plates in hepatocyte culture medium and incubate for 2–4 h at 37 °C and 5 % CO_2 (*see* **Notes 2** and **3**).
10. Remove the cell culture medium and replace with an identical volume of fresh hepatocyte culture medium. Add the same amount of cell culture medium to a well without cells (i.e. control) and place the cell cultures in an incubator at 37 °C and 5 % CO_2.
11. Sample the supernatant and cells after 1, 2, 3 and 5 days. Aliquots of supernatants are immediately frozen in pre-chilled tubes and stored at −80 °C until analysis (*see* **Note 4**).

3.2 Plating and Culture of upcyte® Hepatocytes

Upcyte® hepatocytes are genetically engineered human cell strains derived from single donor primary cells that are able to proliferate, while having a differentiated phenotype exhibiting functional phase 1 and 2 biotransformation activities [17, 18].

1. Thaw the hepatocytes by completely immersing the vial in a waterbath at 37 °C, taking care not to agitate the cells during the thawing process. There should be no ice remaining in the vial.
2. Add the contents of the vial to 50 mL pre-warmed (i.e. 37 °C) hepatocyte culture medium supplemented with 10 % fetal bovine serum. Rinse the vial twice with 1 mL pre-warmed (i.e. 37 °C) hepatocyte culture medium containing 10 % fetal bovine serum and add to the same tube. No fetal bovine serum should be used from this step onwards.
3. Centrifuge the tube for 5 min at $90 \times g$ and room temperature.
4. Aspirate the supernatant without disrupting the pellet. Leave approximately 200–400 μL cell culture medium on the cells.
5. Gently loosen and resuspend the cells without adding any extra medium by agitating and rotating the tube. Do not mix or shake the cells, as this will reduce cell survival.
6. Add an appropriate volume of pre-warmed (i.e. 37 °C) culture medium to the pellet (i.e. approximately 1 mL/million cells thawed) and resuspend the cells. Avoid pipetting the cells up and down.
7. Determine the cell number by trypan blue exclusion or by using a cell counter.
8. Seed 150,000 cells/cm^2 for confluent monolayers in collagen-coated plates in hepatocyte culture medium and incubate for 2–4 h at 37 °C and 5 % CO_2 (*see* **Notes 2** and **3**).
9. Remove the cell culture medium and replace with an identical volume of fresh hepatocyte culture medium. Add the same amount of cell culture medium to a well without cells (i.e. control) and place the cell cultures in an incubator at 37 °C and 5 % CO_2.
10. Sample the supernatant and cells after 1, 2, 3 and 5 days. Aliquots of supernatants are immediately frozen in pre-chilled tubes and stored at −80 °C until analysis (*see* **Note 4**).

3.3 Quantification of Blood Coagulation Factors in Supernatants

3.3.1 1-Stage Clotting Assay

The 1-stage or activated partial thromboplastin time clotting assay is frequently used for the activity measurement of factor VIII, IX, XI and XII, and less frequently for assaying factor II, V and X. The test sample is mixed with the same volume of factor-deficient plasma. Coagulation is then started by adding phospholipids and calcium, followed by recording of the clotting time. The clotting time in the mixture is compared with the clotting time obtained with dilutions of reference plasma. In order to obtain the specific activity, the activity has to be normalized to the antigen level determined by an ELISA assay.

1. Prepare a calibration curve by using the reference plasma according to the manufacturer's instructions. Test at least three dilutions similar to the processing of a sample and establish a concentration–response curve. Express the results as IU/mL corresponding to 100 % plasma concentration.
2. Thaw the samples and reference plasma quickly at 37 °C. Briefly mix before the start of the assay.
3. Test the undiluted samples from the supernatants.
4. Dilute 25 μL of the respective factor-deficient plasma with 25 μL automated activated partial thromboplastin time reagent and 25 μL test sample.
5. Incubate for 3 min at 37 °C.
6. Add 25 μL 25 mM $CaCl_2$ solution and measure the time to clot formation using a fibrometer.

3.3.2 Chromogenic Assay

The activity of factor VIII and IX can also be measured using a 2-stage chromogenic assay. The assay is based on a 2-step reaction consisting of the initial activation of factor X by factor VIII or IX of the sample. Factor IXa in turn converts a chromogenic substrate to a chromophore that is measurable photometrically at 405 nm. For measurement of factor VIII activity, we use the Coatest SP4 chromogenic assay with normal human reference plasma for calibration. The assay is performed according to the manufacturer's instructions for the lower detection range (*see* **Note 5**).

3.3.3 ELISA Assay for the Quantification of Blood Coagulation Factors

Commercial kits are available for most blood coagulation factors. In order to detect low levels of blood coagulation factors produced in culture, calibration curves need to be adapted to detect low values by setting up more dilutions of the reference plasma at a lower level of detection.

1. Coat immunolon plate wells with 100 μL coating buffer with the appropriate coating antibody at a concentration between 1 and 10 μg/mL.
2. Cover the plate and incubate overnight at 4–8 °C.
3. Wash the plate three times with 100 μL wash buffer/well.
4. Add 150 μL blocking buffer to each well and incubate for 60 min at 37 °C.
5. Wash 4 times with 100 μL wash buffer/well.
6. Add 100 μL diluted samples, undiluted samples and standards to the respective wells. Samples and standards should be tested in triplicate.

7. Incubate for 90 min at 37 °C.
8. Wash the plate three times with100 μL wash buffer/well.
9. Add 100 μL diluted detection antibody to each well.
10. Incubate for 1 h at 37 °C.
11. Wash the plate three times with 100 μL wash buffer/well.
12. Add 100 μL substrate solution to each well.
13. Incubate for 30 min at room temperature or until the desired color change is reached.
14. Add 50 μL of the respective stop solution or read the absorbance immediately at the respective wavelength.
15. Mix by gently tapping the plate.
16. Measure the absorbance within 30 min (*see* **Note 6**).

4 Notes

1. Store the cell culture plates in a refrigerator at 4–8 °C in a bag for up to 1 week when not immediately used.
2. For proper functionality, it is critical that the hepatocytes form an evenly distributed confluent cell layer. If not known how many cells are plateable, different cell densities should be tested. Do not rotate the cell culture plates, as this causes uneven distribution. Rather, gently move the cell culture plates back and forth and from left to right.
3. The diameter of the culture dish used mainly depends on the purpose of the assay intended. We recommend the use of 48-well plates for the activity measurement of blood coagulation factors.
4. If the amount of blood coagulation factor secreted is too low to be detected, supernatants can be concentrated up to 400-fold using Vivaspin centrifugal concentrators (Sartorius, Germany). Supernatants are filtered through a 0.22 μm filter before applying to the Vivaspin column with a molecular weight cut-off of 30 kDa. Centrifuge at the recommended speed provided by the manufacturer and check the volume of the concentrate at regular intervals (i.e. every 5 min). Once the desired concentration is reached, harvest by rinsing the filter with the concentrate before aspirating using a pipette.
5. Timing is a critical parameter in this assay. Make sure that the incubation times for each well are identical by always pipetting in the same order and at the same speed.
6. The signal of the ELISA assay can be enhanced by using Supersignal ELISA pico chemi-luminescent substrate (Thermo Scientific, United States of America) and subsequent luminescence measurement.

Acknowledgements

This work was financially supported by the grants of Cosmetics Europe (project HeMiBio, grant agreement No. 266777) and the European Union (FP7) under grant agreement No. 304961 (project Re-Liver).

References

1. Schenone M, Furie BC, Furie B (2004) The blood coagulation cascade. Curr Opin Hematol 11:272–277
2. Monroe DM, Hoffman M (2006) What does it take to make the perfect clot? Arterioscler Thromb Vasc Biol 26:41–48
3. Mannucci PM, Tripodi A (2012) Hemostatic defects in liver and renal dysfunction. Hematol Am Soc Hematol Educ Program 2012:168–173
4. Gordon FH, Mistry PK, Sabin CA et al (1998) Outcome of orthotopic liver transplantation in patients with haemophilia. Gut 42:744–749
5. Ko S, Tanaka I, Kanehiro H et al (2005) Preclinical experiment of auxiliary partial orthotopic liver transplantation as a curative treatment for hemophilia. Liver Transpl 11:579–584
6. Olson JP, Miller LL, Troup SB (1966) Synthesis of clotting factors by the isolated perfused rat liver. J Clin Invest 45:690–701
7. Owen CA Jr, Bowie EJ (1977) Generation of coagulation factors V, XI, and XII by the isolated rat liver. Haemostasis 6:205–212
8. Boost KA, Auth K, Woitaschek D et al (2007) Long-term production of major coagulation factors and inhibitors by primary human hepatocytes *in vitro*: perspectives for clinical application. Liver Int 27:832–844
9. Tatsumi K, Ohashi K, Shima M et al (2008) Therapeutic effects of hepatocyte transplantation on hemophilia B. Transplantation 86:167–170
10. Biron-Andreani C, Raulet E, Pichard-Garcia L et al (2010) Use of human hepatocytes to investigate blood coagulation factor. Methods Mol Biol 640:431–445
11. Do H, Healey JF, Waller EK et al (1999) Expression of factor VIII by murine liver sinusoidal endothelial cells. J Biol Chem 274:19587–19592
12. Follenzi A, Benten D, Novikoff P et al (2008) Transplanted endothelial cells repopulate the liver endothelium and correct the phenotype of hemophilia A mice. J Clin Invest 118:935–945
13. Fomin ME, Zhou Y, Beyer AI et al (2013) Production of factor VIII by human liver sinusoidal endothelial cells transplanted in immunodeficient uPA mice. PLoS One 8:e77255
14. Fahs SA, Hille MT, Shi Q et al (2014) A conditional knockout mouse model reveals endothelial cells as the predominant and possibly exclusive source of plasma factor VIII. Blood 123:3706–3713
15. Nightingale T, Cutler D (2013) The secretion of von Willebrand factor from endothelial cells: an increasingly complicated story. J Thromb Haemost 11:192–201
16. Kane WH, Devore-Carter D, Ortel TL (1990) Expression and characterization of recombinant human factor V and a mutant lacking a major portion of the connecting region. Biochemistry 29:6762–6768
17. Burkard A, Dahn C, Heinz S et al (2012) Generation of proliferating human hepatocytes using upcyte® technology: characterisation and applications in induction and cytotoxicity assays. Xenobiotica 42:939–956
18. Norenberg A, Heinz S, Scheller K et al (2013) Optimization of upcyte® human hepatocytes for the *in vitro* micronucleus assay. Mutat Res 758:69–79

Chapter 24

Functionality Testing of Primary Hepatocytes in Culture by Measuring Urea Synthesis

Jennifer Bolleyn, Vera Rogiers, and Tamara Vanhaecke

Abstract

One of the mechanisms of healthy hepatocytes to detoxify ammonia in the liver consists of converting it into urea. When liver function is impaired, this detoxification capacity decreases and may cause severe pathologies, such as hepatic encephalopathy. Consequently, urea synthesis is a parameter that can be used to monitor liver functionality. In this chapter, a protocol for the measurement of urea synthesis in the culture medium of cultured rat hepatocytes is described. The procedure relies on a chromogenic reagent that specifically forms a colored complex with urea. The latter can be measured colorimetrically and is directly proportional to the urea concentration in the sample.

Key words Ammonia, Urea synthesis, Liver functionality, Primary hepatocyte

1 Introduction

Hepatocytes comprise approximately 80 % of the total liver mass. Hepatocytes are involved in several biological functions, such as synthesis of proteins, cholesterol, phospholipids and bile salts. In addition, detoxification and toxification (i.e. activation) of xenobiotics and endogenous molecules majorly occur in hepatocytes [1, 2]. As such, ammonia, a toxic metabolite of glutamine, is removed by the liver and involves periportal and perivenous hepatocytes. In periportal hepatocytes, ammonia is converted into urea via the urea cycle. In the perivenous hepatocytes, ammonia is metabolized into glutamine by glutamine synthetase [3, 4]. When liver function is impaired, toxins, including ammonia, are more difficult to remove and could be at the origin of severe pathologies, such as hepatic encephalopathy [3]. Thus, urea synthesis represents one of the essential liver functions and can be used to monitor overall liver functionality.

Over the years, several protocols have been used to measure urea in biological samples. In the past, the two most commonly

Mathieu Vinken and Vera Rogiers (eds.), *Protocols in In Vitro Hepatocyte Research*, Methods in Molecular Biology, vol. 1250, DOI 10.1007/978-1-4939-2074-7_24, © Springer Science+Business Media New York 2015

used methods for measuring urea were based either on the measurement of ammonium carbonate produced by hydrolysis with urease or on a measurement of the yellow pigment formed by condensing diacetyl with urea [5–7]. Although these methods were specific and sensitive, several problems, including reagent instability, urease inhibition and inactivation (i.e. urease method) and unstable color development (i.e. diacetyl reaction) occurred [5]. In the current chapter, an improved method is described for the direct measurement of urea in the supernatant of primary hepatocytes in culture. This method is based upon a reaction between urea, *o*-phthalaldehyde and *N*-(1-naphthyl) ethylenediamine forming a colored complex that is directly proportional to the urea concentration which can be measured colorimetrically.

2 Materials

2.1 Establishment of Primary Rat Hepatocyte Cultures and Incubation with NH_4Cl

1. Hepatocyte seeding medium. William's E medium without phenol red containing 7 ng/mL glucagon, 292 mg/mL L-glutamine, 7.33 IE/mL sodium benzyl penicillin, 50 μg/mL kanamycin monosulfate, 10 μg/mL sodium ampicillin, 50 μg/mL streptomycin sulfate, and 10 % fetal bovine serum. Prepare in a laminar air flow cabinet and store for maximum 7 days at 4 °C. Prior to use, place the hepatocyte seeding medium for 30 min in a thermostated bath at 37 °C (*see* **Note 1**).
2. Hepatocyte culture medium. Serum-free hepatocyte seeding medium supplemented with 25 μg/mL hydrocortisone sodium hemisuccinate and 0.5 μg/mL insulin. Prepare in a laminar air flow cabinet and store for maximum 7 days at 4 °C. Prior to use, place the hepatocyte culture medium for 30 min in a thermostated bath at 37 °C (*see* **Note 1**).
3. 6 mM NH_4Cl (Quantichrom urea assay kit, Bioassay Systems, GENTAUR, Belgium).
4. Hank's balanced salt solution (Gibco, Belgium).
5. Incubator (37 °C ± 1 °C, 90 % ± 5 % humidity, 5 % ± 1 % CO_2).
6. Laminar air flow cabinet.
7. Plastic cell culture dishes.

2.2 Quantitative Measurement of Urea Concentration

1. Quantichrom urea assay kit (Bioassay Systems, GENTAUR, Belgium).
2. William's E medium without phenol red.
3. 96-Well plates.
4. Multiplate reader.

3 Methods

3.1 Establishment of Primary Rat Hepatocyte Cultures and Incubation with NH_4Cl

1. Use freshly isolated primary rat hepatocytes [8, 9] (*see* **Note 2**).
2. Plate the hepatocytes evenly on plastic culture dishes at a density of 0.57×10^5 cells/cm^2 in hepatocyte seeding medium (*see* **Note 3**). Place the cell cultures in an incubator for 4 h at 37 °C and 5 % CO_2.
3. Remove the hepatocyte seeding medium and replace by identical volumes of hepatocyte culture medium. Place the cell cultures in an incubator for 24 h at 37 °C and 5 % CO_2.
4. Incubate hepatocytes at various timepoints at 37 °C and 5 % CO_2 for 24 h in William's E medium without phenol red containing 6 mM NH_4Cl (*see* **Note 4**).
5. Collect supernatant after incubation and perform the assay immediately or use liquid nitrogen to freeze the supernatant and store the samples at −80 °C.
6. Rinse the cells once with 1 mL Hank's balanced salt solution buffer and proceed with cell cultivation as initially foreseen.

3.2 Quantitative Measurement of Urea Concentration

1. Place all reagents at room temperature and shake well before use.
2. Prepare the working reagent solution by adding 1 volume of reagent A to 1 volume of reagent B (*see* **Notes 5** and **6**).
3. Prepare a standard curve in phenol red-free medium (Table 1) by adding known amounts of urea from the stock solution present in the kit (i.e. 50 mg/dL) (*see* **Note 7**).

Table 1
Preparation of the urea standard curve

Final urea concentration (mg/dL)	Dilution factor	Stock solution of 50 mg/dL (μL)	Phenol red-free medium (μL)
50.00	1	5.00	'–' or '/'
25.00	2	2.50	2.50
10.00	5	1.00	4.00
7.50	6.66	2.50	14.15
5.00	10	2.50	22.50
2.50	20	2.50	47.50

A standard curve is prepared in phenol red-free medium by adding known amounts of urea from the stock solution present in the kit (i.e. 50 mg/dL)

4. Transfer 5 μL of these dilutions in duplicate into wells of a clear flat bottom 96-well plate (*see* **Note 8**).
5. Transfer 5 μL sample (i.e. supernatant) or ultrapure water (i.e. blank) in duplicate into wells of a clear flat bottom 96-well plate. Freeze the rest of supernatant at −80 °C until further use.
6. Add 200 μL working reagent (i.e. reagent A:reagent B 1:1) to all wells and tap lightly on the plate to gently mix.
7. Incubate 20 min or 50 min for regular or low urea samples, respectively, at room temperature and read the optical density at 520–530 nm. The signal is stable for about 30 min. For low urea concentrations, the optical density can be measured at 430 nm.
8. Calculate the urea concentration expressed in mg/dL as follows:

$$\left(A_{\text{sample}} - A_{\text{Blank}} \,/\, A_{\text{standard}} - A_{\text{blank}}\right) \times n \times \text{urea standard concentration}$$

A = optical density measured at 520–530 nm

Standard = concentration of the urea standard (i.e. 50.00 mg/dL)

N = dilution factor of the samples

4 Notes

1. Check the sterility of the cell culture medium by adding 1 mL of the medium to 25 mL sterile thioglycollate medium and placing this mixture for 2 days in an incubator at 37 °C and 5 % CO_2.
2. Assess cell viability using trypan blue exclusion after hepatocyte isolation, as the enzymatic procedure may cause considerable harm to the cells. Cell viability should be at least 85 %.
3. The type of culture dishes used mainly depends on the purpose of the assay. For the measurement of urea production under the conditions described for the supernatant of hepatocyte cultures, it is recommended to use 3.5 cm diameter cell culture dishes or 6-well culture plates.
4. 6 mM NH_4Cl is used in this assay. This is based upon our own experience with rat hepatocytes, but it is possible to use higher concentrations if necessary, such as for differentiated hepatocyte-like stem cells. For this purpose, prepare an 8 mM NH_4Cl solution by adding 40 mL 10 mM NH_4Cl to 10 mL phenol red-free medium. This solution can be stored at −20 °C for 3 months.

5. Wrap aluminum foil around both the recipients containing reagent B and the working reagent, since reagent B is sensitive to light.
6. Use working reagent within 20 min after mixing.
7. Phenol red-free medium is used to avoid interaction between phenol red and the working reagent, which could lead to false positive results.
8. For colorimetric assays, use clear flat bottom 96-well plates.

Acknowledgements

This work was financially supported by the grants of the European Union (FP7/Cosmetics Europe projects DETECTIVE and HeMiBio).

References

1. Gomez-Lechon MJ, Lahoz A, Gombau L et al (2010) In vitro evaluation of potential hepatotoxicity induced by drugs. Curr Pharm Des 16:1963–1977
2. Goyak KMO, Laurenzana EM, Omiecinski CJ (2010) Hepatocyte differentiation. Methods Mol Biol 640:115–138
3. Butterworth RF, Vaquero J (2009) Hepatic encephalopathy. In: Arias IM, Alter HJ, Boyer JL et al (eds) The liver: biology and pathobiology. Wiley, West Sussex, pp 597–617
4. Damink SWMO, Deutz NEP, Dejong CHC et al (2002) Interorgan ammonia metabolism in liver failure. Neurochem Int 41: 177–188
5. Jung D, Biggs H, Erikson J et al (1975) New colorimetric reaction for endpoint, continuous-flow, and kinetic measurement of urea. Clin Chem 8:1136–1140
6. Henry RJ (1964) Clinical chemistry, principles and techniques. Hoeber Medical Division, Harper & Row, New York
7. Marsh WH, Fingerhut B, Miller H (1965) Automated and manual direct methods for the determination of blood urea. Clin Chem 11:624–627
8. Papeleu P, Vanhaecke T, Henkens T et al (2006) Isolation of rat hepatocytes. Methods Mol Biol 320:229–237
9. Seglen P (1976) Preparation of isolated rat liver cells. Methods Cell Biol 13:29–83

Chapter 25

Assay of Bile Acid Conjugation and Excretion in Human Hepatocytes

Helene Johansson and Ewa C.S. Ellis

Abstract

Primary hepatocytes isolated from human, mouse and rat liver as well as cell lines, such as HepG2 cells, are frequently used in vitro systems in liver research. In regenerative medicine, stem cells are used for differentiation towards hepatocyte-like cells with the goal of creating differentiated functional hepatocytes. It is therefore important to measure the quality and function of highly specialized hepatocyte-specific functions using appropriate methods. In this chapter, we describe an assay to assess conjugation and excretion of labeled bile acids in cultured hepatocytes.

Key words Hepatocytes, Bile acids, Conjugation, Enterohepatic circulation, CDCA, LC-MS/MS

1 Introduction

Bile acid synthesis is highly specific for differentiated hepatocytes and is important for elimination of cholesterol from the body [1, 2]. The human primary bile acids cholic acid (CA) and chenodeoxycholic acid (CDCA) are synthesized mainly by the classical neutral pathway initiated by the rate-limiting microsomal enzyme cholesterol 7α-hydroxylase. The starting point for the alternative acidic pathway is catalyzed by sterol-27-hydroxylation and has no known rate-limiting enzyme [1–3]. The classical pathway accounts for about 90 % of the total bile acid production and the ratio of CA:CDCA is approximately 2:1, while the alternative pathway mostly produces CDCA [4, 5].

Bile acids are conjugated to either glycine or taurine catalyzed by bile acid coenzyme A amino acid *N*-acyltransferase [6–8]. Once conjugated, bile acids are excreted over the apical membrane into the bile canaliculi mainly by the bile salt export pump [5, 9]. Bile acids may undergo further modification when reaching the intestine, such as deconjugation and dehydroxylation by intestinal bacteria, forming the secondary bile acids deoxycholic acid or lithocholic acid [3, 5]. About 95 % of the bile acids are re-absorbed by

Mathieu Vinken and Vera Rogiers (eds.), *Protocols in In Vitro Hepatocyte Research*, Methods in Molecular Biology, vol. 1250, DOI 10.1007/978-1-4939-2074-7_25, © Springer Science+Business Media New York 2015

enterocytes through active transport (i.e. conjugated) or diffusion (i.e. unconjugated) and released to the portal blood [5, 9]. The portal blood transports a mixture of free and conjugated bile acids from the intestine to the liver. In the liver, bile acids are taken up by hepatocytes through active transport (i.e. conjugated) via the sodium-dependent taurocholate co-transporting polypeptide or diffusion (i.e. unconjugated), thus completing the enterohepatic circulation [5, 9, 10]. Deconjugated bile acids are re-conjugated and again transported into the bile together with newly synthesized bile acids [4].

In the assay described in this chapter, the deuterium-labeled free bile acid D_4-CDCA is added to the culture medium of primary human hepatocytes. D_4-CDCA is taken up by the hepatocytes via diffusion followed by conjugation of the free bile acid with glycine (D_4-GCDCA) or taurine (D_4-TCDCA). The conjugated bile acids is unable to diffuse and will be transported out of the hepatocyte via an active transport process mediated by the bile salt export pump. The amount of conjugated bile acids in the medium is analyzed and related to the amount of added D_4-CDCA. Bile acids do not accumulate in normal healthy hepatocytes. Indeed, only low amounts of bile acid intermediates are found within hepatocytes [11].

2 Materials

2.1 Cell Cultures

1. Freshly isolated primary human hepatocytes in cell suspension at 0.75×10^6 cells/mL.
2. Fetal bovine serum.
3. 500 mL William's medium E (Sigma-Aldrich, Sweden) supplemented with 5 mL 200 mM L-glutamine, 10 mL 1 M *N-2-hydroxyethylpiperazine-N*-2-ethane sulfonic acid buffer (Lonza, Sweden), 10 μL 600 μM insulin, 0.5 mL 50 mg/mL gentamicin, 100 μL 250 μg/mL amphotericin B, and 50 μL 1 mM dexamethasone. Store at 4 °C and warm to 37 °C in a waterbath prior to use.
4. 10 mg/mL Matrigel prepared from Engelbreth-Holm-Swarm mouse sarcoma [12]. Thaw on ice.
5. 0.03 mg/mL rat tail collagen type 1 [13]. Collagen should be applied to cell culture plates and kept at 37 °C at least 30 min prior to use.
6. 12-well or 6-well culture plates.
7. 1 g/L D_4-CDCA (Sigma-Aldrich, Sweden) solution in methanol (*see* **Note 1**).
8. Incubator (37 °C ± 1 °C, 90 % ± 5 % humidity, 5 % ± 1 % CO_2).
9. Laminar air flow cabinet.

2.2 Bile Acid Analysis with Liquid Chromatography-Tandem Mass Spectrometry

1. 1 g/L D_4-CDCA (Sigma-Aldrich, Sweden) solution in methanol.
2. Methanol (Solveco, Sweden).
3. Acetonitrile (Sigma-Aldrich, Sweden).
4. Formic acid (Sigma-Aldrich, Sweden).
5. Ammonium acetate (Sigma-Aldrich, Sweden).
6. Glass centrifuge tubes.
7. Vials (Waters, Sweden).
8. Liquid chromatography-tandem mass spectrometry (LC-MS/MS) device MicromassQuattro Micro with C18 reverse-phase column and electrospray ionization in negative mode (Waters, Sweden).

3 Methods

3.1 Cell Cultures

All procedures are carried out at room temperature and in a laminar culture hood unless otherwise specified.

1. Coat 12-well or 6-well culture plates with 1–2 mL rat tail type 1 collagen and incubate for at least 30 min at 37 °C and 5 % CO_2. Alternatively, apply 50–100 μL ice-cold Matrigel to each well and evenly distribute using a sterile rubber cell scraper (*see* **Note 2**). Leave the plates in the laminar hood at room temperature for approximately 30 min to let the Matrigel form a gel.
2. Plate human hepatocytes at a concentration of 0.75×10^6 cells/mL in cold cell culture medium supplemented with 5 % fetal bovine serum. A total volume of 1 and 2 mL should be applied for each well of a 12- and 6-well plate, respectively. When using collagen, aspirate the excess collagen fluid prior to seeding. Incubate at 37 °C and 5 % CO_2 (*see* **Note 3**).
3. Aspirate the cell culture medium after 1–2 h to remove dead and unattached cells. Switch to serum-free medium (*see* **Note 4**) pre-warmed at 37 °C. A total volume of 1 and 1.5 mL should be applied for each well of a 12- and 6-well plate, respectively.
4. When using collagen, wash the wells after 17–24 h by pipetting medium over the well onto the cells to remove dead cells. When using Matrigel, change the medium without washing (*see* **Note 5**).
5. Renew the cell culture medium.
6. Cells are cultured for 5 days (*see* **Note 6**). On day 5, primary human hepatocytes are treated 6 h prior to harvesting, bearing in mind that the optimal treatment time may vary depending on the cell type (*see* **Note 7 and 8**). Change to fresh medium pre-

Table 1
Preparation of medium with D_4-CDCA treatment

Final concentration	5 mL medium (μL)	1.5 mL medium (6-well plate) (μL)	1.0 mL medium (12-well plate) (μL)
0 (blank)	0	0	0
1.5	3	0.9	0.6
3	6	1.8	1.2
5	10	3	2
7	14	4.2	2.8
10	20	6	4
15	30	9	6
20	40	12	8

Mix 5 mL fresh medium with D_4-CDCA or add D_4-CDCA directly to wells containing 1 mL (i.e. 12-well plates) or 1.5 mL (i.e. 6-well plates) fresh medium

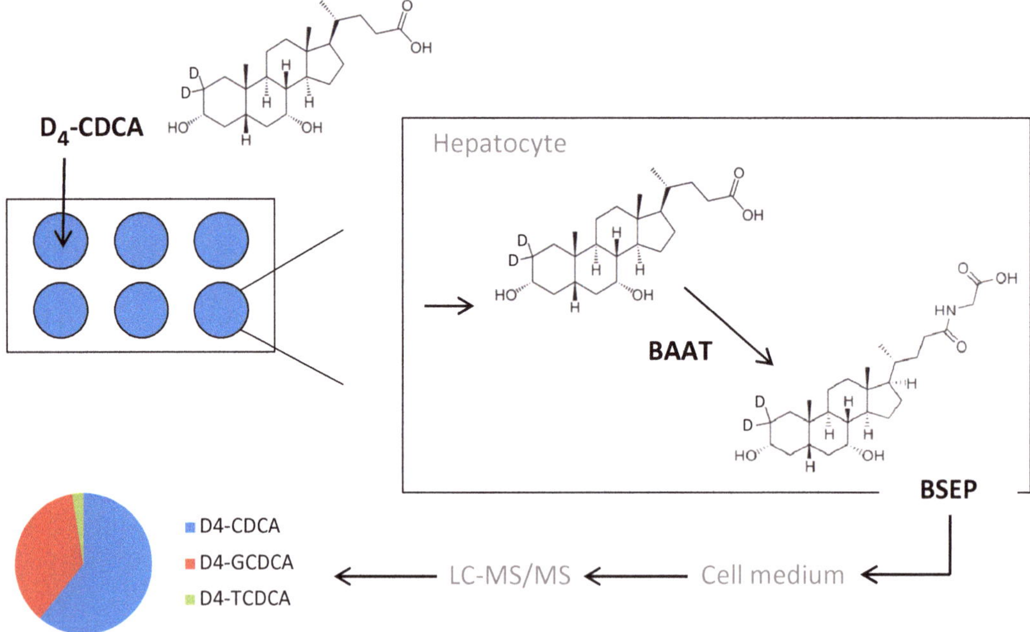

Fig. 1 Schematic overview of the assay. D_4-CDCA is added to the cell culture, taken up by the hepatocytes, conjugated, and again released to the medium where it can be measured with LC-MS/MS

warmed to 37 °C and mixed with D_4-CDCA according to Table 1. Alternatively, add D_4-CDCA directly into the wells with fresh medium. Each concentration point should be made in triplicate (i.e. 3 wells/treatment point) (Fig. 1).

7. Collect the medium from each treatment point and store at −20 to −80 °C until analysis.

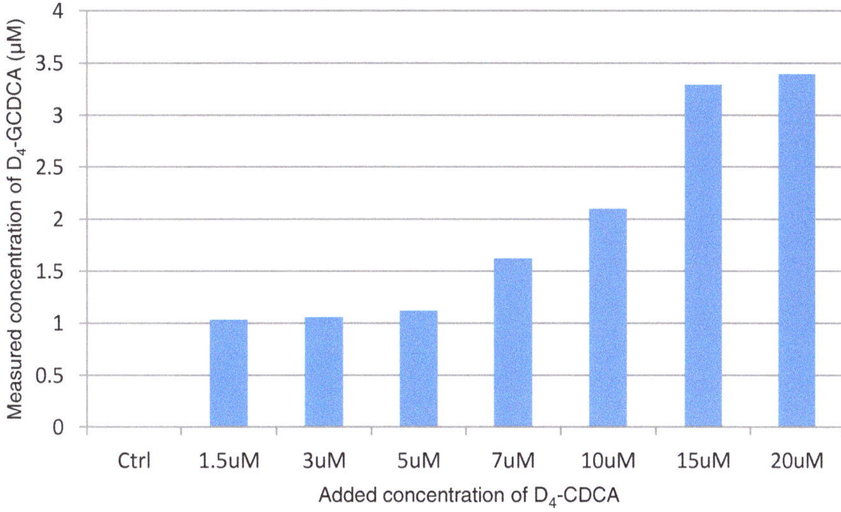

Fig. 2 Formation of GCDCA. 1.5–20 μM D_4-CDCA was added to the cell cultures and 1–3 μM had been absorbed, conjugated, and excreted after 6-h incubation. Small amounts of D_4-TCDCA could also be found (not shown)

3.2 Bile Acid Analysis with LC-MS/MS

1. Dilute 10 μL 1 g/L (i.e. 2.5 mM) D_4-CDCA solution in 100 mL methanol to prepare a 250 nM internal standard solution.
2. Prepare a 40 % methanol solution by mixing with ultrapure water.
3. Prepare a 0.02 % formic acid solution by mixing with ultrapure water.
4. Prepare 10 mM ammonium acetate buffer by dissolving 80 mg ammonium acetate in 100 mL ultrapure water.
5. To extract bile acids from the medium, mix 50 μL medium, 500 μL 250 nM (i.e. 50 ng) internal standard, 40 μL methanol, and 800 μL acetonitril. Centrifuge at 13,000 × *g* at room temperature for 15 min and transfer the upper phase to glass centrifuge tubes. Evaporate the samples under nitric oxide and dissolve in 75 μL methanol, 0.02 % formic acid, and 10 nM ammonium acetate. Transfer to vials and analyze the samples with LC-MS/MS [14].
6. Calculate the percentage of GCDCA, TCDCA, and CDCA from the chromatograms using the chromatogram area (*see* **Note 9**). The control area is first subtracted from the total area of GCDCA, TCDCA, and CDCA. The control area is also subtracted from the area of GCDCA, TCDCA, and CDCA, respectively. The percentage of GCDCA, TCDCA, and CDCA is then calculated by dividing the subtracted area of GCDCA, TCDCA, and CDCA with the subtracted area of the total amount CDCA. The amount of added CDCA (i.e. 1.5–20 μM) that has been conjugated to GCDCA and TCDCA can be calculated (Figs. 2 and 3) (Table 2).

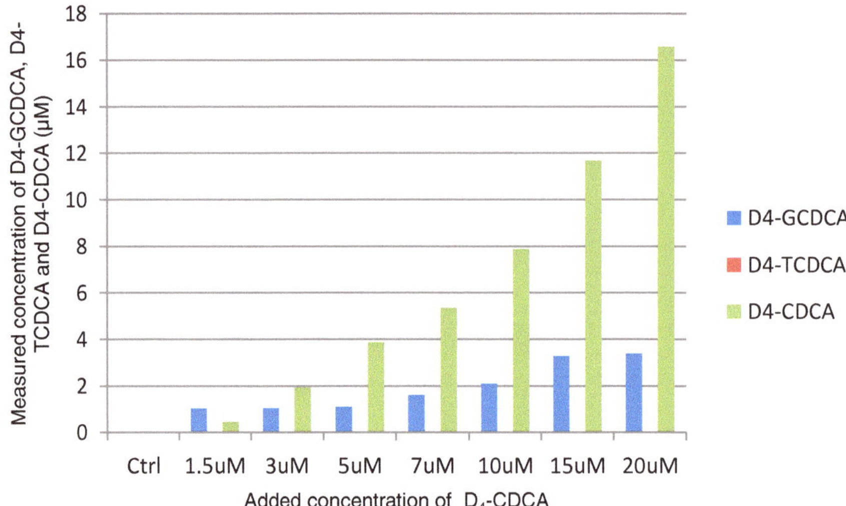

Fig. 3 Composition of D_4-bile acids. Concentration of D_4-GCDCA, D_4-TCDCA and D_4-CDCA expressed in µM calculated from the area of the chromatograms and related to the total amount of added D_4-CDCA

4 Notes

1. Other labeled bile acids, such as deuterated cholic acid or deoxycholic acid, can be used for this assay. Deuterium labels can differ in numbers or positions. Bile acids labeled with C_{13} may also be utilized.
2. Matrigel forms a gel quickly at room temperature. Thaw the Matrigel on ice and keep it on ice at all times while coating the plates. Do not coat too many wells at once, as the Matrigel may start to clot and result in an uneven layer in the well.
3. The morphology of primary human hepatocytes plated on Matrigel and collagen differs. Cells on Matrigel indeed migrate and form groups of cells that may reach out to each other in spider web-like structures, while cells on collagen have a cuboidal appearance (Fig. 4).
4. Serum is only used for the first 2 h to help primary human hepatocytes to attach and can be omitted when using Matrigel. Prolonged use of serum should be avoided, since serum contains bile acids, which will interfere with the bile acid assay.
5. When using Matrigel, cell culture medium should be added cautiously, as hepatocytes may detach if medium is applied directly onto the cells.
6. Cells are cultured for 5 days. However, the optimal time in culture may differ between cell types.
7. In primary human hepatocytes, bile acid production is lower during the first days of culture, but will increase and peak around day 5 followed by a decrease.

Table 2
Calculation of amount of conjugated D_4-CDCA from the chromatogram area

D_4-CDCA	0 μM	1.5 μM	3 μM	5 μM	7 μM	10 μM	15 μM	20 μM
Area								
D_4-GCDCA	20469717	25965782	34211396	37961813	30825512	36290384	36844459	40617783
D_4-TCDCA	0	37783	70004	112038	95263	123325	97719	91928
D_4-CDCA	24680914	27121932	49780737	85302173	58852020	84009327	82829459	123089565
Total area	45150631	53125498	84062137	123376024	89772794	120423036	119771637	163799275
Total area-control area	0	7974866	38911506	78225393	44622163	75272405	74621006	118648644
Area-control area								
D_4-GCDCA	0	5496065	13741679	17492095	10355794	15820666	16374741	20148065
D_4-TCDCA	0	37783	70004	112038	95263	123325	97719	91928
D_4-CDCA	0	2441018	25099823	60621259	34171106	59328413	58148546	98408651
Percentage (%)								
D_4-GCDCA	0	68.9	35.3	22.4	23.2	21.0	21.9	17.0
D_4-TCDCA	0	0.5	0.2	0.1	0.2	0.2	0.1	0.1
D_4-CDCA	0	30.6	64.5	77.5	76.6	78.8	77.9	82.9
Amount (μM)								
D_4-GCDCA	0	1.03	1.06	1.12	1.62	2.10	3.29	3.40
D_4-TCDCA	0	0.01	0.01	0.01	0.01	0.02	0.02	0.02
D_4-CDCA	0	0.46	1.94	3.87	5.36	7.88	11.69	16.59

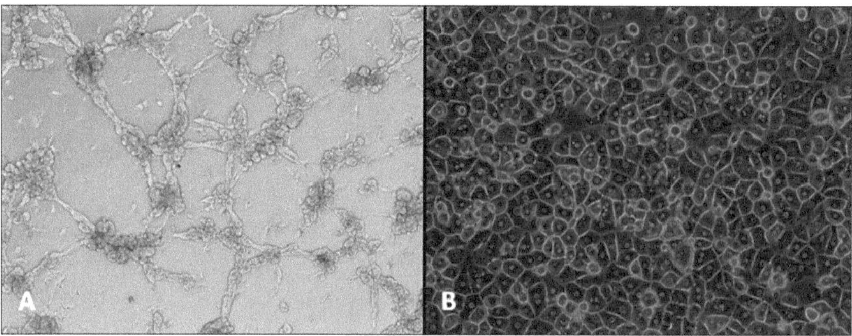

Fig. 4 Morphology of primary human hepatocytes in different culture conditions. Cells plated on (**a**) Matrigel (10× magnification) or (**b**) collagen (10× magnification)

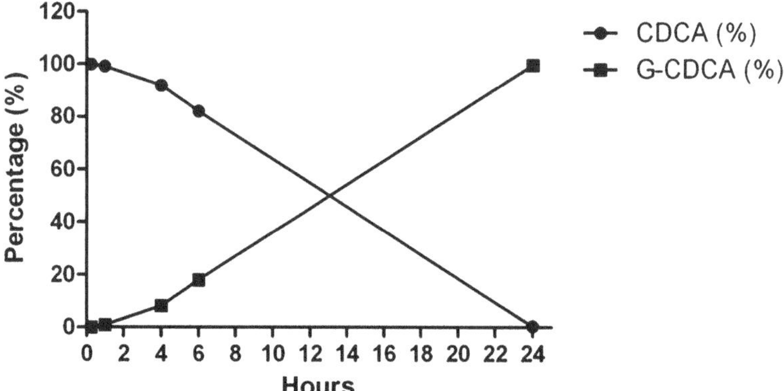

Fig. 5 Time course of GCDCA formation from CDCA. 20 μM CDCA was added to primary human hepatocytes and GCDCA was measured after 15 min, and 1, 4, 6, and 24 h. Approximately 20 % had been absorbed, conjugated, and excreted after 6 h. Almost 100 % of the added CDCA had been conjugated after 24 h

8. The 6 h of incubation with D_4-CDCA was chosen based on a time course where primary human hepatocytes were treated with 20 μM CDCA for 15 min to 24 h. Almost 100 % of the added CDCA had been conjugated to GCDCA after 24 h (Fig. 5). In order not to deplete the substrate, 6 h was chosen for subsequent studies in human primary hepatocytes, but this time scheme may be different for other cell types. Also, if this assay is used to test different treatments that may affect bile acid conjugation, there must be sufficient substrate available to avoid substrate shortage if the conjugation is increased.
9. The amount of conjugated versus free D_4-CDCA is calculated from the percentage of converted CDCA to GCDCA. Absolute values can also be used to calculate conjugation rates. Formation of TCDCA is low (i.e. between 0.1 and 0.5 %) depending on the amount of substrate added.

References

1. Ellis E, Goodwin B, Abrahamsson A et al (1998) Bile acid synthesis in primary cultures of rat and human hepatocytes. Hepatology 27:615–620
2. Einarsson C, Ellis E, Abrahamsson A et al (2000) Bile acid formation in primary human hepatocytes. W J Gastroenterol 6:522–525
3. Ellis E, Axelsson M, Abrahamsson A et al (2003) Feedback regulation of bile acid synthesis in primary human hepatocytes: evidence that CDCA is the strongest inhibitor. Hepatology 38:930–938
4. Glaumann H, Einarsson K, Angelin B (1988) Gallsyrametabolism vid leversjukdom. In: Salde B (ed) Hepatologi, 1st edn. AB Tika, Lund, pp 51–68
5. Li T, Chiang JYL (2011) Bile acid signaling in liver metabolism and diseases. J Lipids 2012:1–9
6. Johnson MR, Barnes S, Kwakye JB et al (1991) Purification and characterization of bile acid-CoA amino acid *N*-acyltransferase from human liver. J Biol Chem 266:10227–10233
7. Solaas K, Ulvestad A, Soreide O et al (2000) Subcellular organization of bile acid amidation in human liver: a key issue in regulating the biosynthesis of bile salts. J Lipid Res 41:1154–1162
8. Falany CN, Johnson MR, Barnes S et al (1994) Glycine and taurine conjugation of bile acids by a single enzyme. J Biol Chem 269:19375–19379
9. Byrne JA, Strautnieks SS, Miele-Vergani G et al (2002) The human bile salt export pump: characterization of substrate specificity and identification of inhibitors. Gastroenterology 123:1649–1658
10. Agellon LB (2002) Metabolism and function of bile acids. In: Vance DE, Vance JE (eds) Biochemistry of lipids, lipoprotein and membranes, 4th edn. Elsevier, Amsterdam, pp 433–448
11. Axelson M, Ellis E, Mork B et al (2000) Bile acid synthesis in cultured human hepatocytes: support for an alternative biosynthetic pathway for cholic acid. Hepatology 31:1305–1312
12. Kleinman HK, McGarvey ML, Hassell JR et al (1986) Basement membrane complexes with biological activity. Biochemistry 25:312–318
13. Strom SC, Michalopoulos G (1982) Collagen as a substrate for cell growth and differentiation. Methods Enzymol 82:544–555
14. Ellis E, Nauglers S, Parini P et al (2013) Mice with chimeric livers are an improved model for human lipoprotein metabolism. Plos One 8:1–10

Chapter 26

General Cytotoxicity Assessment by Means of the MTT Assay

Laia Tolosa, María Teresa Donato, and María José Gómez-Lechón

Abstract

Cytotoxicity assays were among the first in vitro bioassay methods used to predict toxicity of substances to various tissues. In vitro cytotoxicity testing provides a crucial means for safety assessment and screening, and for ranking compounds. The choice of using a particular cytotoxicity assay technology may be influenced by specific research goals. As such, four main classes of assays are used to monitor the response of cultured cells after treatment with potential toxicants. These methods measure viability, cell membrane integrity, cell proliferation, and metabolic activity. In this chapter, we focus on the 3-(4,5-dimethylthiazol-2-yl)-2,5-diphenyltetrazolium bromide tetrazolium reduction colorimetric assay to evaluate detrimental intracellular effects on metabolic activity. This assay is well-characterized, simple to use and remains popular in several laboratories worldwide.

Key words Cytotoxicity, Tetrazolium salt, IC_{50}, Balb/c 3T3 cells, Rat hepatocytes

1 Introduction

Cytotoxicity results from interference with structures or properties that are essential for cell survival, proliferation, or function. Basal (i.e. general) cytotoxicity involves 1 or more of the abovementioned structures or processes when all the cell types studied show similar sensitivities. The safety evaluation of compounds, such as drugs, cosmetics, food additives, pesticides and industrial chemicals, is growing year by year. Therefore, the need for reliable, sensitive, quantitative, easy to handle and fast cytotoxicity tests led to the development of several assays adapted for high-throughput in microtiter plates, now routinely used to detect cytotoxic effects of large number of compounds in cellular systems (Table 1). Assays to measure viability, cell membrane integrity, cell proliferation, and metabolic activity are commonly used to monitor the response of cultured cells to treatment with xenobiotics [1, 2].

Cell viability may be judged by morphological changes and by changes in membrane permeability or the physiological state

Mathieu Vinken and Vera Rogiers (eds.), *Protocols in In Vitro Hepatocyte Research*, Methods in Molecular Biology, vol. 1250, DOI 10.1007/978-1-4939-2074-7_26, © Springer Science+Business Media New York 2015

Table 1
Biological endpoint parameters for detecting cytotoxicity

Endpoint	Experimental parameter	Reference
Cell viability	Vital dye uptake	
	Fluorescein diacetate	[8]
	Hydroethidine	[7]
	Trypan blue	[5]
	Propidium iodide	[6]
	Cellular protein evaluation	
	Sulforhodamine assay	[16, 17]
	Lowry assay	[18]
Cell proliferation	Rate of DNA synthesis	
	^{3}H-Thymidine	[38]
	5-bromo-2′-deoxyuridine	[39]
	5-ethynyl-2′-deoxyuridine	[40]
	Cellular protein evaluation	
	Sulforhodamine assay	[16, 17]
	Lowry assay	[18]
	Carboxyfluorescein diacetate *N*-succinimidyl ester assay	[13]
	Clonogenic assay	[14, 15]
	Metabolic activity (MTT, XTT, WST1, alamar blue)	[2, 19, 20, 22, 23, 25–30, 43]
Plasma membrane damage	Cytosolic enzymes leakage	[11, 41, 42]
	Leakage from pre-loaded cells (vital dye, ^{51}Cr)	
	Protease release	[12]
	5-Carboxyfluorescein diacetate acetoxymethyl ester assay	[9]
Metabolic activity	Co-factor depletion	[27, 28]
	Impairment of mitochondrial function	
	MTT assay	[19, 20]
	3-[4,5-dimethylthiazol-2-yl]-2,5-diphenyl tetrazolium bromide	
	XTT assay	[2, 22]
	4-[3-(4-iodophenyl)-2-(4-nitrophenyl)-2H-5-tetrazolio]-1,3-benzene disulfonate	
	Alamar blue assay	[25, 26]
	WST-1 assay	[23, 43]
	4-[3-(4-iodophenyl)-2-(4-nitrophenyl)-2H-5-tetrazolio]-1,3-benzene disulfonate)	
	Lysomal alteration	
	Neutral red uptake assay	[29, 30]

inferred from the exclusion of certain dyes or the uptake and retention of others [3, 4]. In fact, damage to cells by xenobiotics commonly results in an early cell membrane integrity alteration, leading to changes in permeability. Vital dyes, such as trypan blue [5], propidium iodide [6], hydroethidine [7] and fluorescein diacetate [8], are normally excluded from the inside of healthy cells. However, if cell membrane permeability is compromised, they freely cross the membrane and stain intracellular components [3, 4]. 5-Carboxyfluorescein diacetate acetoxymethyl ester can be converted by non-specific esterases of living cells from a membrane-permeable, non-polar, non-fluorescent substance into a polar and fluorescent dye, namely carboxyfluorescein. Conversion into carboxyfluorescein by cells indicates the integrity of the plasma membrane [9]. Alternatively, membrane integrity can be assessed by monitoring the passage of substances that are normally sequestered from inside cells to the outside. This is typically achieved by measuring constitutive, conserved, and stable enzymatic activities released from dead cells. Among those, lactate dehydrogenase is a favorite marker of cell death for in vitro models. Loss of intracellular lactate dehydrogenase and its release into the cell culture medium is an indicator of irreversible cell death due to cell membrane damage [2, 10, 11]. Leakage of enzymes associated with organelles is also used. Protease biomarkers that enable the measurement of relative numbers of live and dead cells within the same cell population have been identified. The live cell protease is active only in cells that have a healthy cell membrane and activity is lost once the cell is compromised and the protease is exposed to the external environment. Therefore, this assay measures the activity of released intracellular proteases as a result of cell membrane impairment [12].

Assays for cell proliferation may monitor the number of cells over time, the number of cellular divisions, metabolic activity, or DNA synthesis. Carboxyfluorescein diacetate *N*-succinimidyl ester is a popular choice for measuring the number of cellular divisions that a population has undergone [13]. The clonogenic assay or colony formation assay is an in vitro cell survival assay based on the ability of a single cell to grow into a colony. This assay essentially tests every cell in the population for its ability to undergo unlimited division [14, 15]. Cell proliferation can also be measured by sulforhodamine B used for cell density determination based on the cellular protein content measurement [16, 17]. Other colorimetric assays can also be used to determine protein content [18].

Assays that measure metabolic activity are suitable for analyzing proliferation, viability, and cytotoxicity. Reduction of tetrazolium salts, such as 3-(4,5-dimethylthiazol-2-yl)-2,5-diphenyltetrazolium bromide tetrazolium (MTT), 2,3-bis-(2-methoxy-4-nitro-5-sulfophenyl)-2H-tetrazolium-5-carboxanilide (XTT) and water soluble tetrazolium salts (WST1), to colored formazan compounds, or the bioreduction of resazurin by mitochondrial enzyme succinate dehydrogenase with the requirement of cellular nicotinamide adenine dinucleotide occurs only in metabolically active cells. Viable cells reduce the MTT reagent to a colored formazan product, which is water-insoluble, precipitates into cells and that needs to be extracted with organic solvents [2, 10, 19, 20]. The XTT salt yields a soluble product that is secreted into the culture medium.

This method is particularly useful when a cell culture needs monitoring several times to, for example, measure cell proliferation, since this product is non-toxic and, hence allows repetitive testing on a single cell preparation [2, 21, 22]. WST1 was developed as a third-generation tetrazolium salt that exhibits good solubility, is very stable in stock solution and can be used as an alternative to the MTT assay [23]. Another assay, similar to the MTT method, is the alamar blue assay, which measures cellular metabolic activity. The alamar blue assay is based on the reduction of the blue nonfluorescent dye resazurin to pink fluorescent resorufin by a pool of reductase or diaphorase-type enzymes in mitochondria and the cytosol [24–26]. In addition, Alamar Blue shows no cytotoxic effects and the tested cells do not need to be destroyed, thus enabling several tests or kinetic measurements on the same set of cells [26]. In addition to using dyes to indicate the redox potential of cells in order to monitor their viability, researchers have developed assays that use adenosine triphosphate content as a marker of viability [2]. Such adenosine triphosphate-based assays include bioluminescent assays in which adenosine triphosphate is the limiting reagent for the luciferase reaction [27, 28]. The neutral red uptake assay is based on the ability of viable cells to incorporate and bind the supravital dye neutral red in lysosomes [29]. Dead cells or those with membrane damage cannot accumulate the dye. It is thus possible to distinguish viable and damaged or dead cells according to their specific lysosomal capacity for taking up the dye [30].

Cytotoxicity experiments are generally designed to determine IC_{10} and IC_{50} values (i.e. the concentration causing 10 or 50 % cell death, respectively) as well as the maximal non-toxic concentration of a compound (i.e. the highest concentration compatible with cell survival) [1]. Therefore, in vitro cytotoxicity testing provides a crucial means for safety assessment, as well as for labeling and ranking compounds. Although a broad spectrum of assays is currently available for the assessment of cytotoxicity in vitro, relatively few are practical and have sufficient sensitivity, scalability and robustness. The matter of how many toxicity endpoint parameters need to be assessed to identify a potentially cytotoxic compound is worthwhile discussing. In most cases, the different quantitative cytotoxic parameters currently used provide equivalent information on the toxicity of a compound. Proper assay method choice depends not only on the number and type of cells used, but also on the expected outcome. Of the assays with high-throughput utility, considerable practical differences exist as far as their use is concerned. For example, viability assays based on measuring metabolic reductase activities have been successfully used and continue to be a useful cost-effective staple of the industry. MTT is regarded as the first example of a tetrazolium salt used in multi-well viability reductase-based assays for adherent mammalian cells [20] and it is among the easiest cytotoxicity assays to perform. When using MTT salts, the formazan formed is water-insoluble, precipitates into cells and should be extracted with organic solvents [10]. This assay is well-characterized, simple to use, and referenced to this day in the literature. Moreover, it is often considered as a gold standard assay and is routinely used as a benchmark during the development of new cytotoxicity methods.

2 Materials

2.1 Cell Cultures

All solutions, glassware and pipettes must be sterile and all procedures should be carried out under aseptic conditions and in the sterile environment of a laminar flow cabinet.

1. Seeding culture medium for hepatocytes. Ham's F12:William's E medium 1:1 supplemented with 2 % newborn calf serum, 0.2 % bovine serum albumin fraction V, 10 nM insulin, 25 μg/mL transferrin, 0.1 μM sodium selenite, 65.5 μM ethanolamine, 7.2 μM linoleic acid, 7 mM glucose, 6.14 mM ascorbic acid, 0.64 mM *N*-omega-nitro-L-arginine methyl ester, 50 mU/mL penicillin, and 50 μg/mL streptomycin.
2. 24 h serum-free chemically defined medium for hepatocytes. The same culture medium composition as described above, but serum-free and supplemented with 10 nM dexamethasone.
3. Coating mixture for culture plates for hepatocytes. Dissolve 1 mg of human fibronectin in 97 mL of Dulbecco's Modified Eagle's Medium supplemented with 0.1 % bovine serum albumin. Add to the former solution 3 mL 0.1 % collagen from calf skin solution in 0.1 M acetic acid.
4. Culture medium for BALB/c 3T3 cells clone 31 (*see* **Note 1**). Dulbecco's Modified Eagle's Medium supplemented with 4.5 mg/L glucose, 10 % fetal calf serum, 4 mM L-glutamine, 50 U penicillin/mL, and 50 μg streptomycin/mL.
5. Phosphate-buffered saline (PBS). 137 mM NaCl, 2.7 mM KCl, 10 mM $Na_2HPO_4{\cdot}2H_2O$, 1.8 mM KH_2PO_4 in deionized water. Adjust to pH 7.4 and store for maximum 6 months at 4 °C.
6. 0.25 % trypsin/0.02 % ethylenediaminetetraacetic acid.
7. 0.4 % solution trypan blue (Sigma-Aldrich, Spain), sterile-filtered, cell culture tested.

2.2 MTT Assay

1. MTT stock solution. Dissolve 5 mg/mL MTT reagent (Sigma-Aldrich, Spain) in PBS and sterilize by filtration through 0.22 μm filter and store at 4 °C in the dark.
2. MTT incubation media. Mix 1 mL of MTT stock solution and 10 mL of serum-free culture medium.

2.3 Equipment and General Laboratory Ware

1. Plastic culture 96-well microtiter plates and flasks (*see* **Note 2**).
2. Incubator (37 °C ± 1 °C, 90 % ± 5 % humidity, 5 % ± 1 % CO_2).
3. Laminar air flow cabinet.
4. Thermostated waterbath (37 °C).
5. Inverse-phase contrast microscope.
6. Plate reader spectrophotometer equipped with a 550 nm ± 10 nm filter.

3 Methods

3.1 Coating of Culture Plates Procedure for Hepatocytes

1. Use freshly isolated human, rat, or mouse hepatocytes.
2. Coat culture plates with 10 μL/cm^2 coating mixture and leave at 37 °C and 5 % CO_2 for 1 h before hepatocyte seeding.
3. Remove the excess of coating mixture.

3.2 Subculture of BALB/C 3T3 Cells (See Note 3)

1. Decant the medium and wash the cell cultures briefly twice with PBS (i.e. 5 mL/25 cm^2 flask and 15 mL/75 cm^2 flask).
2. Add 0.5 mL trypsin-ethylenediaminetetraacetic acid solution/25 cm^2 flask or 1 mL/75 cm^2 flask to the monolayer for 15–30 s.
3. Remove the excess of the trypsin-ethylenediaminetetraacetic acid solution and incubate the cells at 37 °C and 5 % CO_2. After 2–3 min, lightly tap the flask to detach the cells into a single cell suspension.
4. After detaching the cells, add pre-warmed (i.e. 37 °C) cell culture medium to the flask, count the cells and dilute conveniently in cell culture medium for seeding in 96-well plates.

3.3 Cell Counting and Viability Assessment

1. After resuspension of hepatocytes or BALB/c 3T3 cells in cell culture medium, the viability is determined in a cell aliquot by cell counting using the trypan dye exclusion method.
2. Load an aliquot of a 1:1 mixture of hepatocyte suspension or BALB/c 3T3 cell suspension and 0.4 % trypan blue solution in saline in a counter chamber.
3. Count the viable cells in five different fields under an optical microscope (*see* **Note 4**).

3.4 Cell Seeding

1. Plate the hepatocytes in coated 96-well plates at a cell density of 25×10^3 alive cells/well in seeding culture medium. Incubate at for 1 h 37 °C and 5 % CO_2.
2. Aspire the cell culture medium to remove unattached cells and debris, and add the same volume of fresh seeding culture medium (*see* **Note 5**).

3. Plate the BALB/C 3T3 cells in 96-well plates at a density of 2.5×10^3 alive cells/well in cell culture medium (*see* **Note 6**).
4. Using a multi channel pipette, dispense 100 μL/well of the cell suspension (i.e. hepatocytes or BALB/C 3T3 cells) to each well, except the peripheral wells that only should receive 100 μL/well of culture medium (i.e. blanks). Prepare 1 plate/chemical to be tested (*see* **Note 7**).

3.5 Preparation of Cells for Assays

1. 24 h after cell plating, replace the medium of the hepatocytes by serum-free chemically defined medium just before adding the dilutions of the test chemicals for the MTT assay (*see* **Note 8**).
2. For BALB/c 3T3 cells, incubate the cells for 24 h at 37 °C and 5 % CO_2 in order to form a subconfluent monolayer. This incubation period assures cell recovery, adherence, and progression to the exponential growth phase.

3.6 Preparation of Test Chemicals and Range Finding Experiment (See Note 9)

1. Prepare the stock solution for each test chemical at the recommended highest soluble concentration. The highest test concentration applied to the cells in each range finding experiment should be 0.5 times the highest concentration soluble in culture medium, 1/200 the highest concentration found to be soluble in the solubility test if the chemical was soluble in dimethylsulfoxide or 1/50 the highest concentration found to be soluble in the solubility test if the chemical was soluble in ethanol (*see* **Note 10**).
2. For the range finding experiment, test eight concentrations of each test chemical by diluting the stock solution with a constant factor covering a large range. The initial dilution series should be log dilutions (e.g. 1:10, 1:100, and 1:1,000) (*see* **Note 11**).
3. Test up to eight concentrations of each chemical in a 96-well plate prepared as 2× concentrated solutions in the corresponding cell culture medium (i.e. hepatocytes or BALB/c 3T3 cells).

3.7 Application of Test Chemical

1. At the time of treatment, remove the cell culture medium from the cells by careful inversion of the plate over an appropriate receptacle (*see* **Note 12**).
2. Immediately add 100 μL fresh pre-warmed dilution cell culture medium either for BALB/c 3T3 cells or serum-free chemically defined medium for hepatocytes to all wells of columns 1–10 and 12, including the blanks. Follow the plate configuration in Fig. 1.
3. Transfer 200 μL pre-warmed cell culture medium either for BALB/c 3T3 cells or serum-free chemically defined medium for hepatocytes to the wells of column 11. Follow the plate configuration in Fig. 1.

	1	2	3	4	5	6	7	8	9	10	11	12
A	MTTb	VCb	VCb	C_1b	C_2b	C_3b	C_4b	C_5b	C_6b	C_7b	VCb	VCb
B	MTTb	VC1	C_1	C_2	C_3	C_4	C_5	C_6	C_7	C8	VC2	VCb
C	MTTb	VC1	C_1	C_2	C_3	C_4	C_5	C_6	C_7	C8	VC2	VCb
D	MTTb	VC1	C_1	C_2	C_3	C_4	C_5	C_6	C_7	C8	VC2	VCb
E	MTTb	VC1	C_1	C_2	C_3	C_4	C_5	C_6	C_7	C8	VC2	VCb
F	MTTb	VC1	C_1	C_2	C_3	C_4	C_5	C_6	C_7	C8	VC2	VCb
G	MTTb	VC1	C_1	C_2	C_3	C_4	C_5	C_6	C_7	C8	VC2	VCb
H	MTTb	VCb	VCb	C_1b	C_2b	C_3b	C_4b	C_5b	C_6b	C_7b	VCb	VCb

Fig. 1 96-Well plate configuration for positive controls and test chemicals for the MTT assay. VC1: BALB/c 3T3 cell culture medium or hepatocytes serum-free chemically defined culture medium containing the solvent. VC2: BALB/c 3T3 cell culture medium or hepatocytes serum-free chemically defined culture medium with no solvent. C1 to C8: test chemicals or positive control in eight concentrations (i.e. C1 the highest concentration and C8 the lowest concentration). B: blanks. VCb: VC1 blank. MTTb: blank of the MTT assay. C1b to C7b: blanks. Shaded wells contain no cells

4. Rapidly transfer 100 μL of the 2× concentrated dosing solutions to the appropriate wells of the test plate using a single delivery multi channel pipettor (*see* **Note 13**). Follow the plate configuration in Fig. 1.
5. Incubate cells for 24 h at 37 °C and 5 % CO_2.

3.8 Positive and Negative Controls

1. Set up a separate plate of positive control, in casu sodiumlaurylsulfate (SLS), at eight concentrations for each set of test chemical plates (*see* **Note 14**).
2. If historical means have not yet been established by the laboratory, IC_{50} ± standard deviation (SD) for SLS should be determined. The mean IC_{50} ± 2.5 SD for the SLS acceptable tests (i.e. after the removal of outliers) are the values that will be used as acceptance criteria for test sensitivity for the MTT assay.
3. Use two negative controls, namely control 1 (VC1) consisting of diluted culture medium in which chemicals are added to cells and control 2 (VC2) consisting of culture medium. Follow the plate configuration in Fig. 1.

3.9 MTT Assay

1. Carefully remove the medium with the test chemical and rinse the cells very carefully once with 100 μL pre-warmed PBS. Remove the rinsing solution by gently blotting on paper towels.

2. Add 100 μL MTT incubation medium to each well, including the blanks, and incubate at 37 °C and 5 % CO_2 for 2 h (*see* **Note 15**). Observe the cells briefly during MTT incubation (e.g. between 1 and 2 h for formazan crystal formation).
3. After incubation, remove the MTT incubation medium. Carefully remove the excess of medium in wells by gently blotting on paper towels.
4. Add exactly 100 μL dimethylsulfoxide to all wells, including blanks. Wells of column 1 are the blanks of the assay (MTTb). Follow the plate configuration in Fig. 1.
5. Gently shake the microtiter plate on a microtiter plate shaker for 20–45 s to extract formazan from the cells and to form a homogeneous solution.
6. Leave the plates for at least 10 min at room temperature. Plates should be protected from light by using a cover.
7. Measure the absorption of the resulting colored solution at 550 nm ± 10 nm in a microtiter plate reader using the blanks (MTTb) as a reference (*see* **Note 16**). A low variability (i.e. <15 %), both intra-plate and intra-assay, should be routinely obtained (*see* **Note 17**).

3.10 MTT Test Acceptance Criteria and Intra-assay Variability

The following acceptance criteria must be met for a MTT test to be acceptable:

1. The internal reference compound IC_{50} concentration must be within ± 2.5 SD of the mean established by the laboratory.
2. The mean of VC1 must not differ by more than 15 % from the mean of VC2.
3. A relevant concentration range around the IC_{50} concentration must be covered, preferably with several points of a graded effect, but with a minimum of 2 points, 1 on each side of the estimated IC_{50} value, avoiding too many non-cytotoxic and 100 %-cytotoxic concentrations (*see* **Note 18**).
4. The intra-assay variability in the wells of the same experimental condition (i.e. treatment with each concentration of the test chemicals or controls) must be lower than 15 %.

3.11 Data Analysis

1. Measure MTT optical density values in 4–6 replicates for each of the eight concentrations of the test chemical as well as for VC1 and VC2.
2. Subtract the arithmetic mean of the optical density of the MTTb measurements from all MTT optical density measurements. Express the relative cell viability as percentage by dividing all values by the arithmetic means of the MTT optical density *minus* mean MTTb values that were obtained under

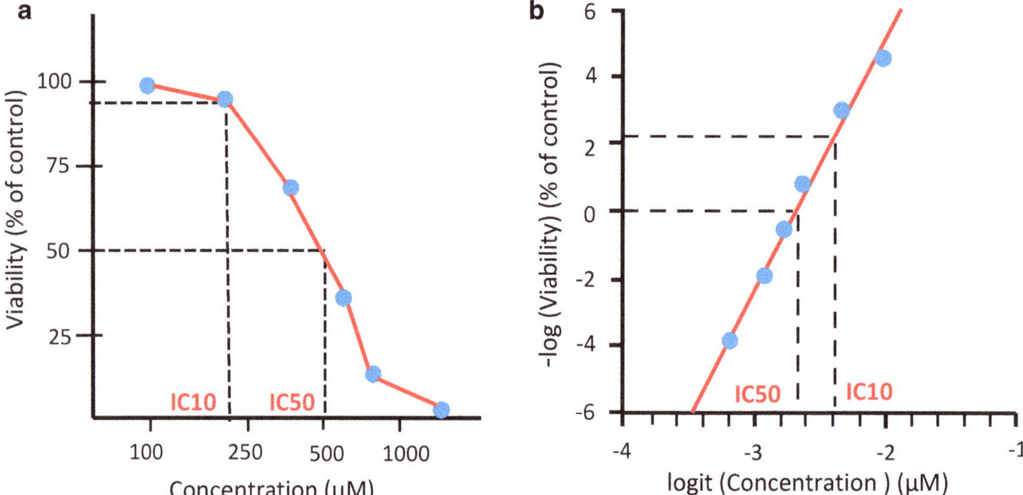

Fig. 2 IC_{10} and IC_{50} estimations. Experiments on cytotoxicity are designed to determine the highest concentration compatible with cell survival. This is roughly estimated for each parameter as the concentration causing only 10 % of the maximal cytotoxic effect (i.e. IC_{10}). The toxic potential of a given compound is expressed as the concentration causing 50 % of cell death (i.e. IC_{50}). Both parameters can be estimated either graphically from the concentration–effect curves (IC_{10} about 220 μM and IC_{50} about 500 μM; *panel A*) or mathematically in the Logit transformed curve (IC_{10} is 231.8 μM and IC_{50} is 461.0 μM; *panel B*)

VC1 and VC2. If achievable, the eight concentrations tested for each chemical will span the range from no effect up to full loss of cell viability.

3. Express toxicity data as IC_{10} and IC_{50} values (Fig. 2). Dose-response assays are generally characterized by a non-linear relationship between the viability and the amount of test compound given. To calculate these IC_{10} and IC_{50} values, Logit transformation of results can be used [1] (*see* **Note 19**). The transformed results fit linear regression. To obtain IC_{10} and IC_{50} values, one should interpolate the desired value in the transformed equation. Other software, such as GraphPadPrism (GraphPad Software, United States of America), is also widely used (*see* **Note 20**).
4. Summarize the IC_{50} values obtained from several experiments that were performed for the same test chemical by computation of the median IC_{50} value. The median should be chosen instead of the mean because it is unaffected by extreme values, whereas the mean strongly depends on extreme values.
5. If, for a specific experiment, the calculated IC_{50} value is larger than the maximum concentration tested, the experiment should be summarized as having no effect since the estimated IC_{50} value cannot be relied on.

3.12 Cytotoxicity Evaluation of Chemicals in Hepatocytes and BALB/c 3T3 Cells Using the MTT Assay

The MTT method is a frequently used cytotoxicity assay. When cells die, they lose the ability to convert MTT into formazan, thus color formation serves as a useful and convenient marker of viable cells. The exact mechanism of the reaction of MTT reduction into formazan is not well understood, but is likely to involve nicotinamide adenine dinucleotide or similar reducing molecules that transfer electrons to MTT by specific mitochondrial enzymes. This has led to the assumption that the MTT assay is actually measuring mitochondrial activity [31]. As an example, cytotoxicity of a set of compounds was assessed in hepatocytes and BALB/c 3T3 cells, including tamoxifen [32], which has been reported to elicit hepatic metabolism-dependent toxicity [33], amitriptyline, lithium sulfate, sodium selenate and pentachlorophenol. A bioactivable compound is expected to cause more toxicity in hepatocytes than in non-metabolizing cells, such as BALB/c 3T3 cells. A compound eliciting similar toxicities in both cell types indicates that it primarily exerts basal cytotoxicity. As shown in Figs. 3 and 4, Table 2, no difference in cytotoxicity is observed between the two cellular models for amitriptyline, lithium sulfate, sodium selenate and pentachlorophenol. Therefore, they are correctly classified as cytotoxic. As significantly lower IC_{10} and IC_{50} values in hepatocytes were obtained for tamoxifen, it was correctly classified as bioactivable [34]. In addition, the IC_{10} and IC_{50} values are mostly used to label and rank the compounds according to their toxic potential. As shown in Table 2, tamoxifen has the lowest IC_{50} value in both

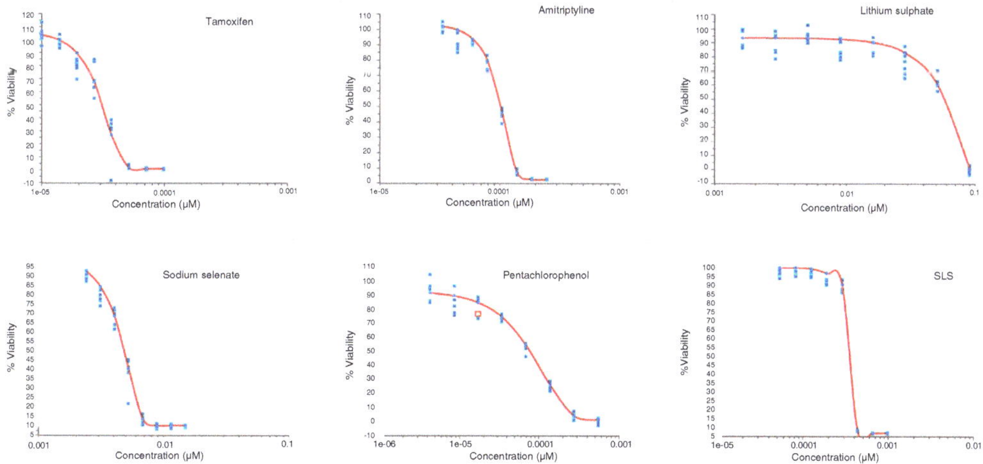

Fig. 3 Cytotoxicity of a set of compounds to rat hepatocytes. Hepatocytes were seeded in 96-well plates and cytotoxicity with the MTT test was assayed after 24 h exposure to a range of compounds concentrations. In total, three independent experiments were performed and SLS was used as the positive control. The values represent the mean and the standard deviation of the IC_{10} and IC_{50} values

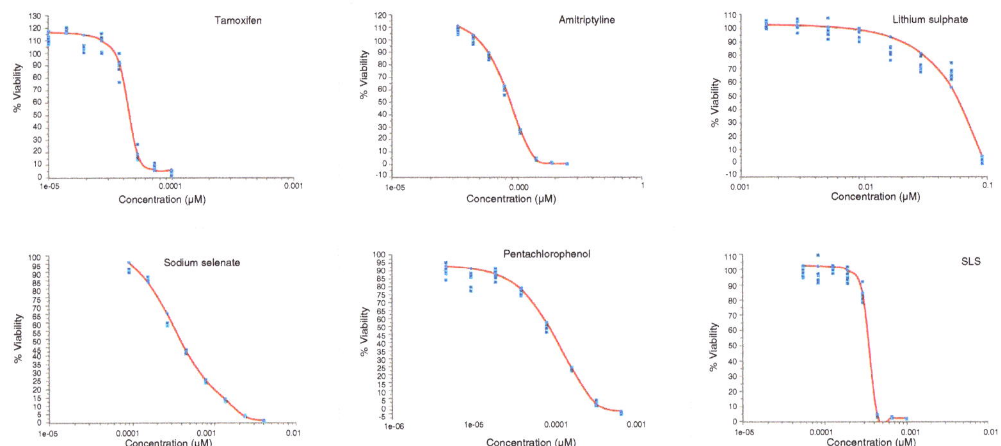

Fig. 4 Cytotoxicity of a set of compounds to BALB/c 3T3 cells. Cells were seeded in 96-well plates and cytotoxicity with the MTT test was assayed after 24 h exposure to a range of compounds concentrations. In total, three independent experiments were performed and SLS was used as the positive control. The values represent the mean and the standard deviation of the IC_{10} and IC_{50} values

Table 2
IC_{10} and IC_{50} values (μM) estimated in rat hepatocytes and BALB c/3T3 cells

	Rat hepatocytes		BALB c/3T3 cells	
Compound	**IC_{10}**	**IC_{50}**	**IC_{10}**	**IC_{50}**
Tamoxifen	5 ± 0.5	22 ± 1.7	7 ± 1	60 ± 5.9
Amitriptyline	56 ± 14	90 ± 11	46 ± 4	79 ± 8
Lithium sulfate	28,250 ± 3,800	46,567 ± 4,000	8,203 ± 1,300	51,963 ± 8,400
Sodium selenate	302 ± 13	525 ± 32	158 ± 23	452 ± 50
Pentachlorophenol	16 ± 5	80 ± 14	24 ± 15	83 ± 10
SLS	271 ± 36	355 ± 6	260 ± 32	349 ± 7

cellular models and is therefore the most cytotoxic, followed by pentachlorophenol and amitriptyline, with similar IC_{10} and IC_{50} values, sodium selenate, SLS and finally by lithium sulfate with the highest IC_{50} value (i.e. least cytotoxic).

4 Notes

1. BALB/c 3T3 cells clone 31 can be supplied by the European Collection of Cell Cultures (United Kingdom) or by the American Type Culture Collection (United States of America).

2. 96-Well plates should be checked to ensure that they adequately support the attachment and spreading of hepatocytes.
3. When BALB/C 3T3 cells reach 50–80 % confluence, they should be removed from the flask by trypsinization and subcultured.
4. Hepatocyte preparations with viability below 70 % are not adequate for further cultivation and use in the MTT assay. Viability of BALB/c 3T3 cells to be subcultured after detaching usually is more than 90 %.
5. The attachment efficiency of hepatocyte suspensions to fibronectin/collagen-coated plates 1–2 h after cell seeding is usually 70–80 %.
6. The seeding density should be noted to ensure that the cells in the control wells are not overgrown after 24 h incubation during exposure to test chemicals.
7. Examine each plate under a phase-contrast microscope after seeding to assure that cell density is relatively even across the microtiter plate. This also allows to check for experimental and systemic cell seeding errors.
8. It is known that the activity of cytochrome P450 enzymes in hepatocytes decreases with culture time [35–37]. In case the toxicity of test chemicals is related to a bioactivation process, test compounds can be added to hepatocytes preferably at the time of medium renewal 1–2 h after cell seeding. In this case, seeding medium for preparing dilutions of test chemicals should be used.
9. Prepare test chemicals ex tempore. Test chemical solutions should not be prepared in bulk for use in subsequent tests. The solutions must not have noticeable precipitations. Each stock dilution should have at least 1–2 mL total volume to ensure adequate solution. For chemicals dissolved in dimethylsulfoxide, the final concentration of the latter for application to the cells must be 0.5 % in the dilution medium either in the controls or in the eight test concentrations. For ethanol, a maximal concentration of 2 % can be employed.
10. Each chemical should be tested using freshly prepared dilutions on each occasion. Preparation under red or yellow light is recommended to preserve chemicals that degrade upon exposure to light.
11. Do not use automatic plate washers for this procedure nor vacuum aspiration. It is recommended to gently blot the plate on a sterile paper towel so that the monolayer is minimally disrupted.
12. If a range finder experiment does not generate enough cytotoxicity, higher doses should be attempted. If cytotoxicity is

limited by solubility, more stringent solubility procedures to increase the stock concentration should be employed.

13. The test chemical dosing solutions should be transferred from the lowest to the highest concentration so that the same pipette tips on the multichannel pipettor can be used for the whole plate.
14. Two or three compounds as internal reference controls in each assay can be used. The concentration range should be determined experimentally.
15. William's E medium can interfere with formazan color formation. Therefore, at this stage, replace William's E medium by any other culture medium, including all supplements, and use it to prepare the MTT incubation media.
16. Absorbance readings can only be accepted when there is no noticeable turbidity in the working dilutions in the culture medium due to the precipitation of the compound.
17. Lack of reproducibility could be due to the physico-chemical properties of the test compounds, many of them showing hydrophobicity and poor water solubility. In addition, insolubility when diluting in culture medium to prepare working concentrations may occur. Adsorption to plastic could also occur. Volatile compounds show great variability among experiments.
18. Experiments revealing less than 1 cytotoxic concentration on each side of the IC_{50} value shall be repeated, whenever possible with a smaller dilution factor. Each experiment should have at least 1 cytotoxicity value higher than 0 and 50 % viability and at least 1 cytotoxicity value higher than 50 % and lower than 100 % viability.
19. This notion of the IC_{50} value can be generalized further such that ICX is the concentration at which the response is X % between minimal and maximal activity.
20. In the Logit transformation, the results of concentration (x) and viability (y) are expressed as $x' = -\log(x)$ and $y' = \text{Logit}(y) = \ln((y - y_{oo})/(y_o - (y - y_{oo}))$. In this equation, y_{oo} is the asyndote of an infinite concentration (i.e. almost 0) and y_o is the asyndote of a concentration of 0 (i.e. 100 %). The transformed results fit a linear regression. To obtain a given IC_{10} or IC_{50} value, one should interpolate the desired value in the transformed equation.

Acknowledgements

The authors acknowledge financial support from the ALIVE Foundation. Laia Tolosa was a recipient of a Sara Borrell Contract from the "Instituto de Salud Carlos III" of the Spanish Ministry of Economy and Competitiveness.

References

1. Gomez-Lechon MJ, Ponsoda X, Castell JV (2000) In vitro toxicity testing. In: Doyle A, Griffiths JB (eds) Cell and tissue culture for medical research. Wiley, Chinchester, pp 402–419
2. Niles AL, Moravec RA, Riss TL (2008) Update on in vitro cytotoxicity assays for drug development. Expert Opin Drug Discov 3:655–669
3. Yang H, Acker J, Chen A et al (1998) In situ assessment of cell viability. Cell Transplant 7:443–451
4. Stoddart MJ (2011) Cell viability assays: introduction. Methods Mol Biol 740:1–6
5. Freshney R (1987) Culture of animal cells: a manual of basic technique. Alan R. Liss, Inc., New York
6. Krishan A (1975) Rapid flow cytofluorometric analysis of mammalian cell cycle by propidium iodide staining. J Cell Biol 66:188–193
7. Bucana C, Saiki I, Nayar R (1986) Uptake and accumulation of the vital dye hydroethidine in neoplastic cells. J Histochem Cytochem 34: 1109–1115
8. Hutz RJ, DeMayo FJ, Dukelow WR (1985) The use of vital dyes to assess embryonic viability in the hamster, *Mesocricetus auratus*. Stain Technol 60:163–167
9. Schreer A, Tinson C, Sherry JP et al (2005) Application of alamar blue/5-carboxyfluorescein diacetate acetoxymethyl ester as a noninvasive cell viability assay in primary hepatocytes from rainbow trout. Anal Biochem 344:76–85
10. Castell JV, Gomez-Lechon MJ, Ponsoda X et al (1997) In vitro investigation of the molecular mechanisms of hepatotoxicity. Arch Toxicol 19:313–321
11. Ponsoda X, Jover R, Castell JV et al (1991) Measurement of intracellular LDH activity in 96-well cultures: a rapid and automated assay for cytotoxicity studies. J Tissue Cult Meth 13:21–24
12. Cho MH, Niles A, Huang R et al (2008) A bioluminescent cytotoxicity assay for assessment of membrane integrity using a proteolytic biomarker. Toxicol In Vitro 22: 1099–1106
13. Quah BJ, Parish CR (2010) The use of carboxyfluorescein diacetate succinimidyl ester (CFSE) to monitor lymphocyte proliferation. J Vis Exp 44:2259
14. Franken NA, Rodermond HM, Stap J et al (2006) Clonogenic assay of cells in vitro. Nat Protoc 1:2315–2319
15. Plumb JA (1999) Cell sensitivity assays: clonogenic assay. Methods Mol Med 28:17–23
16. Skehan P, Storeng R, Scudiero D et al (1990) New colorimetric cytotoxicity assay for anticancer-drug screening. J Natl Cancer Inst 82:1107–1112
17. Vichai V, Kirtikara K (2006) Sulforhodamine B colorimetric assay for cytotoxicity screening. Nat Protoc 1:1112–1116
18. Sapan CV, Lundblad RL, Price NC (1999) Colorimetric protein assay techniques. Biotechnol Appl Biochem 29:99–108
19. Fotakis G, Timbrell JA (2006) In vitro cytotoxicity assays: comparison of LDH, neutral red, MTT and protein assay in hepatoma cell lines following exposure to cadmium chloride. Toxicol Lett 160:171–177
20. Mosmann T (1983) Rapid colorimetric assay for cellular growth and survival: application to proliferation and cytotoxicity assays. J Immunol Methods 65:55–63
21. Brosin A, Wolf V, Mattheus A et al (1997) Use of XTT-assay to assess the cytotoxicity of different surfactants and metal salts in human keratinocytes (HaCaT): a feasible method for in vitro testing of skin irritants. Acta Derm Venereol 77:26–28
22. Ponsoda X, Gomez-Lechon MJ, Castell JV (1998) Toxicity and cell density monitoring in monolayer and three-dimmensional cultures with the XTT assay. ATLA 26:331–342
23. Ishiyama M, Shiga M, Sasamoto K et al (1993) A new sulfonated tetrazolium salt that produces a highly water-soluble formazan dye. Chem Pharm Bull 41:1118–1122
24. Gonzalez RJ, Tarloff JB (2001) Evaluation of hepatic subcellular fractions for alamar blue and MTT reductase activity. Toxicol In Vitro 15:257–259
25. McMillian MK, Li L, Parker JB et al (2002) An improved resazurin-based cytotoxicity assay for hepatic cells. Cell Biol Toxicol 18: 157–173
26. O'Brien J, Wilson I, Orton T et al (2000) Investigation of the alamar blue (resazurin) fluorescent dye for the assessment of mammalian cell cytotoxicity. Eur J Biochem 267:5421–5426
27. Crouch SP, Kozlowski R, Slater KJ et al (1993) The use of ATP bioluminescence as a measure of cell proliferation and cytotoxicity. J Immunol Methods 160:81–88
28. Lundin A, Hasenson M, Persson J et al (1986) Estimation of biomass in growing cell lines by adenosine triphosphate assay. Methods Enzymol 133:27–42
29. Borenfreund E, Puerner JA (1985) Toxicity determined in vitro by morphological alterations

and neutral red absorption. Toxicol Lett 24:119–124

30. Jones PA, King AV (2003) High-throughput screening (HTS) for phototoxicity hazard using the in vitro 3T3 neutral red uptake assay. Toxicol In Vitro 17:703–708
31. van Meerloo J, Kaspers GJ, Cloos J (2011) Cell sensitivity assays: the MTT assay. Methods Mol Biol 731:237–245
32. Sridar C, D'Agostino J, Hollenberg PF (2012) Bioactivation of the cancer chemopreventive agent tamoxifen to quinone methides by cytochrome P4502B6 and identification of the modified residue on the apoprotein. Drug Metab Dispos 40:2280–2288
33. Park BK, Laverty H, Srivastava A et al (2011) Drug bioactivation and protein adduct formation in the pathogenesis of drug-induced toxicity. Chem Biol Interact 192:30–36
34. Gomez-Lechon MJ, Tolosa L, Castell JV et al (2010) Mechanism-based selection of compounds for the development of innovative in vitro approaches to hepatotoxicity studies in the LIINTOP project. Toxicol In Vitro 24:1879–1889
35. Binda D, Lasserre-Bigot D, Bonet A et al (2003) Time course of cytochromes P450 decline during rat hepatocyte isolation and culture: effect of *L*-NAME. Toxicol In Vitro 17:59–67
36. Rodriguez-Antona C, Donato MT, Boobis A et al (2002) Cytochrome P450 expression in human hepatocytes and hepatoma cell lines: molecular mechanisms that determine lower expression in cultured cells. Xenobiotica 32:505–520
37. Suolinna EM (1982) Isolation and culture of liver cells and their use in the biochemical research of xenobiotics. Med Biol 60:237–254
38. Fallot P, Girgis A, Laine-Boeszoermenyi M et al (1965) Use of liquid scintillation spectrometer for the direct measurement of the radioactivity of tritium incorporated in the cells of mammals cultivated in vitro. Int J Appl Radiat Isot 16:349–356
39. Dolbeare F, Gratzner H, Pallavicini MG et al (1983) Flow cytometric measurement of total DNA content and incorporated bromodeoxyuridine. Proc Natl Acad Sci U S A 80:5573–5577
40. Ranall MV, Gabrielli BG, Gonda TJ (2010) Adaptation and validation of DNA synthesis detection by fluorescent dye derivatization for high-throughput screening. Biotechniques 48:379–386
41. Korzeniewski C, Callewaert DM (1983) An enzyme-release assay for natural cytotoxicity. J Immunol Methods 64:313–320
42. Sewell RB, Ling TS, Yeomans ND (1986) Ethanol-induced cell damage in cultured rat antral mucosa assessed by chromium-51 release. Dig Dis Sci 31:853–858
43. Tan AS, Berridge MV (2000) Superoxide produced by activated neutrophils efficiently reduces the tetrazolium salt, WST-1 to produce a soluble formazan: a simple colorimetric assay for measuring respiratory burst activation and for screening anti-inflammatory agents. J Immunol Methods 238:59–68

Chapter 27

Measurement of Apoptotic and Necrotic Cell Death in Primary Hepatocyte Cultures

Michaël Maes, Tamara Vanhaecke, Bruno Cogliati, Sara Crespo Yanguas, Joost Willebrords, Vera Rogiers, and Mathieu Vinken

Abstract

Hepatotoxicity, including drug-induced liver injury, is frequently accompanied by cell death. The latter is typically driven by apoptosis or necrosis, which substantially differ based upon biochemical and morphological criteria. This chapter describes two commonly used methods to probe apoptotic and necrotic activities in adherent monolayer cultures of primary hepatocytes. The apoptosis assay uses a prototypical substrate of caspase 3, the main executor of apoptotic cell death, which can be cleaved, yielding a product that can be measured fluorimetrically. The second assay relies on the disruption of the cell plasma membrane, which typically occurs in necrotic cell death and that results in the extracellular release of cytoplasmic enzymes, such as lactate dehydrogenase. The latter can be indirectly assessed by spectrophotometrically measuring the consumption of reduced nicotinamide adenine dinucleotide.

Key words Primary hepatocyte culture, Apoptosis, Caspase 3, Necrosis, Lactate dehydrogenase

1 Introduction

Drug-induced liver injury is a ubiquitous issue in clinical settings and pharmaceutical industry. Recently, a unifying mechanistic model of drug-induced liver injury has been introduced. According to this model, drug-induced hepatotoxicity relies on three consecutive steps, namely an initial cellular insult that leads to the occurrence of mitochondrial permeability transition, which in turn ultimately burgeons into the onset of cell death. The latter may be manifested as apoptosis or necrosis [1–3]. Apoptosis is a genetically programmed and well-orchestrated process that can be triggered via two routes, namely through the mitochondria-mediated intrinsic cascade or through the death receptor-mediated extrinsic pathway. Both routes rely on the proteolytic activity of cystein proteases, the caspases, which are expressed as inactive pro-enzymes

Mathieu Vinken and Vera Rogiers (eds.), *Protocols in In Vitro Hepatocyte Research*, Methods in Molecular Biology, vol. 1250, DOI 10.1007/978-1-4939-2074-7_27, © Springer Science+Business Media New York 2015

that become activated upon proteolytic cleavage. As such, two subsets of caspases can be distinguished, namely the initiator caspases and the effector caspases. The former, including caspase 8, caspase 9 and caspase 10, mediate the cleavage and thus the activation of effector caspases. Effector caspases, such as caspase 3, caspase 6 and caspase 7, subsequently cleave a large number of cellular proteins, like major cytoplasmic and nuclear elements, which forms the biochemical basis of the apoptotic morphological phenotype. The extrinsic apoptotic pathway is initiated by binding of a specific subset of ligands, such as Fas ligand, to their corresponding receptors at the cell plasma membrane surface, resulting in the proteolytic cleavage and auto-activation of procaspase 8. Activated caspase 8 induces caspase 3, which subsequently cleaves its cellular targets. In the intrinsic apoptotic pathway, cytochrome C is released from mitochondria, a process that is mediated by members of the B-cell lymphoma-2 protein family. Cytochrome C triggers caspase 9 activation, which then induces caspase 3 [4–9]. Necrosis, as opposed to apoptosis, is a rather passive and unorganized process that is caused by a plethora of external stress factors, including extremely high concentrations of xenobiotics. It usually starts with the loss of ion homeostasis, which eventually evokes cell swelling, loss of cell plasma membrane integrity, and cell lysis [4, 5, 9–12].

This chapter describes two methods to measure apoptotic and necrotic cell death in primary hepatocyte cultures at the activity level, using an established in vitro model of Fas-mediated cell death [13]. The apoptosis activity method is based upon the use of a synthetic caspase 3 substrate called acetyl-aspartic acid-glutamic acid-valine-aspartic acid-7-aminotrifluoromethylcoumarin (Ac-DEVD-AFC) (Fig. 1) (*see* **Note 1**). When the AFC label is attached to the peptide substrate, it produces a blue fluorescence upon exposure to ultraviolet light. Caspase 3 cleaves this tetrapeptide between the second aspartic acid and the AFC moiety. Released AFC produces a yellow-green fluorescence at 505 nm when exposed to ultraviolet light,

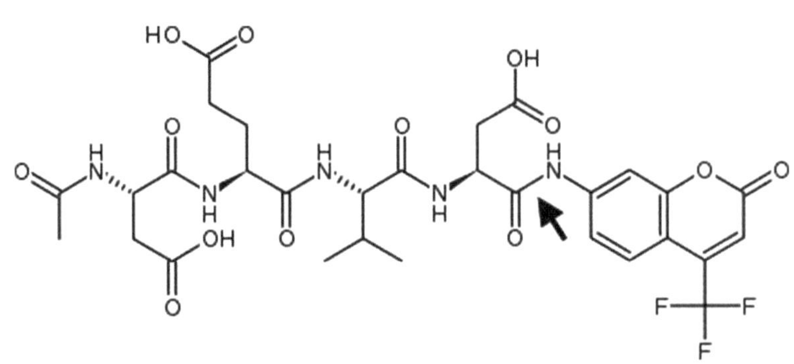

Fig. 1 Chemical structure acetyl-aspartic acid-glutamic acid-valine-aspartic acid-7-aminotrifluoromethyl-coumarin (Ac-DEVD-AFC). The *arrow* indicates the cleavage site for caspase 3

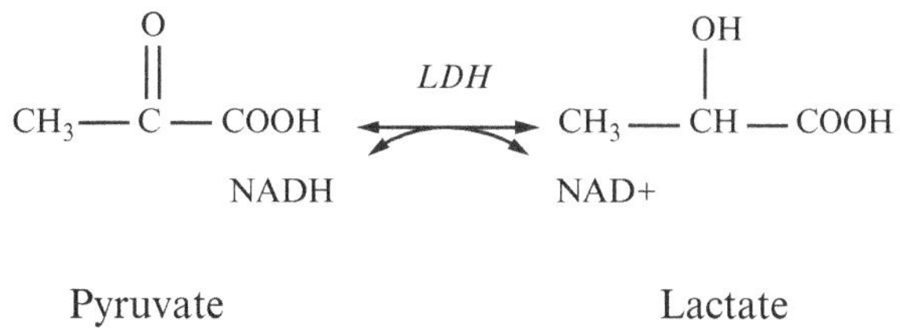

$$CH_3-\overset{\overset{\Large O}{||}}{C}-COOH \underset{NADH \quad NAD+}{\overset{LDH}{\rightleftharpoons}} CH_3-\overset{\overset{\Large OH}{|}}{CH}-COOH$$

Pyruvate Lactate

Fig. 2 Catalytic function of lactate dehydrogenase (LDH). LDH catalyzes the interconversion of pyruvate and lactate with concomitant interconversion of reduced and oxidized nicotinamide adenine dinucleotide

which can be measured fluorimetrically. The caspase 3 activity is proportional to the amount of free AFC produced [8, 14]. The assay to assess necrotic cell death relies on the disruption of the cell plasma membrane, which results in the extracellular release of cytoplasmic enzymes. Among those is lactate dehydrogenase (LDH), a stable enzyme that leaks from the cell upon cell plasma membrane damage in relatively high amounts. LDH catalyzes the interconversion of pyruvate and lactate with concomitant interconversion of reduced (NADH) and oxidized nicotinamide adenine dinucleotide (Fig. 2). The principle of the assay described in this chapter is based upon this reaction. In particular, the consumption of NADH is spectrophotometrically assessed and serves as a measure that is proportional to LDH activity [15].

2 Materials

2.1 Establishment of the In Vitro Model of Fas-Mediated Hepatocellular Cell Death

1. Hepatocyte seeding medium. William's medium E containing 7 ng/mL glucagon, 292 mg/mL L-glutamine, 7.33 IE/mL sodium benzyl penicillin, 50 μg/mL kanamycin monosulfate, 10 μg/mL sodium ampicillin, 50 μg/mL streptomycin sulfate, and 10 % fetal bovine serum. Prepare in a laminar air flow cabinet and store for maximum 7 days at 4 °C. Prior to use, the hepatocyte seeding medium should be placed for 30 min in a thermostated bath at 37 °C.
2. Hepatocyte culture medium. Serum-free hepatocyte seeding medium supplemented with 25 μg/mL hydrocortisone sodium hemisuccinate and 0.5 μg/mL insulin. Prepare in a laminar air flow cabinet and store for maximum 7 days at 4 °C. Prior to use, the hepatocyte culture medium should be placed for 30 min in a thermostated bath at 37 °C.
3. Hepatocyte cell death medium. Hepatocyte culture medium supplemented with 200 ng/mL Fas ligand (Alexis, Switzerland) and 2 μg/mL cycloheximide (Sigma-Aldrich, Belgium). Prepare ex tempore in a laminar air flow cabinet. Prior to use,

the hepatocyte cell death medium should be placed for 30 min in a thermostated bath at 37 °C.

4. Incubator (37 °C ± 1 °C, 90 % ± 5 % humidity, 5 % ± 1 % CO_2).
5. Laminar air flow cabinet.
6. Thermostated bath.

2.2 Measurement of Caspase 3 Activity in Cultured Primary Hepatocytes

1. Phosphate-buffered saline (PBS). 137 mM NaCl, 2.7 mM KCl, 10 mM $Na_2HPO_4{\cdot}2H_2O$, 1.8 mM KH_2PO_4 in deionized water. Adjust to pH 7.4, sterilize by passing through a 0.22 μm filter and store for maximum 6 months at 4 °C. Prior to use, PBS should be placed for 30 min at room temperature.
2. 25× Concentrated reaction buffer. 250 mM 1,4-piperazinediethanesulfonic acid, 125 mM DL-dithiothreitol, 50 mM ethylenediaminetetraacetic acid tetrasodium salt dehydrate in double distilled water. Adjust pH of this solution to 7.4 (*see* **Note 2**) and store in aliquots of 2.5 mL for maximum 6 months at −20 °C.
3. Reaction buffer. 1 volume part 25× concentrated reaction buffer and 24 volume parts double distilled water. Prepare this buffer ex tempore and place at least 10 min at 37 °C prior to use.
4. *N*-[2-hydroxyethyl]piperazine-*N′*[2-ethanesulfonic acid] solution. 50 mM *N*-[2-hydroxyethyl]piperazine-*N′*[2-ethanesulfonic acid] in double distilled water. Sterilize this solution by passing through a 0.22 μm filter and store for maximum 6 months at 4 °C.
5. Ethylenediaminetetraacetic acid tetrasodium salt dihydrate solution. 500 mM ethylenediaminetetraacetic acid tetrasodium salt dihydrate solution in double distilled water. Adjust the pH of this solution to 7.4. Sterilize by passing through a 0.22 μm filter and store for maximum 6 months at room temperature.
6. Phenylmethylsulfonyl fluoride solution. 100 mM phenylmethylsulfonyl fluoride in isopropanol. Protect this solution from light and store for maximum 2 weeks at −20 °C.
7. Pepstatin A solution. 1.5 mM pepstatin A in dimethyl sulfoxide. Protect this solution from light and store for maximum 4 weeks at −20 °C.
8. Aprotinin solution. 0.15 mM aprotinin in PBS. Protect this solution from light and store for maximum for 6 months at −20 °C.
9. Leupeptin solution. 2.1 mM leupeptin in PBS. Protect this solution from light and store for maximum 4 weeks at −20 °C.
10. Lysis buffer. 1 mg/mL 3-[(3-cholamidopropyl)dimethylammonio]-1-propanesulfonate, 5 mM DL-dithiothreitol, 20 % *N*-[2-hydroxyethyl]piperazine-*N′*[2-ethanesulfonic acid] solution,

0.4 % ethylenediaminetetraacetic acid tetrasodium salt dihydrate solution, 1 % phenylmethylsulfonyl fluoride solution, 1 % pepstatin A solution, 1 % aprotinin solution, 2 % leupeptin solution in double distilled water. Prepare this buffer ex tempore (*see* **Note 3**).

11. Ac-DEVD-AFC solution. 1 mg/mL Ac-DEVD-AFC in dimethyl sulfoxide. Protect this solution from light and store for maximum 6 months at −20 °C.
12. AFC solution. 200 mM AFC in dimethyl sulfoxide. Protect this solution from light and store for maximum 2 weeks at room temperature.
13. Amber microcentrifuge tubes (VWR, Belgium).
14. Black 96-well plates (VWR, Belgium).
15. Multiplate reader with temperature control.

2.3 Measurement of the Extracellular Release of Lactate Dehydrogenase in Cultured Primary Hepatocytes

1. LDH reaction buffer (DiaSys, Germany). 4 volume parts of reagent 1 (i.e. NADH dissolved in 1 mM Good's buffer, pH 9.6) and 1 part of reagent 2 (i.e. pyruvate in 800 μM phosphate buffer, pH 7.5). Prepare this buffer ex tempore and place in a thermostated bath at 25 °C before use.
2. Plastic microcuvettes.
3. Spectrophotometer, computer and software.
4. Ultrasonicator.

3 Methods

3.1 Establishment of a Monolayer Culture of Primary Hepatocytes and Induction of Cell Death

1. Use freshly isolated primary rat hepatocytes [16].
2. Evenly plate the hepatocytes on plastic culture dishes at a density of 0.56×10^5 cells/cm^2 in hepatocyte seeding medium. Place the cell cultures in an incubator at 37 °C and 5 % CO_2 for 4 h.
3. Remove the hepatocyte seeding medium and replace by identical volumes of hepatocyte culture medium. Place the cell cultures in an incubator at 37 °C and 5 % CO_2 for 20 h.
4. Replace the hepatocyte culture medium. Place the cell cultures in an incubator at 37 °C and 5 % CO_2 for 20 h.
5. Remove the hepatocyte culture medium and replace by identical volumes of hepatocyte cell death medium. Place the cell cultures in an incubator at 37 °C and 5 % CO_2.
6. Sample at the start of cell death induction and 2, 4, and 6 h thereafter.

3.2 Measurement of Caspase 3 Activity in Cultured Primary Hepatocytes

3.2.1 Sampling from Cultured Primary Hepatocytes

1. Remove the hepatocyte cell death medium from 10 cm diameter culture dishes and wash the cells twice with 10 mL PBS (*see* **Note 4**).
2. Add 5.5 mL PBS to the culture dish and scrape the cells (*see* **Note 5**).
3. Transfer the cell suspension to a 15 mL centrifuge tube on ice.
4. Repeat **steps 2** and **3**.
5. Centrifuge for 5 min at 4,332 × *g* and 4 °C.
6. Gently remove the supernatant and resuspend the pellet in 1 mL ice-cold PBS.
7. Transfer the cell suspension to a 2 mL microcentrifuge tube on ice.
8. Centrifuge for 5 min at 2,040 × *g* and 4 °C.
9. Remove the supernatant and directly merge the microcentrifuge tube into liquid nitrogen.
10. Stock the microcentrifuge tubes, containing the dry pellets, at −80 °C.

3.2.2 Preparation of Cell Homogenates

1. Defrost the microcentrifuge tubes on ice.
2. Add 100 μL lysis buffer to the microcentrifuge tubes and resuspend by pipetting up and down (*see* **Note 5**).
3. Leave the microcentrifuge tubes on ice for 20 min.
4. Merge the microcentrifuge tubes in liquid nitrogen and subsequently thaw at 37 °C in a thermostated bath for 2 min.
5. Repeat **step 4** twice without mixing the microcentrifuge tube between the freeze–thaw cycles.
6. Centrifuge for 30 min at 16,065 × *g* and 4 °C.
7. Transfer 5 μL of the supernatant to another microcentrifuge tube, add 95 μL double distilled water and determine the protein concentration (*see* **Note 6**).
8. Transfer the remaining supernatant to a new microcentrifuge tube, merge into liquid nitrogen and store at −80 °C.

3.2.3 Measurement of Caspase 3 Activity in the Standards

1. Prepare the reaction buffer and the AFC solution.
2. Prepare a series of AFC standards (Table 1) in 2 mL amber microcentrifuge tubes protected from light.
3. Add 240 μL of each standard to a corresponding well in a black 96-well plate.
4. Repeat **step 3** for the same standard.
5. Repeat **steps 3–4** for the remaining standards.
6. Place the black 96-well plate in the multiplate reader and measure the fluorescence at 37 °C with λ_{ex} 390 nm and λ_{em} 500 nm (*see* **Note 7**).

Table 1
Preparation of the 7-amino-trifluoromethylcoumarin (AFC) standards

Standard number	Concentration AFC (nM)	Reaction buffer (μL)	Solution (μL)
0 (blank)	0	1,000	0
1	1,200	994	6 (AFC solution)
2	600	500	500 (standard 1)
3	300	500	500 (standard 2)
4	150	500	500 (standard 3)
5	75	500	500 (standard 4)
6	37.5	500	500 (standard 5)
7	18.75	500	500 (standard 6)
8	9.375	500	500 (standard 7)

3.2.4 Measurement of Caspase 3 Activity in the Samples

1. For each of the samples, calculate the volume that corresponds with 50 μg protein (x μL).
2. Mix 230-x μL double distilled water with 10 μL 25× concentrated reaction buffer and x μL sample in a well of a black 96-well plate.
3. Repeat **step 2** twice for the same sample.
4. Repeat **steps 2** and **3** for the remaining samples.
5. In a well of the same black 96-well plate, mix 230 μL double distilled water with 10 μL 25× concentrated reaction buffer.
6. Repeat **step 5** twice for the blank.
7. Incubate the black 96-well plate for 2 min at room temperature under mild shaking (i.e. 120 oscillations per minute).
8. Add 10 μL Ac-DEVD-AFC solution to each well using a dispenser pipette.
9. Slightly tap the black 96-well plate and place into the multiplate reader.
10. Place the black 96-well plate in the multiplate reader and perform 99 measurements of the fluorescence at 37 °C with λ_{ex} 390 nm and λ_{em} 500 nm over a total time frame of 60 min (*see* **Note** 7).

3.2.5 Processing of the Results (Fig. 3) (See Note 8)

1. Collect the results of the measurement of the caspase 3 activity in the standards and calculate the mathematical equation of the best-fit line:

$$y = ax + b$$

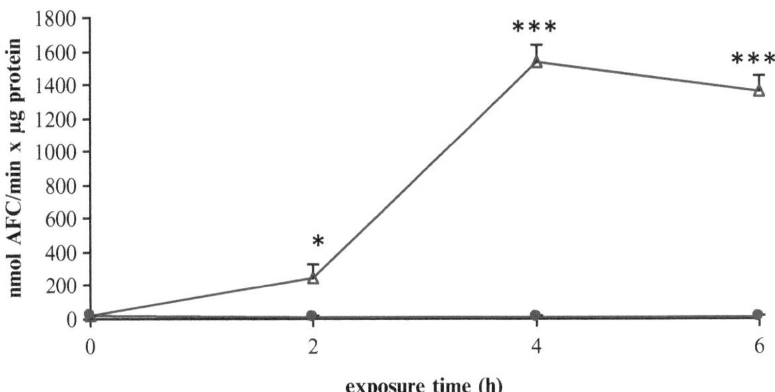

Fig. 3 Measurement of caspase 3-like activity. Freshly isolated rat hepatocytes were cultivated in a monolayer configuration and were exposed to 200 ng/mL Fas ligand and 2 µg/mL cycloheximide, starting at 44 h postplating. Samples were taken at the start of the exposure (0 h), and after 2, 4, and 6 h. Caspase 3-like activity was determined and results are expressed as nmol aminotrifluoromethylcoumarin (AFC)/min × µg protein. Results represent the means ± standard deviation of three independent experiments. Results were evaluated by 1-way analysis of variance followed by post hoc Bonferroni tests. *Filled circle* control, *open square* Fas ligand/cycloheximide; *$p<0.05$; ***$p<0.001$. Reproduced in modified form with permission from [13]

2. Calculate the caspase 3 activity in the samples, expressed as nmol produced AFC/minute x µg protein, by completing the following equation:

$$\left[\left(\Delta F_{\text{sample}} - b\right) \times 250\right] / \left[a \times X \times P \times 60\right]$$

With parameters:

$$\Delta F_{\text{sample}} = F_{\text{sample}} - F_{\text{blank}}$$

where b intercept of the best-fit line, 250 = total reaction volume (µL)/well, a slope of best-fit line, X volume (µL) sample added/well, P protein concentration (µg/µL) of the sample, 60 = reaction time (min).

3.3 Measurement of the Extracellular Release of Lactate Dehydrogenase in Cultured Primary Hepatocytes

3.3.1 Sampling from Cultured Primary Hepatocytes

1. Take 500 µL cell culture medium from a 3.5 cm diameter culture dish and transfer into a recipient. Snapfreeze in liquid nitrogen and store at −20 °C (*see* **Note 9**).
2. Scrape the cells in the remaining hepatocyte cell death medium using a policeman and transfer the cell suspension into a recipient. Snapfreeze in liquid nitrogen and store at −20 °C.

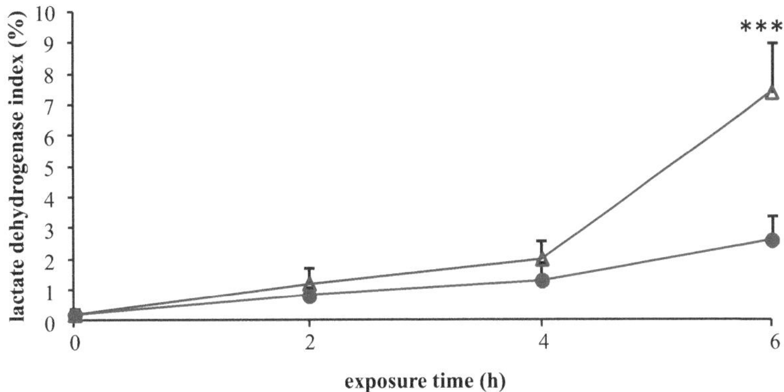

Fig. 4 Measurement of lactate dehydrogenase leakage. Freshly isolated rat hepatocytes were cultivated in a monolayer configuration and were exposed to 200 ng/mL Fas ligand and 2 μg/mL cycloheximide, starting at 44 h postplating. Samples were taken at the start of the exposure (0 h), and after 2, 4, and 6 h. Lactate dehydrogenase activity was measured and the lactate dehydrogenase index was calculated as [(100 × lactate dehydrogenase activity in supernatant)/(lactate dehydrogenase activity in cell suspension)]. Results represent the means ± standard deviation of three independent experiments. Results were evaluated by 1-way analysis of variance followed by post hoc Bonferroni tests. *Filled circle* control, *open square* Fas ligand/cycloheximide; ***$p < 0.001$. Reproduced in modified form with permission from [13]

3.3.2 Measurement of the Lactate Dehydrogenase Activity

1. Defrost the microcentrifuge tubes on ice.
2. Homogenize the content of the cell culture medium and the cell suspension by ultrasonication with 20 pulses and 2 × 20 pulses with repeat duty cycle 0.5 s, respectively (*see* **Note 10**).
3. Mix 1 mL LDH reaction buffer with 20 μL homogenized cell suspension in a plastic microcuvette and place in the spectrophotometer (*see* **Note 11**).
4. After exactly 1 min, measure the decrease in absorption (ΔC/min) at 340 nm for a total of 1 min against an air blank (*see* **Note 12**).
5. Repeat **steps 3** and **4** twice.
6. Repeat **steps 3–5** for the corresponding homogenized cell culture medium.

3.3.3 Processing of the Results (Fig. 4) (See Note 8)

1. Collect and average the 3 ΔC/min values for both the cell suspension (i.e. total LDH activity) and the cell culture medium (i.e. extracellular LDH activity).
2. Calculate the LDH index by assessing the relative extracellular release of LDH, using the following equation:

$$\text{LDH index}\,(\%) = \left[(\text{extracellular LDH activity}) / (\text{total LDH activity})\right] \times 100$$

4 Notes

1. With respect to the specificity of the apoptosis assay, it should be kept in mind that Ac-DEVD-AFC not only is a substrate for caspase 3, but also for other caspases, including caspase 1, caspase 4, caspase 6, caspase 7, caspase 8 and caspase 10 [8]. Hence, the enzymatic activity measured is not restricted to caspase 3 and thereby should be rather considered as caspase 3-like activity.
2. The pH of the reaction buffer is a crucial parameter and should be checked regularly. Deviation of the pH can result in crystallization of the compounds in the reaction buffer, which in turn will affect the outcome of the assay (Table 2).
3. Dissolve the 3-[(3-cholamidopropyl)dimethylammonio]-1-propanesulfonate and the DL-dithiothreitol in 746 μL double distilled water when preparing a volume of 1 mL lysis buffer and subsequently add the different solutions.
4. This washing step could remove dead cells and therefore may be omitted.
5. The diameter of the culture dish used mainly depends on the purpose of the assay intended. For measurement of caspase 3 activity, it is strongly recommended to use 10 cm diameter culture dishes with hepatocytes plated at a density of 0.57×10^5 cells/cm^2 (i.e. a total of 4.4×10^6 hepatocytes). When using 6 cm diameter culture dishes (i.e. a total of 1.6×10^6 hepatocytes), harvesting of the cells and preparation of cell homogenates should be performed with 3 mL PBS and 50 μL lysis buffer, respectively.
6. Among the different assays to assess protein concentration in biological samples, the Bradford method [17] is the most commonly used one. In this assay, which has been commercialized (Bio-Rad, Belgium), a protein sample is added to a solution of the dye Coomassie Brilliant Blue G-250 in phosphoric acid and ethanol. Under the acid conditions, the dye normally has

Table 2
Troubleshooting of the caspase 3-like activity measurement procedure

Problem	Cause	Solution
Negative testing results	Decomposed reaction buffer and solutions	Prepare new reaction buffer and solutions and protect from light
	Air bubbles in the wells of the black 96-well plate	Tap the black 96-well plate before incubation in the multiplate reader
Divergent testing results	Incomplete lysis of the cells	Prepare new lysis buffer and strongly pipette up and down when lyzing cells
	Variation in temperature	Check temperature in the multiplate reader and ensure a constant temperature of 37 °C
	Variation in pH and crystallization of the reaction buffer	Check the pH of the reaction buffer and adjust to 7.4

a brownish color, but upon binding to the protein, the blue form of the dye is produced. The optical absorbance of the solution is measured at a wavelength of 595 nm and is proportional to the amount of protein present and bound to the dye. It is a relative method, whereby bovine serum albumin is a typical standard used. Clearly, other assays, such as the bicinchoninic acid-based method [18] and the Lowry assay [19], can also be addressed to assess protein concentration.

7. Temperature is a critical parameter in this assay, as it strongly influences protein stability and hence caspase 3 activity. Thus, it is of the utmost importance to ensure a constant temperature of 37 °C in the cabinet of the multiplate reader before and during incubation of the black 96-well plate (Table 2).

8. From a statistical point of view, it is strongly recommended to perform the experiments on hepatocytes isolated from the livers of three different animals. Results can be processed and evaluated by 1-way analysis of variance followed by post hoc Bonferroni tests.

9. The diameter of the culture dish used mainly depends on the purpose of the assay intended. For measurement of the extracellular release of LDH, it is sufficient to use 3.5 cm diameter culture dishes, though culture dishes with a higher diameter can also be used.

10. It is of the utmost importance to continuously keep samples on ice during ultrasonication, since this procedure produces heat, which may cause proteolysis and hence loss of enzyme activity.

11. Temperature is a critical parameter in this assay, as it strongly influences enzyme activity. Therefore, it is important to continuously store the LDH reaction buffer at 25 °C and to regularly check the temperature. The procedure may also be performed at 37 °C, though in this case only 10 μL homogenized sample must be mixed with 1 mL LDH reaction buffer and a correction factor of 16,030 must be handled during processing of the results (*see* **Note 12**).

12. The assessment of the enzymatic activity relies on modifications in NADH concentration as a function of time (ΔC/min). Upon implementation of the Beer–Lambert law, this yields the equation $(\Delta A/\text{min})/(\varepsilon \times l)$, whereby the parameters ΔA, ε and l represent the decrease in absorption, the molecular absorption coefficient ($\text{cm}^{-1}\ \text{mol}^{-1}\ \text{l}$) and the optical path (1 cm), respectively. When applied to LDH activity, this gives $[(\Delta A/\text{min}) \times 10^6]/(\varepsilon \times l)$. In this equation, LDH activity is expressed in units per liter, whereby 1 unit corresponds with the quantity of LDH that catalyzes the conversion of 1 μmol NADH/min at 25 °C. This formula can be simplified by taken into account an established correction factor of 8,095, as experimentally established by the manufacturer. Consequently, LDH activity

Table 3
Troubleshooting of the lactate dehydrogenase (LDH) assay

Problem	Cause	Solution
Negative testing results	Decomposed LDH reaction buffer Air bubbles in the microcuvette	Prepare new LDH reaction buffer and store for a maximum of 5 days or 8 h at 4 °C or 25 °C, respectively Shake the microcuvette upon mixing the LDH reaction buffer and the sample
Divergent testing results	Incomplete homogenization of samples Variation in temperature	Subject samples to an additional series of 20 ultrasonication pulses Check temperature and store the LDH reaction buffer at 25 °C

can be calculated as $(\Delta A/\text{min}) \times 8{,}095$. The LDH activity is automatically calculated by the computer software upon measurement of the samples (Table 3).

Acknowledgements

This work was financially supported by the grants of the Agency for Innovation by Science and Technology in Flanders (IWT), the University, Hospital of the Vrije Universiteit Brussel—Belgium (Willy GeptsFonds UZ-VUB), the University of Sao Paulo—Brazil (USP), the Sao Paulo Research Foundation—Brazil (FAPESP), the Fund for Scientific Research—Flanders (FWO-Vlaanderen), the European Research Council (project CONNECT), the European Union (FP7) and Cosmetics Europe (projects DETECTIVE and HeMiBio).

References

1. Vinken M, Maes M, Vanhaecke T et al (2013) Drug-induced liver injury: mechanisms, types and biomarkers. Curr Med Chem 20: 3011–3021
2. Russmann S, Kullak-Ublick GA, Grattagliano I (2009) Current concepts of mechanisms in drug-induced hepatotoxicity. Curr Med Chem 16:3041–3053
3. Vinken M, Maes M, Oliveira AG et al (2014) Primary hepatocytes and their cultures in liver apoptosis research. Arch Toxicol 88: 199–212
4. Jaeschke H, Gores GJ, Cederbau AI et al (2002) Mechanisms of hepatotoxicity. Toxicol Sci 65:166–176
5. Raffray M, Cohen GM (1997) Apoptosis and necrosis in toxicology: a continuum or distinct modes of cell death? Pharmacol Ther 75: 153–177
6. Gomez-Lechon MJ, O'Connor E, Castell JV et al (2002) Sensitive markers used to identify compounds that trigger apoptosis in cultured hepatocytes. Toxicol Sci 65:299–308
7. Gill GH, Dive D (2000) Apoptosis: basic mechanisms and relevance to toxicology. In: Roberts R (ed) Apoptosis in toxicology. Taylor & Francis, London, pp 1–20
8. Yin X-M, Dong Z (2003) Essentials of apoptosis: a guide for basic and clinical research. Humana Press, Totowa, NJ
9. Feldmann G (1997) Liver apoptosis. J Hepatol 26:1–11
10. Decrock E, Vinken M, De Vuyst E et al (2009) Connexin-related signaling in cell

death: to live or let die? Cell Death Differ 16:524–536
11. Malhi H, Gores GJ (2008) Cellular and molecular mechanisms of liver injury. Gastroenterology 134:1641–1654
12. Schulze-Bergkamen H, Untergasser A, Dax A et al (2003) Primary human hepatocytes: a valuable tool for investigation of apoptosis and hepatitis B virus infection. J Hepatol 38:736–744
13. Vinken M, Decrock E, De Vuyst E et al (2009) Biochemical characterization of an in vitro model of hepatocellular apoptotic cell death. Altern Lab Anim 37:209–218
14. Schwartz LM, Ashwell JD (2001) Methods in cell biology volume 66: apoptosis. Academic, San Diego
15. Bergmeyer HU (1974) Lactate dehydrogenase. In: Bergmeyer HU (ed) Methods of enzymatic analysis. Academic, New York, pp 574–579
16. Papeleu P, Vanhaecke T, Henkens T et al (2006) Isolation of rat hepatocytes. Methods Mol Biol 320:229–237
17. Bradford MM (1976) A rapid and sensitive method for the quantitation of microgram quantities of protein utilizing the principle of protein-dye binding. Anal Biochem 72:248–254
18. Smith PK, Krohn RI, Hermanson GT et al (1985) Measurement of protein using bicinchoninic acid. Anal Biochem 150:76–85
19. Lowry OH, Rosebrough NJ, Farr AL et al (1951) Protein measurement with the Folin phenol reagent. J Biol Chem 193:265–275

Chapter 28

Critical Factors in the Assessment of Cholestatic Liver Injury In Vitro

Benjamin L. Woolbright and Hartmut Jaeschke

Abstract

Cholestasis is a common pathological component of numerous liver diseases. The initiating event during cholestatic liver injury is widely believed to be the accumulation of bile acids in hepatocytes and the hepatic parenchyma. As bile acids are considered the primary toxic compounds in the injury, numerous in vitro models of bile acid-induced injury and bile acid-induced changes in gene expression have been developed to attempt to better define cholestasis at a cellular level. This chapter focuses on the establishment of a system for determining the effects of cholestatic concentrations of bile acids on hepatocytes using primary hepatocytes or hepatoma cell lines. Moreover, this chapter addresses significant differences in the response of different species to bile acid exposure and novel information on the relevance of treating hepatocytes with concentrations of specific bile acids.

Key words Apoptosis, Necrosis, In vitro model, Bile acid, Primary hepatocyte, Cholestasis, Inflammation, Liver, Bile salt, Human hepatocytes, Rodent models, Sodium taurocholate, Co-transporting polypeptide, Bile salt export pump, HepG2, Huh7, Transfected hepatocytes, Neutrophils, Methods

1 Introduction

Cholestasis results from either blockade of any portion of the biliary tracts or inhibition of bile acid export from hepatocytes. When left untreated, chronic cholestasis proceeds to significant liver damage and cirrhosis. Common in vivo models for study include the bile duct ligation (BDL) model [1], which results in extrahepatic cholestasis, administration of lithocholic acid (LCA) [2] or alpha-naphthylisothiocyanate [3], which result in intrahepatic cholestasis, or use of the multi-drug resistance two transporter knock-out mouse [4] or inhibition of the bile salt export pump (BSEP) [5], which mimic inhibition of export of bile acids from hepatocytes. As the accumulation of bile acids in hepatocytes and the hepatic parenchyma is considered to be the initiator of the injury process [6], considerable effort has gone into the establishment of a suitable

Mathieu Vinken and Vera Rogiers (eds.), *Protocols in In Vitro Hepatocyte Research*, Methods in Molecular Biology, vol. 1250, DOI 10.1007/978-1-4939-2074-7_28, © Springer Science+Business Media New York 2015

in vitro model for directly studying the effects of increased bile acid levels on hepatocytes. In general, the administration of bile acids to cultured hepatocytes is a simple process. However, recent advances have illustrated numerous potential pitfalls that can be encountered during otherwise routine cell culture experiments. As a preface, current models for the study of cholestatic liver injury are discussed in this chapter, along with potential pitfalls that must be considered before initiation of experimentation.

1.1 Differential Effects of Specific Bile Acids on Hepatocytes

Bile acids are the major constituents of bile. Numerous species are present in the body and are derived from the two primary bile acids, namely cholic acid (CA) and chenodeoxycholic acid (CDCA). Dehydroxylation reactions result in secondary bile acids, namely deoxycholic acid (DCA) and lithocholic acid (LCA) from cholic acid and chenodeoxycholic acid, respectively. A majority of these bile acids are then conjugated to either of the amino acids taurine or glycine before they enter enterohepatic circulation. Recent methodology has allowed for a quantitative assessment of individual bile acid concentrations in multiple species and in multiple tissue compartments during cholestasis [7–11]. Perhaps the most important aspect of these new data is the elucidation of vast differences in individual bile acid concentrations in rodents and humans, both in control patients and during cholestatic liver injury [1]. While numerous studies [6, 12, 13] have focused on administering micromolar levels of toxic bile acids or their conjugate salt, such as glycochenodeoxycholate (GCDC) or taurolithocholate (TLC), to hepatocytes in vitro, newer data based on mass spectrometric bile acid analysis questions the use of micromolar concentrations of these bile acids in rodent cultures, as they likely do not accumulate to similar values in any compartment in vivo [8, 10]. Moreover, murine hepatocytes are highly resistant to TLC-induced injury in vitro, suggesting mechanisms of toxicity specific to the rat [10, 14]. Additionally, the use of unconjugated bile acids to investigate pathophysiology seems unwarranted as more than 99 % of bile acids in liver, plasma and bile are conjugated to taurine or glycine in multiple species [7–9, 11]. While intrahepatic concentrations of bile acids are considered to be integral to the injury, recent reports have asserted that the characteristic areas of necrosis seen after BDL or after administration of 1 % LCA via the diet are due to infarction of the small biliary tracts and the ensuing release of bile into the hepatic parenchyma, suggesting that biliary levels of bile acids may be driving the hepatic accumulation of bile acids during some models of cholestatic liver injury [2, 10]. Cholestasis models, such as BDL, display a lack of zonation in the area of injury. Instead, the necrosis correlates specifically with areas of biliary leakage, implicating leakage of biliary constituents onto the hepatic parenchyma as an initiator of injury [2, 4, 10, 11]. However, even in bile, a majority of the rodent bile acid milieu is largely composed of hydrophilic bile acids, such as taurocholic acid (TCA), that remain non-toxic up to millimolar concentrations [8, 10]. While the hydrophilic bile acids, like TCA, do not cause direct hepatotoxicity [8, 10], exposing these bile acids to hepatocytes results in dramatic increases in pro-inflammatory genes, such as macrophage inflammatory protein 2, mouse keratinocyte-derived chemokine, intercellular adhesion molecule 1, early growth response protein 1 and others [15, 16]. Many of these genes play a significant role in the pathophysiology of cholestatic liver injury, as intercellular adhesion molecule 1 and early growth response protein 1 knockout mice are protected against BDL injury [17, 18]. This reduction in injury is due to attenuation of neutrophil function and recruitment, as BDL injury is mediated by cytotoxic neutrophils [17, 19, 20]. In contrast to the BDL model, only in models where the bile is toxified, such as through administration of LCA via the diet, direct hepatotoxicity can be seen when hepatocytes are exposed to rodent bile [2, 10]. As such, experiments where mouse or rat hepatocytes are exposed to 50–100 μM GCDC or TLC may not be relevant to the pathophysiology of cholestatic liver injury in mice or man. Furthermore, as hepatocytes are exposed to multiple bile acids simultaneously during cholestasis, exposure of hepatocytes to single bile acids may obscure the anti-apoptotic effects that hydrophilic bile acids can induce through activation of nuclear factor kappa beta via upregulation of cytokines [21]. Thus, while the simplicity of adding a single bile acid to hepatocytes can produce interesting and repeatable results, the relevance of exposure of hepatocytes to a single toxic bile acid may bear little resemblance to the in vivo pathophysiology. An increased focus is instead called for on how a complete bile acid milieu derived from pathophysiological assessment of levels in the appropriate species affects hepatocytes both in vitro and in vivo. Recent studies using complete mixtures of serum or biliary bile acid values suggest that hepatocytes may be less susceptible on a mole to mole basis to a total bile acid mixture than to direct exposure to a single bile acid [11, 22]. It is therefore imperative to carefully assess what bile acids and what models are being used to investigate biological effects of bile acids to ensure pathophysiological relevance.

1.2 Species Differences in In Vitro Bile Acid Toxicity

Bile acid levels in serum, liver, and bile, and the subsequent cellular response vary substantially between commonly used laboratory species and man. These differences have resulted in significant controversy in the field with regard to which method of cell death, apoptosis or necrosis occurs during cholestasis [2, 6, 23–25]. Part of this discussion revolves around differences seen during in vitro exposure to bile acids, particularly pro-apoptotic bile acids, such as GCDC. Humans produce a relatively larger amount of the more toxic glycine-conjugated bile acids, particularly GCDC, whereas mice and rats produce larger quantities of the less toxic taurine-conjugated bile acids [7–9, 11, 26]. Numerous studies expose rat hepatocytes or transfected human hepatoma lines to about 50 μM GCDC, which then respond with prototypical apoptosis including nuclear condensation, cellular shrinkage, nuclear fragmentation, and activation of caspases [27]. Complete rescue is possible by pretreatment with the pan-caspase inhibitor z-VAD-fmk, indicating the injury is predominately apoptosis [28]. Even given this well-defined and established response, the validity of applying bile acids commonly observed in humans to rat hepatocytes does not accurately model the pathophysiology, especially as levels of these bile acids either do not reach or far exceed 50 μM during cholestasis [7, 12]. In contrast, human hepatocytes show little evidence for apoptosis at doses consistent with the pathophysiology and die primarily through necrosis, although only when exposed to higher concentrations of GCDC or a complete biliary bile salt mixture [11]. Interestingly, primary human hepatocytes are particularly resistant to GCDC-induced injury requiring near millimolar levels of specific toxic bile acids before the onset of toxicity [11, 29, 30]. Species differences also exist between mice and humans, as humans do not exhibit the same increases in cytokine production that mice do when stimulated with millimolar concentrations of TCA [11, 15]. As human bile is considerably more concentrated with glycine-conjugated toxic bile acids, this may be a by-product of the direct toxicity produced by human bile in comparison to the relatively less harmful murine bile. While inflammation is a noted part of multiple human cholestatic disorders, cytokine production likely results from another source. A final important note is the marked difference in transporter expression levels over time between primary human and primary rodent hepatocytes. While human hepatocytes retain a majority of their transporter activity (e.g. BSEP) over the first 24 h of culture [31], rodent hepatocytes rapidly dedifferentiate and lose transporter expression [32, 33]. Loss of transporter activity rapidly decreases bile acid function in vitro. Thus, primary hepatocytes from rodents need to be used immediately after adherence for them to be useful. Bile salt efflux is considerably different between humans and rats as well. While humans have essentially similar levels of basolateral and canalicular efflux, rats rely predominantly on basolateral transport, increasing their resistance to canalicular BSEP inhibition-induced cholestasis [34]. Thus, critical differences are present between species in regard to bile acid toxicity and metabolism that merit consideration before experimentation in this field.

1.3 BSEP Inhibition In Vitro

BSEP is responsible for export of bile salts out of the hepatocytes into the biliary tracts [35]. Troglitazone, a peroxisome proliferator-activated receptor activator, and bosentan, an endothelin receptor antagonist, are currently the most widely recognized BSEP inhibitors [36]. Troglitazone was removed from the market due to hepatotoxicity and bosentan currently carries a black box warning label that requires monthly monitoring for liver damage associated with taking the drug. Drug-induced BSEP inhibition results in severe cholestasis, making it a pharmacological and a clinical issue [37]. While not directly cytotoxic, drugs that inhibit BSEP result in increases in serum transaminases and serum bile acid levels, suggesting they may kill cells through loss of homeostasis of native bile acid levels and subsequent liver injury [38]. Long-term drug-induced cholestasis also causes both a ductular reaction and hepatitis that may worsen the injury [36]. The mechanisms behind how BSEP inhibition produces hepatotoxicity are not well understood, although intrahepatic accumulation of bile acids and oxidative stress are implicated [39]. Both functional [40] and gene expression assays [41] have some predictive value in determining the degree by which a drug can inhibit BSEP. Thus, while these drugs are not directly cytotoxic, drug-induced cholestasis is a separate yet important issue. Multiple efforts are underway currently to establish a reproducible and appropriate model for drug-induced cholestasis, as it relates to liver injury in vitro [25].

1.4 Primary Cells Versus Hepatoma Cell Lines

Due to the difficulty in acquiring primary cells from animals or humans, hepatoma cell lines are sometimes used [42, 43]. Most commonly used hepatoma lines, such as the Huh7 line and HepG2 line, do not natively express the sodium taurocholate co-transporting polypeptide (NTCP), the primary bile acid uptake transporter, and must be transfected to express the protein, whereupon they will actively take up bile acids. Thus, non-transfected HepG2 or Huh7 cells make a poor model for bile acid-induced liver injury or cholestatic liver injury. Multiple reports have used HepG2 cells transfected with NTCP to assess the effects of bile acids [43–45]. In general, these cells recapitulate responses seen in the rat and not the human, which suggests that they may not be a suitable surrogate for primary human hepatocytes. In contrast, the recently isolated HepaRG cell line functionally expresses both canalicular and basolateral transporters [31, 46] and may serve as a novel human cell line for in vitro bile acid-induced toxicity. While hepatocytes from this line do not express transporters to the same degree primary human hepatocytes do, some functional activity is retained [31, 39]. As this is a bipotential cell line that actively differentiates into both cholangiocytes and hepatocytes, it is imperative to establish bile acid toxicity without cholangiocyte toxicity when assessing cell death in this model.

Given these potential pitfalls, and the current state of knowledge of the field, a model for in vitro assessment of cholestatic liver injury is proposed herein.

2 Materials

1. Hepatocyte seeding and culture medium. William's medium E containing 10 % fetal bovine serum (FBS), suitable antibiotics and 2 ng/mL insulin. Media should be warmed to 37 °C before use (*see* **Note 1**).
2. Phosphate-buffered saline (PBS). 137 mM NaCl, 2.7 mM KCl, 10 mM $Na_2HPO_4{\cdot}2H_2O$, 1.8 mM KH_2PO_4 in deionized water. Adjust to pH 7.4 and store for maximum 6 months at 4 °C.
3. Depending on the animal model used, cell death is best measured by lactate dehydrogenase (LDH) release into media and caspase-3 activation. For the LDH assay, potassium phosphate buffer is composed of 60 mM KH_2PO_4 and 60 mM K_2HPO_4 along with 100 mM pyruvate and 250 μM nicotinamide adenine dinucleotide (NADH), pH 7.5, in deionized water. This solution can be stored for up to 1 month at 4 °C. The fluorogenic substrate for caspase-3 assay, namely acetyl-aspartic acid-glutamic acid-valine-aspartic acid-7-aminotrifluoromethylcoumarin (Ac-DEVD-AMC) is commercially available (Enzo Life Sciences, United States of America).
4. Depending on the animal model used, reverse transcriptase polymerase chain reaction (RT-PCR) analysis and immunoblotting can be used to assess changes in expression of pro-inflammatory genes and proteins, such as adhesion molecules, after bile acid exposure. Reagents and protocols for these assays are widely available from relevant manufacturers (Life Technologies, United States of America).
5. Incubator (*see* **Note 2**).
6. Spectrophotometer capable of reading at 340 nm.
7. Plate reader capable of exciting at 380 nm and reading emission at 460 nm.
8. Bicinchoninic acid (BCA) protein quantification reagents (Sigma-Aldrich, United States of America).
9. Kits and reagents for RNA extraction and RT-PCR analysis (Invitrogen, United States of America).

3 Methods

3.1 Cell Culture and Bile Acid Administration

1. Use freshly isolated primary human, rat, or mouse hepatocytes (*see* **Note 3**). Cell viability is assessed by means of trypan blue exclusion [47].
2. Plate the hepatocytes on 6-well plastic culture dishes at a density of $0.5\text{–}1.0 \times 10^6$ cells/well (*see* **Note 4**). Hepatocytes

should be allowed to adhere for 3–4 h before the onset of experimentation (*see* **Note 5**).

3. Wash cells with 1 mL PBS and replace 1 mL culture medium.
4. Add the bile acid of interest to the appropriate concentration and incubate as desired (*see* **Notes 6** and 7).

3.2 Measurement of LDH and Alanine Aminotransferase Activity (See Note 8)

1. Treat cells with bile acid of interest.
2. Remove 1 mL medium and store in a microcentrifuge tube at −80 °C. Upon thawing, centrifuge at 14,000 × *g* and room temperature for 3 min. Wash with 1 mL PBS.
3. Lyze cells in 300 μL appropriate protein buffer, scrape and store in a microcentrifuge tube at −80 °C. Upon beginning the assay, sonicate cellular mixture 3 times for 3 s at a low level to ensure complete disruption of membranes. Centrifuge at 14,000 × *g* and room temperature for 10 min.
4. LDH activity can be measured by analyzing the oxidation of NADH in an LDH buffer via a loss in absorbance at 340 nm on a spectrophotometer. LDH activity in both medium and cells should be analyzed. Approximate LDH release can be measured using the equation:

$$U / L = \left[(dA / dT) \times (1 / \varepsilon) \times \text{dilution}\right] \text{with } \varepsilon_{340} = 6.3$$

$$\text{Death ratio} = (U / L)^{\text{Media}} / \left[(U / L)^{\text{Media}} + (U / L)^{\text{Lysate}}\right] \times 100\%$$

5. To analyze LDH activity, dilute media 1:7 into LDH buffer and analyze the oxidation of NADH (i.e. loss of absorbance at 340 nm) in the LDH buffer with a spectrophotometer over a 2 min period. In a matched sample, dilute cellular contents 1:30 and analyze the oxidation of NADH in the LDH buffer with a spectrophotometer over a 2 min period. Use the above equation to calculate the death ratio and report as percentage LDH release. A sample figure is presented showing increasing cell death in glycocholic acid treated mouse hepatocytes in Fig. 1.
6. Alanine aminotransferase (ALT) activity can be measured similarly and involves a 2-step process that also measures the oxidation of NADH via a loss in absorbance at 340 nm on a spectrophotometer. In some cases, measurement of ALT can be superior to measurement of LDH, as loss of LDH activity may occur during extended cell culture. To measure ALT instead of LDH, complete the steps above, but substitute the LDH buffer with an ALT buffer (Pointe Scientific, United States of America). All calculations and steps are otherwise identical as is the equation for calculation.

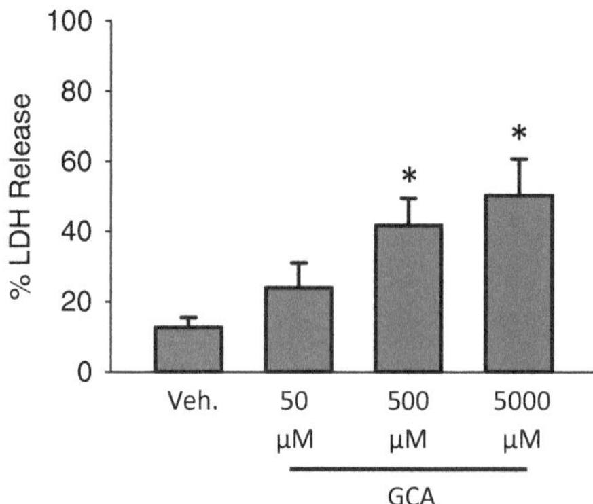

Fig. 1 Bile acid-induced necrotic cell death in murine hepatocytes. Hepatocytes were isolated from livers of C57Bl/6 mice and exposed to various concentrations of glycocholic acid (GCA) for 6 h after an initial 3 h attachment period. LDH was measured as an indicator of cell death ($n=3$; *$p<0.05$ vs. control by analysis of variance)

3.3 Measurement of Caspase-3 Activity (See Note 9)

1. Treat cells with bile acid of interest.
2. Remove media and wash cells with 1 mL PBS.
3. Lyze cells in 150 μL of appropriate protein buffer, scrape, and store in a microcentrifuge tube at −80 °C if necessary. Upon thawing or upon starting the assay, centrifuge at 14,000 × *g* and room temperature for 10 min.
4. Dilute Ac-DEVD-AMC substrate to 2 mM. Dilute z-VAD-fmk to 100 μM (*see* **Note 10**).
5. The assay is best completed in a 96-well plate on a plate reader. Generate 2 master mixes, for 1 caspase-3 activity and 1 for z-VAD-fmk inhibitable activity:
   - Mix A: 50 μL protein homogenizing buffer and 10 μL diluted substrate.
   - Mix B: 40 μL protein homogenizing buffer, 10 μL diluted substrate, and 10 μL diluted z-VAD-FMK.
6. Add 50 μL of cell lysate to each well and then pipette 50 μL of either mix A or mix B in with the samples. This yields 2 wells for each sample, namely mix A well (i.e. caspase-3 activity) and mix B well (i.e. z-VAD-inhibitable caspase activity).
7. Analyze activity over 1 h using an excitation of 380 nm and emission of 460 nm. Activity should be recalculated as caspase-3 activity *minus* z-VAD-inhibitable caspase-3 activity.
8. Measure protein quantity via the BCA assay as per manufacturer's instructions and express as relative fluorescence/mg protein/min.

3.4 RNA Isolation and RT-PCR Analysis (See Note 11)

1. Treat cells with bile acid of interest.
2. Remove media and wash cells twice with PBS.
3. Lyse cells in Trizol buffer for RNA isolation.
4. Extract RNA and reverse transcribe to cDNA.
5. Use PCR to assess gene levels (*see* **Note 12**). Numerous protocols are available for simple quantitative PCR techniques. Following the manufacturer's recommended instructions for the apparatus is advisable for this assay.

3.5 Immunoblot Analysis

1. Treat cells with bile acids of interest.
2. Remove media and wash cells twice with PBS.
3. Lyze and scrape cells in 100 μL of appropriate protein isolation buffer.
4. Measure protein and assess protein levels by immunoblot analysis (*see* **Note 12**). Numerous protocols are available for this process. Following the manufacturer's recommended instructions for the apparatus and antibodies in use is advisable for this assay.

4 Notes

1. Fetal bovine serum can be omitted, but may affect results in some assays. As cells are exposed to serum constantly in vivo, use of fetal bovine serum is recommended in all assays. If fetal bovine serum is not used, a direct comparison between cells given fetal bovine serum and cells not given fetal bovine serum should be done for the indicated assay.
2. Use of room air can have profound effects on hepatocyte culture [48, 49]. Physiological levels of oxygen are protective against GCDC-induced apoptosis in rat hepatocytes [49]. Primary hepatocytes are typically incubated in ambient levels of oxygen (i.e. about 20 %), whereas hepatic oxygen levels can be as low as 4 %, which results in substantial increases in available oxygen for reactive oxygen species generation in GCDC-treated cultured hepatocytes. Reactive oxygen species production likely mediates the apoptosis, although the source of the reactive oxygen species is not well established. Experiments may want to be repeated under conditions of both atmospheric oxygen and physiological hypoxia to verify changes observed during in vitro culture.
3. The animal model used has profound effects on the results obtained.
4. Typically, assays done in our laboratory are done in a 6-well plate. It is recommended to consult commercial literature on

cell density numbers for other sizes of plates. About 80 % confluence is typical for these assays. For all assays listed, 1 well of a 6-well plate at the given cell density will be sufficient to complete the assay.

5. NTCP is rapidly downregulated after culture in murine and rat hepatocytes [32]. To avoid loss of effects, experiments should be carried out immediately at this point. Avoid cultures that do not express NTCP, as there is a serious deficit of bile acid internalization and thus quite likely, bile acid function. These cell lines may be prone to generating results with limited relevance for the human pathophysiology.
6. Longer incubation times will result in loss of NTCP function and alternations in read-out. Assays are best run over 4–6 h to avoid these problems. If longer time periods are needed, stably transfected cells or primary human hepatocytes, which do not lose NTCP activity as quickly, can be used [11, 31]. While the shorter timeframe likely does not mimic the consequences of long-term cholestasis, the loss of functional transporter activity is a more serious concern, as observed effects may be secondary to bile acid function.
7. Bile acids have varied effects. Toxicity and gene induction can be acquired through multiple different bile acids or administration of a single bile acid. In general, the more hydrophobic bile acids are more toxic and also require hydrophobic solvents, such as dimethyl sulfoxide or methanol. Concentrations of solvent below 1 % are suggested to reduce solvent effects. Vehicle solvents should always be added to experiments as a control for the solvent effects [50].
8. LDH is best used to measure necrosis, which may be secondary to apoptosis. The simultaneous use of LDH and caspase-3 is attainable in a single well from a 6-well plate and provides the best overall image of cell death using enzyme kinetics. Additional potential assays include staining necrotic nuclei with the cell permeable reagent propidium iodide.
9. Caspase-3 activity is a specific and direct marker of apoptosis. As caspase-7 generally has the same substrate specificity, some groups will label assays as caspase-3/7 activity, which is also correct. Confirmation of caspase-3 activity can be obtained via immunoblotting for pro-caspase-3 and its cleaved active form. The antibody used in our laboratory is from Cell Signaling (United States of America). While antibodies specific for cleaved caspase-3 are available, acknowledgement of the pro-band in immunoblot imaging is encouraged, as its compensatory decrease can be seen as a confirmation of antibody specificity. Figure 2 depicts the proper representation of an immunoblot for caspase-3 that contains both the pro-form

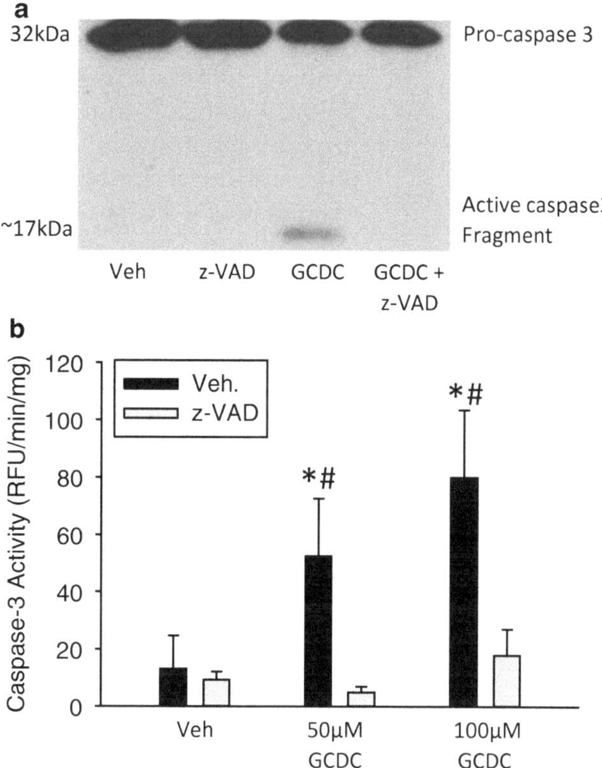

Fig. 2 Bile acid-induced apoptosis in rat hepatocytes. Hepatocytes were isolated from rats and exposed to vehicle (i.e. 0.1 % dimethyl sulfoxide), 50 μM, or 100 μM GCDC for 6 h with and without a 30 min pretreatment with 10 μM of the pan-caspase inhibitor z-VAD-FMK. Caspase-3 activation after 50 μM GCDC was measured by immunoblot analysis (**a**) and through a fluorogenic caspase activity assay (**b**) (*$p < 0.05$ vs. control; #$p < 0.05$ vs. matched z-VAD-fmk treated sample)

band and its cleaved form. Multiple cleavage products are sometimes formed from the 32 kDa precursor between 19 and 14 kDa.

10. As there may be endogenous protease activity, the use of a pan-caspase inhibitor (e.g. z-VAD-fmk) control ensures that the assay is specific for cellular caspase activity and excludes caspase-independent protease activities. This is a vital control, as some cell lysates will contain a great deal of endogenous protease activity.
11. For both RNA and protein, most changes occur within a similar timepoint as the toxicity, thus prolonged exposure results in less activity and not more, presumably due to loss of bile acid uptake.
12. Commonly assessed genes and proteins for assessing inflammation include macrophage inflammatory protein 2, mouse

keratinocyte-derived chemokine, intercellular adhesion molecule 1, early growth response protein 1 and others [15, 16], although any gene of interest could be measured. Using RT-PCR analysis to assess caspase-3 levels is not a useful way to measure apoptosis, as caspase function is dependent upon posttranslational cleavage events largely by other caspases. Use of immunoblot analysis to assess caspase-3 cleavage is essential for demonstration of active apoptosis.

Acknowledgements

Work in this laboratory was supported in part by a CTSA grant from NCRR and NCATS awarded to the University of Kansas Medical Center for Frontiers: The Heartland Institute for Clinical and Translational Research # UL1RR033179 which is now at NCATS # UL1TR000001, grants from the National Institutes of Health (R01 DK070195 and R01 AA12916) (to H.J.), and from the National Center for Research Resources (5P20RR021940) and the National Institute of General Medical Sciences (8 P20 GM103549) of the National Institutes of Health. B.L.W. was supported by the "Training Program in Environmental Toxicology" T32 ES007079-26A2 from the National Institute of Environmental Health Sciences.

References

1. Woolbright BL, Antoine DJ, Jenkins R et al (2013) Plasma biomarkers of liver injury and inflammation demonstrate a lack of apoptosis during obstructive cholestasis in mice. Toxicol Appl Pharmacol 273:524–531
2. Fickert P, Fuchsbichler A, Marschall HU et al (2006) Lithocholic acid feeding induces segmental bile duct obstruction and destructive cholangitis in mice. Am J Pathol 168:410–422
3. Krell H, Metz J, Jaeschke H et al (1987) Drug-induced intrahepatic cholestasis: characterization of different pathomechanisms. Arch Toxicol 60:124–130
4. Fickert P, Zollner G, Fuchsbichler A et al (2002) Ursodeoxycholic acid aggravates bile infarcts in bile duct-ligated and Mdr2 knockout mice via disruption of cholangioles. Gastroenterology 123:1238–1251
5. Rodrigues AD, Lai Y, Cvijic ME (2014) Drug-induced perturbations of the bile acid pool, cholestasis, and hepatotoxicity: mechanistic considerations beyond the direct inhibition of the bile salt export pump. Drug Metab Dispos 42:566–574
6. Malhi H, Guicciardi ME, Gores GJ (2010) Hepatocyte death: a clear and present danger. Physiol Rev 90:1165–1194
7. Trottier J, Białek A, Caron P et al (2012) Metabolomic profiling of 17 bile acids in serum from patients with primary biliary cirrhosis and primary sclerosing cholangitis: a pilot study. Dig Liver Dis 44:303–310
8. Zhang Y, Hong JY, Rockwell CE et al (2012) Effect of bile duct ligation on bile acid composition in mouse serum and liver. Liver Int 32: 58–69
9. Dilger K, Hohenester S, Winkler-Budenhofer U et al (2012) Effect of ursodeoxycholic acid on bile acid profiles and intestinal detoxification machinery in primary biliary cirrhosis and health. J Hepatol 57:133–140
10. Woolbright BL, Li F, Xie Y et al (2014) Lithocholic acid feeding results in direct hepatotoxicity independent of neutrophil function in mice. Toxicol Lett 228:56–66
11. Woolbright BL, Li F, Gholami P et al (2014) Human pathophysiological concentrations of bile salts induce necrosis in primary human

hepatocytes (abstract). FASEB J 28(Suppl 1): 398.1

12. Marrero I, Sanchez-Bueno A, Cobbold PH (1994) Taurolithocholate and taurolithocholate 3-sulphate exert different effects on cytosolic free Ca^{2+} concentration in rat hepatocytes. Biochem J 300:383–386
13. Botla R, Spivey JR, Aguilar H et al (1995) Ursodeoxycholate (UDCA) inhibits the mitochondrial membrane permeability transition induced by glycochenodeoxycholate: a mechanism of UDCA cytoprotection. J Pharmacol Exp Ther 272:930–938
14. Sokol RJ, Devereaux M, Khandwala R et al (1993) Evidence for involvement of oxygen free radicals in bile acid toxicity to isolated rat hepatocytes. Hepatology 17:869–881
15. Allen K, Jaeschke H, Copple BL (2011) Bile acids induce inflammatory genes in hepatocytes: a novel mechanism of inflammation during obstructive cholestasis. Am J Pathol 178:175–186
16. O'Brien KM, Allen KM, Rockwell CE et al (2013) IL-17A synergistically enhances bile acid-induced inflammation during obstructive cholestasis. Am J Pathol 183:1498–1507
17. Gujral JS, Liu J, Farhood A et al (2004) Functional importance of ICAM-1 in the mechanism of neutrophil-induced liver injury in bile duct-ligated mice. Am J Physiol Gastrointest Liver Physiol 286:G499–G507
18. Kim ND, Moon JO, Slitt AL et al (2006) Early growth response factor-1 is critical for cholestatic liver injury. Toxicol Sci 90:586–595
19. Gujral JS, Farhood A, Bajt ML et al (2003) Neutrophils aggravate acute liver injury during obstructive cholestasis in bile duct-ligated mice. Hepatology 38:355–363
20. Yang M, Ramachandran A, Yan HM et al (2014) Osteopontin is an initial mediator of inflammation and liver injury during obstructive cholestasis after bile duct ligation in mice. Toxicol Lett 224:186–195
21. Schoemaker MH, Gommans WM, Conde de la Rosa L et al (2003) Resistance of rat hepatocytes against bile acid-induced apoptosis in cholestatic liver injury is due to nuclear factor-kappa B activation. J Hepatol 39:153–161
22. Jang JH, Rickenbacher A, Humar B (2012) Serotonin protects mouse liver from cholestatic injury by decreasing bile salt pool after bile duct ligation. Hepatology 56:209–218
23. Gujral JS, Liu J, Farhood A et al (2004) Reduced oncotic necrosis in Fas receptor-deficient C57BL/6J-lpr mice after bile duct ligation. Hepatology 40:998–1007
24. Fickert P, Trauner M, Fuchsbichler A et al (2005) Oncosis represents the main type of cell death in mouse models of cholestasis. J Hepatol 42:378–385
25. Vinken M, Landesmann B, Goumenou M et al. (2013) Development of an adverse outcome pathway from drug-mediated bile salt export pump inhibition to cholestatic liver injury. Toxicol Sci 136:97–106
26. Chatterjee S, Bijsmans IT, van Mil SW et al (2014) Toxicity and intracellular accumulation of bile acids in sandwich-cultured rat hepatocytes: role of glycine conjugates. Toxicol In Vitro 28:218–230
27. Patel T, Bronk SF, Gores GJ (1994) Increases of intracellular magnesium promote glycodeoxycholate-induced apoptosis in rat hepatocytes. J Clin Invest 94:2183–2192
28. Jones B, Roberts PJ, Faubion WA et al (1998) Cystatin A expression reduces bile salt-induced apoptosis in a rat hepatoma cell line. Am J Physiol 275:G723–G730
29. Galle PR, Theilmann L, Raedsch R et al (1990) Ursodeoxycholate reduces hepatotoxicity of bile salts in primary human hepatocytes. Hepatology 12:486–491
30. Gonzalez R, Cruz A, Ferrin G et al (2011) Nitric oxide mimics transcriptional and post-translational regulation during α-tocopherol cytoprotection against glycochenodeoxycholate-induced cell death in hepatocytes. J Hepatol 55:133–144
31. Szabo M, Veres Z, Baranyai Z et al (2013) Comparison of human hepatoma HepaRG cells with human and rat hepatocytes in uptake transport assays in order to predict a risk of drug induced hepatotoxicity. PLoS One 8: e59432
32. Liang D, Hagenbuch B, Stieger B et al (1993) Parallel decrease of Na(+)-taurocholate cotransport and its encoding mRNA in primary cultures of rat hepatocytes. Hepatology 18:1162–1166
33. Rippin SJ, Hagenbuch B, Meier PJ et al (2001) Cholestatic expression pattern of sinusoidal and canalicular organic anion transport systems in primary cultured rat hepatocytes. Hepatology 33:776–782
34. Jemnitz K, Veres Z, Vereczkey L (2010) Contribution of high basolateral bile salt efflux to the lack of hepatotoxicity in rat in response to drugs inducing cholestasis in human. Toxicol Sci 115:80–88
35. Kullak-Ublick GA, Meier PJ (2000) Mechanisms of cholestasis. Clin Liver Dis 4: 357–385

36. Padda MS, Sanchez M, Akhtar AJ et al (2011) Drug-induced cholestasis. Hepatology 53: 1377–1387
37. Kubitz R, Droge C, Stindt J et al (2012) The bile salt export pump (BSEP) in health and disease. Clin Res Hepatol Gastroenterol 36:536–553
38. Fattinger K, Funk C, Pantze M et al (2001) The endothelin antagonist bosentan inhibits the canalicular bile salt export pump: a potential mechanism for hepatic adverse reactions. Clin Pharmacol Ther 69:223–231
39. Antherieu S, Bachour-El Azzi P, Dumont J et al (2013) Oxidative stress plays a major role in chlorpromazine-induced cholestasis in human HepaRG cells. Hepatology 57:1518–1529
40. Stieger B, Beuers U (2011) The canalicular bile salt export pump BSEP (ABCB11) as a potential therapeutic target. Curr Drug Targets 12:661–670
41. Garzel B, Yang H, Zhang L et al (2014) The role of bile salt export pump gene repression in drug-induced cholestatic liver toxicity. Drug Metab Dispos 42:318–322
42. Rust C, Bauchmuller K, Bernt C et al (2006) Sulfasalazine reduces bile acid induced apoptosis in human hepatoma cells and perfused rat livers. Gut 55:719–727
43. Muhlfeld S, Domanova O, Berlage T et al (2012) Short-term feedback regulation of bile salt uptake by bile salts in rodent liver. Hepatology 56:2387–2397
44. Denk GU, Maitz S, Wimmer R et al (2010) Conjugation is essential for the anticholestatic effect of NorUrsodeoxycholic acid in taurolithocholic acid-induced cholestasis in rat liver. Hepatology 52:1758–1768
45. Denk GU, Kleiss CP, Wimmer R et al (2012) Tauro-β-muricholic acid restricts bile acid-induced hepatocellular apoptosis by preserving the mitochondrial membrane potential. Biochem Biophys Res Commun 424: 758–764
46. Le Vee M, Jigorel E, Glaise D et al (2006) Functional expression of sinusoidal and canalicular hepatic drug transporters in the differentiated human hepatoma HepaRG cell line. Eur J Pharm Sci 28:109–117
47. Bajt ML, Knight TR, Lemasters JJ et al (2004) Acetaminophen-induced oxidant stress and cell injury in cultured mouse hepatocytes: protection by N-acetyl cysteine. Toxicol Sci 80: 343–349
48. Yan HM, Ramachandran A, Bajt ML et al (2010) The oxygen tension modulates acetaminophen-induced mitochondrial oxidant stress and cell injury in cultured hepatocytes. Toxicol Sci 117:515–523
49. Hohenester S, Vennegeerts T, Wagner M et al (2014) Physiological hypoxia prevents bile salt-induced apoptosis in human and rat hepatocytes. Liver Int 34(8):1224–1231
50. Woolbright BL, Ramachandran A, McGill MR et al (2012) Lysosomal instability and cathepsin B release during acetaminophen hepatotoxicity. Basic Clin Pharmacol Toxicol 111:417–425

Chapter 29

In Vitro Cell Culture Models of Hepatic Steatosis

Gahl Levy, Merav Cohen, and Yaakov Nahmias

Abstract

The liver is the systemic hub of lipid metabolism. The excessive accumulation of lipids in hepatocytes, steatosis, is a major clinical concern, whose progressive forms lead to end-stage liver disease. Currently, animal studies are the gold standard in toxicological risk assessment. Fueled by an integration of modern omics technologies, in silico models and in vitro system optimization, a new paradigm in the basis for toxicological risk assessment is emerging away from the use of animals. In recent years, in vitro assays have been developed for the early screening of the steatogenic potential of compounds. The present chapter describes an assay for the intracellular detection of lipids, a high-content screen for the distinction between steatosis and phospholipidosis, a multiparametric high-content screen for steatogenic potential and a liver X receptor reporter cell line.

Key words Liver, Steatosis, Phospholipidosis, Nile red, High-content screening, Liver X receptor

1 Introduction

Central to mammalian energy homeostasis, the liver is the systemic hub of lipid metabolism. The liver itself is capable of amassing abundant quantities of lipids under stress conditions associated with prolonged surplus energy consumption or impaired fatty acid metabolism. The excessive accumulation of lipids in hepatocytes, steatosis, is a major clinical concern, whose progressive forms lead to end-stage liver disease [1, 2]. Drug-induced hepatic steatosis is of particular pharmaceutical and regulatory concern, potentially resulting in disrupted clinical trials, postmarketing withdrawal or restricted use guidelines [3]. Currently, animal studies are the gold standard in toxicological risk assessment [4]. Yet, animal studies fail to identify approximately half of tested hepatotoxic compounds [5, 6]. Fueled by an integration of modern omics technologies, in silico biokinetic and toxicodynamic models, and in vitro system optimization, a new paradigm in the basis for toxicological risk assessment is emerging. The shift to non-animal test methods requires the development of well-standardized and reliable semi-throughput or high-throughput

Mathieu Vinken and Vera Rogiers (eds.), *Protocols in In Vitro Hepatocyte Research*, Methods in Molecular Biology, vol. 1250, DOI 10.1007/978-1-4939-2074-7_29, © Springer Science+Business Media New York 2015

in vitro drug screening protocols. In recent years, in vitro assays have been developed for the early screening of the steatogenic potential of compounds [7–9].

Drug-induced hepatic steatosis can result from several mechanisms. Many drugs directly inhibit mitochondrial beta-oxidation enzymes (e.g. tetracycline) or transiently sequester cofactors involved in the metabolic pathway (e.g. valproic acid) [10, 11]. Some drugs reduce beta-oxidation by severely impairing mitochondrial function. In certain cases, reduced mitochondrial respiration is due to the direct inhibition of electron transfer within the respiratory chain (e.g. amiodarone) or by damage to mitochondrial DNA (e.g. zidovudine) [12]. Reduction in the regeneration of oxidized cofactors (i.e. nicotinamide adenine dinucleotide and flavin adenine dinucleotide) needed for beta-oxidation is another mechanism of mitochondrial function impairment [8]. Other potential mechanisms include impairment of the formation of very-low density lipoproteins [13].

Given the variety of mechanisms of drug-induced hepatic steatosis, steatogenic potential screening strategies require the analysis of multiple parameters. This chapter focuses on recent developments in in vitro assays of hepatic steatosis. An assay is described for the detection of intracellular lipid accumulation utilizing the lipophilic dye Nile red. Previous studies have shown that drugs which cause hepatic steatosis increase Nile red binding at as low as a tenfold lower concentrations than required for cytotoxicity [14]. We describe a high-content screen for the distinction between steatosis and phospholipidosis utilizing LipidTOX, a commercially available kit. We also describe a multiparametric high-content screen for drug-induced steatosis, which concomitantly analyzes lipid accumulation, mitochondrial membrane potential and oxidative stress. This fluorescent-based approach enables the use of 96-well formats, thus increasing the assay throughput. This assay has been shown to correctly identify positive and negative steatosis inducers [9]. Finally, we describe the development and use of a reporter cell line for liver X receptor (LXR), whose activation has been considered as the molecular initiating event in steatotic adverse outcome pathways [15].

2 Materials

2.1 HepG2 Culture Medium

1. 10,000 units–10 mg/mL penicillin-streptomycin solution (Sigma, Israel). Aliquot to 5 mL and store at −20 °C.
2. European grade heat-inactivated fetal bovine serum (Biological Industries, Israel).
3. HepG2 culture medium. Under sterile conditions, add 5 mL penicillin-streptomycin solution and 50 mL fetal bovine serum

to 500 mL of high glucose Dulbecco's Modified Eagle's Medium. HepG2 culture medium should be stored at 4 °C and used within a few weeks (*see* **Note 1**).

4. Incubator (37 °C ± 1 °C, 90 % ± 5 % humidity, 5 % ± 1 % CO_2).
5. Laminar air flow cabinet.
6. 96-Well culture plates.

2.2 Nile Red Intracellular Lipid Detection

1. Phosphate-buffered saline (PBS). 137 mM NaCl, 2.7 mM KCl, 10 mM $Na_2HPO_4{\cdot}2H_2O$, 1.8 mM KH_2PO_4 in ultrapure water. Adjust to pH 7.4 and store for maximum 6 months at 4 °C.
2. 1 mg/mL Nile red solution in acetone. Aliquot to 100 μL and store at −20 °C.
3. 10 mg/mL Hoechst 33342 solution (Sigma-Aldrich, Israel) in ultrapure water. Store at 4 °C.
4. Nile red working solution. Utilize a 100 μg/mL intermediate acetone dilution to prepare a final working solution of 320 ng/mL Nile red in PBS. Immediately add 10 mg/mL Hoechst 33342 solution to a final concentration of 1 μg/mL (*see* **Note 2**).
5. Bovine serum albumin (AMRESCO, Israel).
6. 50 mM oleate acid stock solution prepared in culture medium containing 1 % bovine serum albumin. Store at 4 °C.
7. 50 mM palmitate (Sigma-Aldrich, Israel) stock solution prepared in culture medium containing 1 % bovine serum albumin. Store at 4 °C.
8. Lipid-overloading induction medium. Prepare a 50 μM mixture of oleate and palmitate in a 2:1 ratio in culture medium utilizing the 50 mM stock solutions (*see* **Note 3**).
9. Bradford reagent (Sigma-Aldrich, Israel).
10. 0.1 % sodiumdodecylsulfate lysis buffer: dilute sodiumdodecylsulfate (Sigma-Aldrich, Israel) to a concentration of 0.1 % using double deionized water.

2.3 LipidTOX Steatosis and Phospholipidosis High-Content Screen

1. 10 mM cyclosporin A steatosis positive control stock solution (HCS LipidTOX kit, Invitrogen, USA). Vigorously mix cyclosporin A and dimethylsulfoxide (DMSO). Store at −20 °C for up to 1 month.
2. 10 mM propanolol phospholipidosis positive control stock solution (HCS LipidTOX kit, Invitrogen, USA). Vigorously mix propranolol and DMSO. Store at −20 °C for up to 1 month.
3. 20 % paraformaldehyde (PFA) stock solution: Dissolve PFA electron microscopy-grade (Polysciences, USA) in PBS and NaOH adjusted to 1 N by heating up to 60 °C on a magnetic

stirrer. Filter the solution with a 0.22 µm filter. After cooling on ice, adjust the pH to 7.2 using 1 M HCl. Aliquot 2 mL and store up to 6 months at −20 °C.

4. PFA 4 % solution. Dilute the 20 % PFA stock solution to 4 % by adding PBS and heating in a 37 °C waterbath until dissolved. This solution must be used immediately.
5. 10 mg/mL Hoechst 33342 solution in ultrapure water. Store at 4 °C.

2.4 Multiparametric High-Content Screen for Drug Induced Steatosis

1. 50 mM oleate acid stock solution prepared in culture medium containing 1 % bovine serum albumin. Store at 4 °C.
2. 50 mM palmitate stock solution prepared in culture medium containing 1 % bovine serum albumin. Store at 4 °C.
3. Lipid-overloading induction medium. Prepare a 50 µM mixture of oleate and palmitate in a 2:1 ratio in culture medium utilizing the 50 mM stock solutions.
4. Boron-dipyrromethene (BODIPY) 493/503 stock solution. Dissolve 1 mg BODIPY 493/503 (Sigma, Israel) in 1 mL of DMSO. Store at −20 °C for up to 6 months and protected from light (*see* **Note 4**).
5. 1 mg/mL 2′,7′-dichlorofluorescein diacetate (DHCF-DA) in DMSO. Store at −20 °C for up to 6 months and protected from light.
6. 1 mg/mL tetramethyl rhodamine methyl ester (TMRM) (Sigma-Aldrich, Israel) in DMSO. Store at −20 °C for up to 6 months and protected from light.
7. 10 mg/mL Hoechst 33342 solution in ultrapure water. Store at 4 °C.
8. 1 mg/mL propidium iodide Sigma-Aldrich, Israel stock solution in double deionized water. Store at 4 °C.
9. Intracellular lipid working solution. Prepare PBS containing 5 ng/mL BODIPY 493/503, 2 µg/mL propidium iodide and 2 µg/mL Hoechst 33342.
10. Mitochondrial function working solution. Prepare PBS containing 75 µg/mL TMRM, 2 µg/mL DHCF-DA, 2 µg/mL propidium iodide and 2 µg/mL Hoechst 33342.
11. Drug treatment working solution. Prepare HepG2 culture medium with the desired concentration of a treatment drug as well as controls containing only the solvent used for dissolving the drug in the appropriate concentration.
12. Automated fluorescence microscope and image analysis software ZEN ASSAYBuilder (Carl Zeiss, Germany).

2.5 LXR Reporter Cell Line Construction

1. LXR green fluorescent protein lentiviral reporter construct (SBI Biosciences, Israel).
2. Lentiviral construction medium additive. Dilute 6 μg of LXR green fluorescent protein reporter construct DNA, 4 μg of pGAG-pol and 2 μg of pVSVG in 275 μL Optimem (Invitrogen, USA) and 25 μL polyethylenimine (Sigma-Aldrich, Israel). This step should be done immediately prior to use (*see* **Notes 5** and **6**).
3. HEK293T (ATCC, USA) culture medium. Under sterile conditions, add 5 mL 10,000 units-10 mg/mL penicillin-streptomycin solution and 50 mL fetal bovine serum to 500 mL of high glucose Dulbecco's Modified Eagle's Medium. Culture medium should be stored at 4 °C and used within a few weeks.
4. HEK293T infection medium. Under sterile conditions, add 50 mL of high glucose Dulbecco's Modified Eagle's Medium. Culture medium should be stored at 4 °C and used within 1 week.
5. Polybrene (Sigma, Israel).
6. T0901317 (Sigma, Israel).
7. Naringenin (Sigma, Israel).
8. 6-Well and 10 cm diameter culture dishes.

3 Methods

3.1 Nile Red Intracellular Lipid Detection

3.1.1 HepG2 Cell Culture (See Note 7)

1. Seed HepG2 cells in a 96-well plate in HepG2 culture medium at 37 °C and 5 % CO_2 and allow cultures to grow to 75 % confluency. 100 μL medium per well is needed.
2. For subculturing purposes, detach cells by treatment with 0.25 % trypsin/0.02 % ethylenediaminetetraacetic acid at 37 °C for 3 min.

3.1.2 HepG2 Lipid-Overloading Induction

1. Allow the HepG2 cells at least 24 h after seeding before initiating the lipid-overloading induction.
2. Remove the HepG2 culture medium and gently wash with PBS.
3. Add lipid-overloading induction medium to each well and incubate at 37 °C and 5 % CO_2 for 12 h (*see* **Note 8**).
4. Replace the lipid-overloading induction medium with standard HepG2 culture medium containing different concentrations of test compounds and incubate at 37 °C and 5 % CO_2 for 24 h (*see* **Note 9**).
5. Remove the HepG2 culture medium containing the test compounds and gently wash with PBS (*see* **Note 10**).

6. Add 100 μL of Nile red working solution to each well and incubate for 15 min at 37 °C and 5 % CO_2 covered from light.
7. Remove the Nile red working solution and wash with PBS.
8. Add 100 μL PBS.
9. Image the plate directly with a fluorescence microscope with excitation/emission maxima 488/550 nm for Nile red and 350/461 nm for Hoechst 33342 stain. Alternatively, Nile red measurements can also be conducted without Hoechst 33342 utilizing a microplate reader using the same excitation/emission maxima (*see* **Note 11**). In the case a microplate reader is utilized, perform a Bradford protein content assay.

3.1.3 Bradford Protein Content Assay

1. Weigh 5 mg of bovine serum albumin and dissolve in double deionized water.
2. Make bovine serum albumin dilutions in concentrations 0, 2.5, 1.25, 0.625, 0.3125, 0.156 mg/mL. This should be done using serial dilution.
3. After microplate Nile red reading is complete, remove PBS and wash twice with PBS (*see* **Note 12**).
4. Add 50 μL 0.1 % sodiumdodecylsulfate lysis buffer to each well. Mix and pipette up and down to make sure all cells have been lyzed (*see* **Note 13**).
5. Pipette 7 μL of each sample and standard curve preparation to a new 96-well plate. Make sure to thoroughly mix prior to sampling.
6. Add 150 μL undiluted Bradford reagent at room temperature to each well and mix (*see* **Note 14**).
7. Cover the plate with aluminum foil to protect from light and incubate at room temperature for 15 min.
8. Measure the absorbance at 585 nm in a microplate reader.

3.1.4 Processing the Results

1. Normalize the Nile red staining intensity to the Hoechst 33342 staining intensity.
2. For plate reader measurements, normalize Nile red absorbance to Bradford assay results and present per mg protein.
3. For Bradford assay data analysis, create a standard curve by plotting the 595 nm values in the *y*-axis versus their concentration in μg/mL in the *x*-axis. Determine the unknown sample concentration using the standard curve. If the samples were diluted (*see* **Note 13**), adjust the final concentration of the unknown samples by multiplying by the dilution factor used. To obtain total mg protein, multiply by the total mL of lysate (i.e. 0.05 mL).

3.2 LipidTOX Steatosis and Phospholipidosis High-Content Screen

3.2.1 Phospholipid Staining (See Note 15)

1. Thaw the vials containing the LipidTOX stains at room temperature.
2. Dilute the 1,000× concentrated LipidTOX red phospholipid stain 1:500 in cell culture medium to obtain a 2× concentrated stain solution. A total volume of 50 μL per well is required.
3. Filter the 2× concentrated stain solution through a 0.22 μm filter.
4. Prepare a 2× concentrated solution of the test compounds in cell culture medium. A total volume of 50 μL per well is required.
5. Remove the growth medium from the cell culture plates.
6. Add 50 μL of the 2× concentrated LipidTOX red phospholipid stain per well.
7. Add 50 μL cell culture medium containing the 2× concentrated test compound or 50 μL control medium.
8. Incubate the cells at 37 °C and 5 % CO_2 for 48 h (*see* **Note 16**).

3.2.2 Fixation

1. Remove the incubation medium.
2. Immediately add 100 μL 4 % PFA solution to each well.
3. Incubate for 10 min at room temperature.
4. Remove the 4 % PFA solution and gently wash twice with PBS.

3.2.3 Neutral Lipid Staining

1. Dilute the 1,000× concentrated LipidTOX green neutral lipid stain 1:1,000 in PBS. A volume of 100 μL per well is required.
2. Remove the PBS from the wells.
3. Add 100 μL 1× concentrated LipidTOX green neutral lipid stain to each well.
4. Wrap the plates in aluminum foil to protect from light.
5. Incubate the plates at room temperature for 45 min.
6. Remove the 1× concentrated LipidTOX green neutral lipid stain solution. Do not wash.
7. Dilute the Hoechst 33342 stock solution 1:10,000 for a final concentration of 1 μg/mL in PBS.
8. Incubate cells in the Hoechst 33342 solution for 2 min at room temperature protected from light.
9. Remove the Hoechst 33342 solution and wash twice with PBS. Add 100 μL PBS to each well.
10. Image the plate directly with a fluorescence microscope with excitation/emission maxima 595/615 nm for LipidTOX red phospholipid stain, 495/505 nm for LipidTOX green neutral lipid stain and 350/461 nm for Hoechst 33342 stain.
11. Plates can be stored at 4 °C for not longer than 1 week before imaging and protected from light.

3.2.4 Processing the Results

Normalize absolute red or green intensity to Hoechst 33342 intensity (*see* **Note 17**). Alternatively, count red or green particles and normalize to blue particles utilizing high-content screen software. Analyze the particles with ImageJ software.

3.3 Multiparametric High-Content Screen for Drug-Induced Steatosis

3.3.1 HepG2 Cell Culturing

Seed HepG2 cells in 2 96-well plates and incubate with 100 μL of HepG2 culture medium per well at 37 °C and 5 % CO_2. Allow cultures to grow to 75 % confluency. Cells should be given ample time to attach to the plate (i.e. 12–24 h) (*see* **Note 18**).

3.3.2 HepG2 Lipid-Overloading Induction

1. Remove the HepG2 culture medium and gently wash with PBS.
2. Add lipid-overloading induction medium to each well and incubate at 37 °C and 5 % CO_2 for 12 h.
3. Replace the lipid-overloading induction medium with drug treatment working solution and incubate at 37 °C and 5 % CO_2 for 24 h. Repeat the identical treatments in the second 96-well plate.
4. Remove the HepG2 culture medium containing the test compounds and gently wash with PBS.
5. Incubate cells in 1 of the 96-well plates with the intracellular lipid working solution for 30 min at 37 °C and protected from light.
6. Incubate cells in the other 96-well plate with the mitochondrial function working solution for 30 min at 37 °C and protected from light.
7. After incubation, remove probe mixtures and wash once with PBS. Add 100 μL PBS to the wells.
8. Analyze the plates directly with a fluorescence microscope with excitation/emission maxima 493/503 nm for BODIPY 493/503, 540/608 nm for propidium iodide, 548/574 nm for TMRM, 360/461 nm for Hoechst 33342 and 517/527 nm for DHCF-DA.

3.3.3 Image and Statistical Analysis

1. Collect images with the 10× objective for the distinct fluorescence channels using automated fluorescence microscopy.
2. Using the automated image analysis software, predefine the nuclear and cytoplasmic compartments in the images obtained.
3. Utilize Hoechst 33342 fluorescence to identify the nuclear region.
4. Use the propidium iodide fluorescence intensity in the predefined nuclear region to determine cell viability. Positive staining for propidium iodide is indicative of cell death.

5. Use TMRM fluorescence intensity in the cytoplasmic region to define the mitochondrial membrane potential.
6. Use DHCF fluorescence intensity in the cytoplasmic region to define reactive oxygen species production.
7. Use BODIPY 493/503 fluorescence intensity in the cytoplasmic region to assess lipid deposits.
8. Perform each measurement for individual cells and then average.
9. To correct for any potential fluorescence emitted from the treating compounds themselves, subtract the measured fluorescence of cells treated with tested compounds, but not with dyes.
10. Normalize the drug-treatment fluorescence intensity results for each parameter, such as lipid content, to the solvent only treated cells assumed to be a 100 % value.

3.4 LXR Reporter Cell Line Construction

3.4.1 HEK293T Cell Culture

Seed HEK293T cells in a 10 cm diameter cell culture plate at low density. Allow cells to grow to approximately 40 % confluency.

3.4.2 Lentivirus Preparation

1. Once cells reach approximately 40 % confluence, replace the culture medium with HEK293T infection medium.
2. Under sterile conditions, prepare lentivirus construction medium additive. Mix and allow to stand at room temperature for 5 min.
3. Add the additive to the HEK293T cells dropwise and incubate at 37 °C and 5 % CO_2 for 24 h.
4. Replace the medium with fresh HEK293T infection medium and incubate at 37 °C and 5 % CO_2 for 24 h. (*see* **Note 19**).
5. Collect the medium and replace with fresh medium.
6. Filter the collected medium through a 0.45 μm syringe filter and add polybrene to a final concentration of 8 μg/mL.

3.4.3 HepG2 LXR Lentivirus Infection

1. Add the collected medium with polybrene to HepG2 cells grown on a 6-well plate at 30 % confluence.
2. After 48 h, replace the medium on the HepG2 cell plate with more collected medium with polybrene.
3. At the end of infection, test the reporter cell line utilizing T0901317 or naringenin (Fig. 1). Cells can now be subcultured and expanded.
4. Treat the cells with various concentrations of drugs for 24 to 72 h and image for green fluorescent protein. Utilize Hoechst 33342 for normalization.

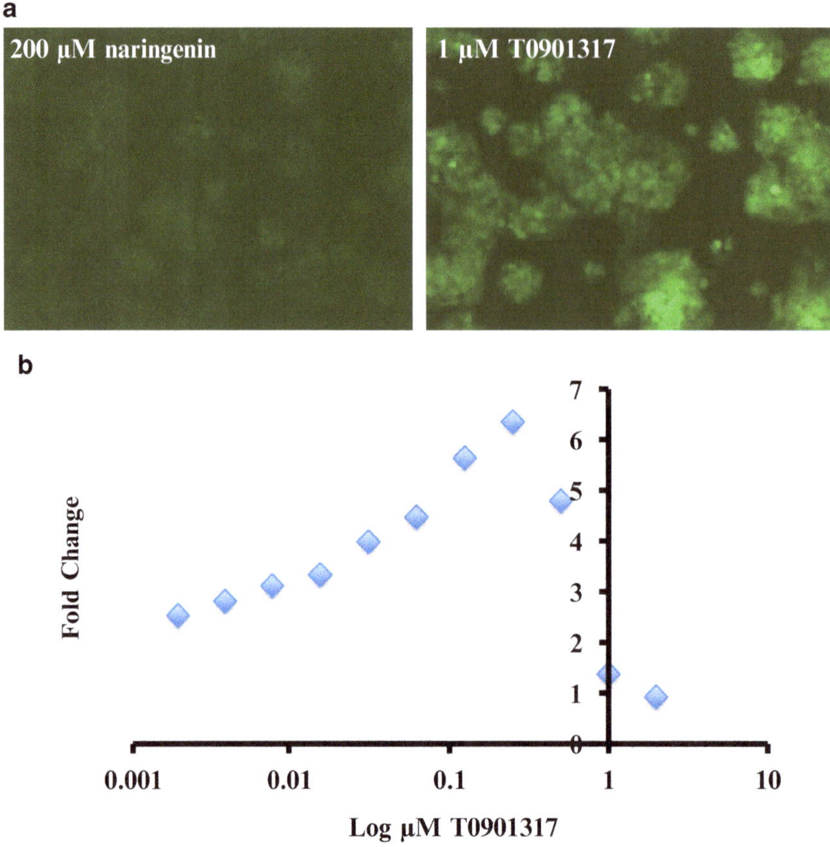

Fig. 1 Measurement of green fluorescent protein intensity in LXR reporter cell lines. (**a**) LXR green fluorescent protein reporter HepG2 cells were incubated with 1 μM of the LXR agonist T0901317 or 200 μM of the LXR antagonist naringenin for 24 h and imaged. (**b**) Fold change is reported as fluorescence intensity of culture treated with T0901317 relative to control LXR green fluorescent protein reporter HepG2 cells

4 Notes

1. Phenol red is known to be a weak estrogen [16]. Change to phenol red-free basal medium if this is an experimental concern.
2. Depending on the cell type, you may need to increase or decrease the concentration of Nile red. We recommend that a test stain should be performed where feasible, utilizing several concentrations of Nile red to determine the optimal concentration.
3. The proportions of oleate and palmitate can be critical. It has been previously shown that a high proportion of palmitic acid promotes an acute harmful effect [17]. Thus, different proportions can be utilized to study different steatotic injury.

4. Do not substitute BODIPY 493/503 with Nile red. The emission of BODIPY 493/503 was chosen specifically because of the 503 nm emission.
5. Before you begin, contact your institution's biosafety office to receive permission and instructions on working with lentiviruses. Follow strict safety procedures and the required biosafety level 2 suitable for handling HIV-derivative viruses.
6. The process requires expansion of the LXR reporter construct by *Escherichia coli* transformation and the subsequent isolation of DNA. If your laboratory does not have this capability, complete kits can be purchased from SBI Biosciences (Israel).
7. Other cell types can be easily adapted to this and the other protocols. We used Upcyte hepatocytes [18], and primary human hepatocytes. Several companies offer high-quality human hepatocytes in a collagen-coated 6-well plate format, including BD Bioscience (Israel) and Cambrex (Israel).
8. This step is optional. It may not be necessary to induce lipid-overloading. We have had success in identifying steatogenic compounds with this protocol without this induction (Fig. 2).
9. Determining the concentration of drugs is a critical step in this process. You may want to perform a cell viability assay at different concentrations. Common assays for viability are

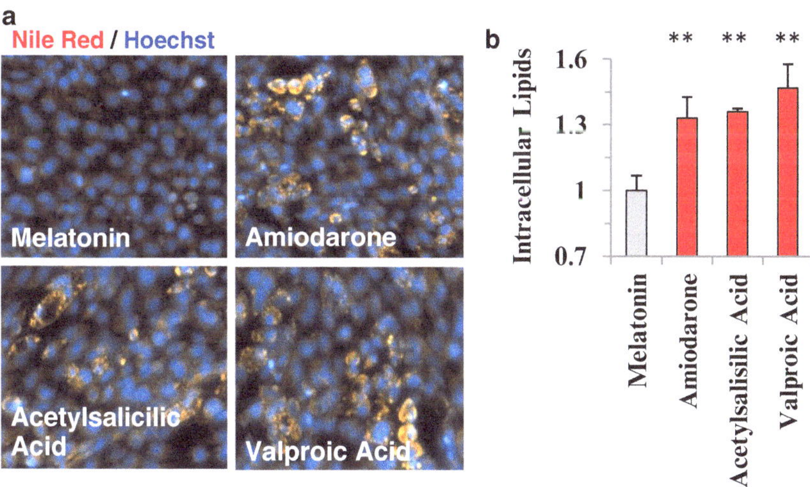

Fig. 2 Toxicological endpoint assay utilizing Nile red intracellular lipid detection. (**a**) Nile red fluorescence micrographs of hepatocytes following 48 h exposure to 2.5 μM amiodarone, 200 μM acetylsalicylic acid, 500 μM valproic acid or melatonin as control. (**b**) Quantification of intracellular lipids was performed utilizing Nile red and Hoechst 33342 stainings. Drug concentrations utilized were TC_{20} concentrations as determined by means of the LIVE/DEAD assay. Results represent the means ± standard deviation of four experimental replicates as staining intensity (i.e. Nile red/Hoechst 33342) normalized to melatonin (i.e. negative control). Results were evaluated by student *t*-test. $^{**}p < 0.01$

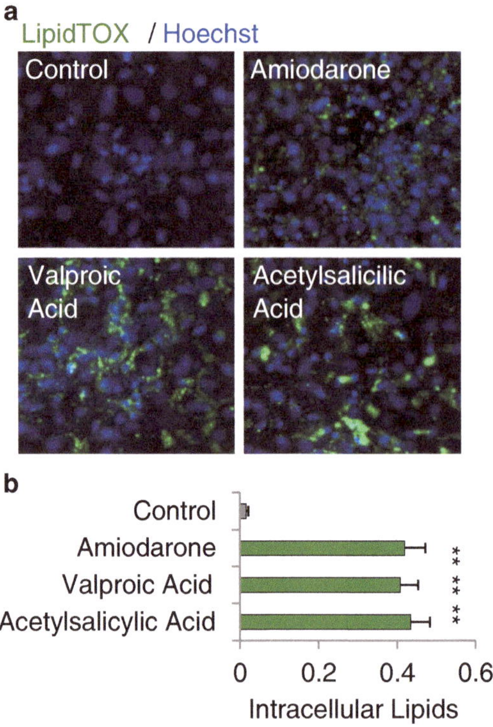

Fig. 3 Toxicological endpoint assay utilizing LipidTOX neutral lipid stain. (**a**) LipidTOX fluorescence micrographs of embryonic stem cell-derived hepatocytes following 24 h exposure to 135 μM amiodarone, 2 mM acetylsalicylic acid or 60 μM valproic acid. Cells were treated with 1 μM LipidTOX and 1 μg/mL Hoechst 33342. Drug concentrations utilized were TC_{20} concentrations as determined by means of the LIVE/DEAD assay. (**b**) Results represent the means ± standard deviation as staining intensity (i.e. LipidTOX/Hoechst 33342) normalized to negative control. Results were evaluated by student *t*-test. $^{**}p < 0.01$

the 3-(4,5-dimethylthiazol-2-yl)-2,5-diphenyltetrazolium bromide tetrazolium test [19] and LIVE/DEAD assay (Life Technologies, Israel). Once these assays are performed at different concentrations, a dose-dependent toxicity curve is utilized to select the drug concentrations for the experiment. We have used TC_{20} concentrations (i.e. concentrations that cause 20 % of the cells to die) successfully in our experiments (Figs. 2 and 3).

10. It is crucial that the washing is done gently. Depending on the condition of the cells, washing can disrupt the culture and wash away material. We make habit of observing the cells under the microscope to determine, by morphology, the condition of the culture. In certain cases, it may be necessary to avoid washing altogether. In such a case, be sure to remove the incubation medium completely and gently add PBS.

11. The use of a plate reader is tempting due to the high-throughput ability. We find that the microscope is preferred. Often, you will find points in the data that may need a closer look. The micrographs can provide a tremendous amount of information.
12. Washing here is very important. Be sure to wash any medium from the rim and the upper portion of each well. Any left-over medium will greatly taint the sample.
13. In the case that the lysate is too viscous, add more of the lysis buffer to all the wells. Make sure that all the cells have been lyzed, especially when the lysate is highly viscous.
14. We find that when tilting the plate while adding the Bradford reagent and pipetting reduces bubble formation.
15. You can perform the two portions of this assay separately. We have utilized the LipidTOX green neutral lipid stain (Life Technologies, Israel) only for experiments (Fig. 3).
16. For some drugs, less than 48 h may be necessary. Other drugs may require longer periods.
17. To correct for any potential fluorescence emitted from the treating compounds themselves, subtract the measured fluorescence of cells treated with tested compounds, but not with dyes.
18. If an experiment needs to be performed sooner, seeding HepG2 cells on collagen will allow the cells to attach faster. We also find that for microscopy, seeding HepG2 cells on collagen allows for better imaging. This could be very useful when performing this assay for the first time, as it will allow for better identification of cellular compartments.
19. The medium removed and all pipettes utilized in the process should be placed directly into a sealable receptacle containing bleach and allowed to stand for at least 45 min before discarding.

Acknowledgements

This work was funded by a European Research Council Starting Grant TMIHCV (project number 242699), a Marie Curie Reintegration Grant microLiverMaturation (project number 248417), the Israel-Japan Ministry of Science (award number 9645), the British Council BIRAX Regenerative Medicine award (number 33BX12HGYN) and HeMiBio, a jointly funded consortium by the European Commission and Cosmetics Europe as part of the SEURAT-1 cluster (project number HEALTH-F5-2010-266777).

References

1. Begriche K, Igoudjil A, Pessayre D et al (2006) Mitochondrial dysfunction in NASH: causes, consequences and possible means to prevent it. Mitochondrion 6:1–28
2. Mantena SK, King AL, Andringa KK et al (2008) Mitochondrial dysfunction and oxidative stress in the pathogenesis of alcohol- and obesity-induced fatty liver diseases. Free Radic Biol Med 44:1259–1272
3. Kleiner DE, Gaffey MJ, Sallie R et al (1997) Histopathologic changes associated with fialuridine hepatotoxicity. Mod Pathol 10: 192–199
4. Bhattacharya S, Zhang Q, Carmichael PL et al (2011) Toxicity testing in the 21st century: defining new risk assessment approaches based on perturbation of intracellular toxicity pathways. PloS One 6:e20887
5. Blomme EA, Yang Y, Waring JF (2009) Use of toxicogenomics to understand mechanisms of drug-induced hepatotoxicity during drug discovery and development. Toxicol Lett 186:22–31
6. Laverty HG, Antoine DJ, Benson C et al (2010) The potential of cytokines as safety biomarkers for drug-induced liver injury. Eur J Clin Pharmacol 66:961–976
7. Janorkar AV, King KR, Megeed Z et al (2009) Development of an in vitro cell culture model of hepatic steatosis using hepatocyte-derived reporter cells. Biotechnol Bioeng 102: 1466–1474
8. Donato MT, Martinez-Romero A, Jimenez N et al (2009) Cytometric analysis for drug-induced steatosis in HepG2 cells. Chem Biol Interact 181:417–423
9. Donato MT, Tolosa L, Jimenez N et al (2012) High-content imaging technology for the evaluation of drug-induced steatosis using a multiparametric cell-based assay. J Biomol Screen 17:394–400
10. Fromenty B, Pessayre D (1997) Impaired mitochondrial function in microvesicular steatosis. Effects of drugs, ethanol, hormones and cytokines. J Hepatol 26:43–53
11. Labbe G, Pessayre D, Fromenty B (2008) Drug-induced liver injury through mitochondrial dysfunction: mechanisms and detection during preclinical safety studies. Fundam Clin Pharmacol 22:335–353
12. Chariot P, Drogou I, De Lacroix-Szmania I et al (1999) Zidovudine-induced mitochondrial disorder with massive liver steatosis, myopathy, lactic acidosis, and mitochondrial DNA depletion. J Hepatol 30:156–160
13. Letteron P, Sutton A, Mansouri A et al (2003) Inhibition of microsomal triglyceride transfer protein: another mechanism for drug-induced steatosis in mice. Hepatology 38:133–140
14. Mcmillian MK, Grant ER, Zhong Z et al (2001) Nile red binding to HepG2 cells: an improved assay for in vitro studies of hepatosteatosis. In Vitr Mol Toxicol 14:177–190
15. Vinken M (2013) The adverse outcome pathway concept: a pragmatic tool in toxicology. Toxicology 312:158–165
16. Berthois Y, Katzenellenbogen JA, Katzenellenbogen BS (1986) Phenol red in tissue culture media is a weak estrogen: implications concerning the study of estrogen-responsive cells in culture. Proc Natl Acad Sci U S A 83:2496–2500
17. Gomez-Lechon MJ, Donato MT, Martinez-Romero A et al (2007) A human hepatocellular in vitro model to investigate steatosis. Chem Biol Interact 165:106–116
18. Burkard A, Dahn C, Heinz S et al (2012) Generation of proliferating human hepatocytes using Upcyte technology: characterisation and applications in induction and cytotoxicity assays. Xenobiotica 42:939–956
19. Mosmann T (1983) Rapid colorimetric assay for cellular growth and survival: application to proliferation and cytotoxicity assays. J Immunol Methods 65:55–63

Chapter 30

Assessment of Liver Fibrotic Insults In Vitro

Luis Perea, Mar Coll, and Pau Sancho-Bru

Abstract

In vitro systems are required to evaluate potential liver fibrogenic effects of drugs and compounds during drug development and toxicity screening, respectively. Upon liver injury or toxicity, hepatic stellate cells are activated, thereby acquiring a myofibroblastic phenotype and participating in extracellular matrix deposition and liver fibrosis. The most widely used in vitro models to investigate liver fibrogenesis are primary cultures of hepatic stellate cells, which can be isolated from healthy human livers. Currently, there are no effective methods to maintain hepatic stellate cells in vitro in a quiescent phenotype. Therefore, when cells are plated, they spontaneously become activated in few days. Most in vitro studies in this area have been performed with monocultures of hepatic stellate cells in order to assess the direct effects of a given factor on hepatic stellate cell activation or the induction of inflammatory and fibrogenic responses. In this chapter, focus is put on basic protocols to isolate hepatic stellate cells from human tissue and to maintain them in culture as well as on common in vitro assays to evaluate their response to profibrogenic factors.

Key words Hepatic stellate cells, Liver fibrosis, Hepatic stellate cell isolation, Hepatic stellate cell culture

1 Introduction

In vitro systems to evaluate liver fibrogenesis are required for the development of new antifibrogenic drugs and for the assessment of potential liver fibrogenic effects of drugs and compounds during drug development and toxicity screening [1]. Although not standardized, the most widely used in vitro models to assess the response to liver fibrotic insults and fibrogenesis consist of cultured primary hepatic stellate cells (HSCs). HSCs can be isolated from human and animal livers and cultured in a monoculture setting or in coculture with other liver cell types [2, 3]. In vivo, HSCs are located within the perisinusoidal space of Disse, which is the area between the hepatocytes and the liver sinusoidal endothelial cells. In normal liver, HSCs have a quiescent phenotype and their principal function is the storage of vitamin A in the form of retinyl palmitate in lipid droplets within the cytoplasm as well as the control of liver extracellular matrix homeostasis [4].

Mathieu Vinken and Vera Rogiers (eds.), *Protocols in In Vitro Hepatocyte Research*, Methods in Molecular Biology, vol. 1250, DOI 10.1007/978-1-4939-2074-7_30, © Springer Science+Business Media New York 2015

Hepatic fibrosis is defined as the excessive accumulation of collagen and other extracellular matrix molecules in the liver parenchyma. During liver injury, a complex relationship exists between the different hepatic cell types [5]. However, the main event in liver fibrogenesis is the activation of HSCs, which become activated from a quiescent phenotype to myofibroblast-like cells [5, 6]. During HSC activation, there is a change in the composition of the extracellular matrix produced by HSCs as well as an increase in the overall production of fibrillar components and a reduction of the degradation of the extracellular matrix. Moreover, once activated, HSCs acquire contractility and chemotaxis capacities, and increase the expression of cytoskeleton components, such as alpha smooth muscle actin, a number of cytokine receptors, growth factors, cytokines and inflammatory mediators [4, 5].

In liver injury, HSCs become activated to support liver regeneration by producing extracellular components and growth factors. If the insult is acute, HSCs return to their quiescent state or disappear by apoptosis [7]. However, if the insult and inflammation persist, there is a perpetuation of HSC activation and an excessive accumulation of extracellular matrix, which leads to fibrogenesis [8, 9] (Fig. 1). Progression of fibrogenesis may trigger the development of cirrhosis and subsequent associated complications [10, 11].

Primary quiescent HSCs can be isolated from normal human liver based on their content in lipid droplets, which confers them a lower density than other liver cells. Another method to isolate quiescent HSCs is by taking advantage of the autofluorescence of vitamin A present in lipid droplets [7, 12]. Vitamin A produces a green-blue fluorescence when it is excited with 330–360 nm ultraviolet light. By means of this fluorescent signal, HSCs can be sorted out from a mixture of cells using flow cytometry.

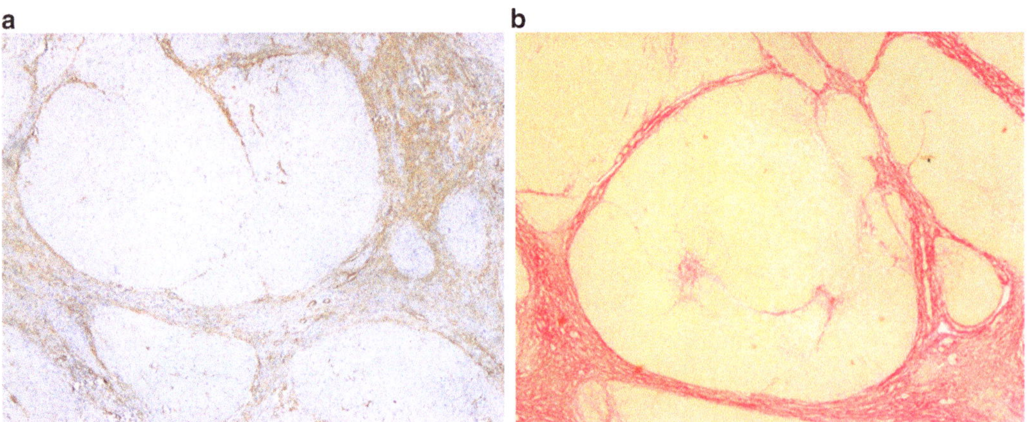

Fig. 1 Inmunohistochemistry staining of fibrosis in human liver tissue. (**a**) Representative images of alpha smooth muscle actin staining in liver tissue from a cirrhotic patient (200× magnification). (**b**) representative images of Sirius red staining in liver tissue from a cirrhotic patient (200× magnification)

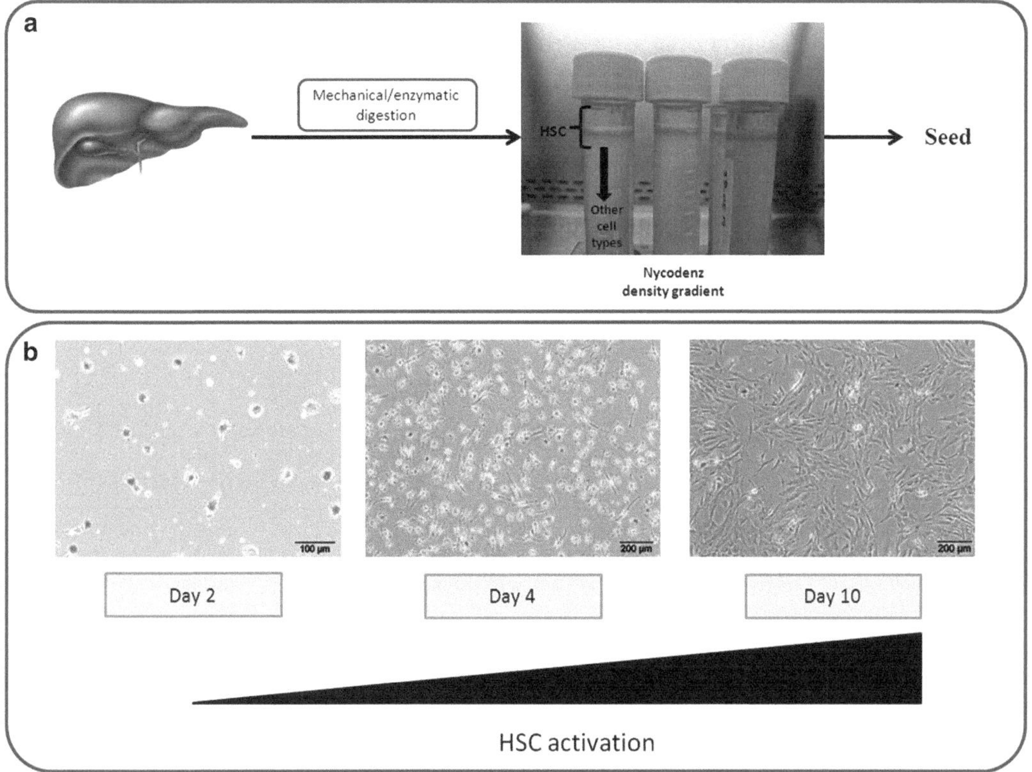

Fig. 2 Isolation of human quiescent HSCs. (**a**) HSCs are isolated from human tissue in 2 steps, namely a mechanical and enzymatic digestion in order to disaggregate the tissue, and a posterior Nycodenz density gradient. (**b**) Once isolated, HSCs are cultured on plastic with growth medium. After few days in culture, cells lose their lipid droplets and activate from a quiescent phenotype to myofibroblast-like cells

When plated on plastic, HSCs undergo an activation process that, although not identical, resembles the process that takes place during liver fibrogenesis [12]. Due to the surface stiffness and the presence of serum, plated HSCs start to activate, losing their lipid droplets and acquiring myofibroblastic properties. Currently, there are no methods to maintain HSCs in vitro in their quiescent phenotype. Hence, when cells are plated, they become activated within a few days (Fig. 2).

Although primary HSCs can be maintained in vitro for 10–12 doublings and can be frozen and thawed in early passages, high-throughput assays may require a renewable cell source as a culture model. As such, two widely used cell lines provide a source of immortalized human HSCs, namely LX-1 and LX-2. These cell lines keep essential features of HSCs, although they do not reproduce all the characteristics of primary HSCs [13]. Most in vitro studies in the context of liver fibrogenesis have been performed with HSC monocultures in order to assess the direct effects of a given factor on HSC activation or the induction of inflammatory and fibrogenic responses.

Table 1
Basic in vitro assays to study hepatic stellate cells response

Effects	Methods	Key markers
Fibrogenesis/inflammation	qRT-PCR	COL1A1, ACTA2, LOX, TIMP1, MCP-1, IL-8
Inflammation	ELISA	MCP1, IL-8, CCL20
Signaling pathways/protein expression	Immunoblot	ERK, AKT, p38, COL1A1, ACTA2, LOX, TIMP1
Migration	Boyden chamber Wound healing assay	
Proliferation, viability, cytotoxicity	Colorimetric assay	
Extracellular matrix degradation	Zymography	MMP2, MMP7, MMP9

ACTA2 actin alpha 2 smooth muscle, *AKT* protein kinase B, *CCL20* chemokine (C-C motif) ligand 20, *COL1A1* collagen type I alpha 1, *ERK* extracellular signal-regulated kinase, *IL-8* interleukin 8, *LOX* lysyl oxidase, *MCP-1* monocyte chemotactic protein 1, *MMP2/7/9* matrix metallopeptidase 2/7/9, *p38* mitogen-activated protein kinase 14, *TIMP1* metallopeptidase inhibitor 1

Cocultures that incorporate hepatocytes as well as other non-parenchymal liver cells may be useful to obtain information about the crosstalk between liver cell types. Several studies have demonstrated that cocultures of HSCs with hepatocytes maintain hepatocyte functionality, while preventing HSC activation [14, 15]. Moreover, in order to achieve a more physiological environment, several matrices as well as three-dimensional culture conditions of liver cells have been developed [16].

In vitro HSC response to fibrogenic insults can be evaluated at the gene expression, cell response, protein synthesis, and functional level (Table 1). Early cell response is usually assessed by evaluating the induction of key signaling pathways involved in HSC activation, such as mitogen-activated protein kinase, protein kinase B or p38. In order to determine the downstream effects of a fibrogenic insult, gene and protein expression analyses as well as the secretion of inflammatory mediators can be evaluated (Table 1). To confirm findings at the transcriptional level, most studies evaluate the expression of key genes at the translational level by immunoblotting or immunostaining. The functional changes associated with HSC activation can be monitored in cultured HSCs by assessing their viability, proliferation, migration and contractile response upon a fibrogenic insult. In this chapter, the wound healing assay is highlighted as a functional assay for monitoring HSC activation. This method reflects key aspects of HSC activation, including viability, proliferation and migration and as such may be considered as an assay to assess HSC activation and potential profibrogenic responses. This chapter equally describes protocols to isolate HSCs from human tissue and to maintain them in culture.

2 Materials

2.1 Isolation of Human Quiescent HSCs

1. 9.5 mM Dulbecco's phosphate-buffered saline without calcium and magnesium (Lonza, United States of America).
2. Enzymatic solution 1. Gey's balanced salt solution (Sigma, United States of America) with 3.15 mg/mL pronase from *Streptomyces griseus* (Roche, Germany), 0.38 mg/mL collagenase A (0.240 U/mg) from *Clostridium histolyticum* (Roche, Germany), 0.01 mg/mL DNAse I from bovine pancreas (Roche, Germany) and 1 % penicillin-streptomycin solution (Life Technologies, Spain). This enzymatic solution should be prepared ex tempore.
3. Enzymatic solution 2. Gey's balanced salt solution with 0.6 mg/mL pronase from *Streptomyces griseus*, 0.38 mg/mL collagenase A (i.e. 0.240 U/mg) from *Clostridium histolyticum*, 0.01 mg/mL DNAse I from bovine pancreas and 1 % penicillin-streptomycin solution. This enzymatic solution should be prepared ex tempore.
4. 18 % Nycodenz solution (Sigma, United Stated of America) in Gey's balanced salt solution (*see* **Note 1**).
5. Centrifuge with swinging tube bucket rotor (Thermo Scientific, United States of America).

2.2 HSC Cultivation

1. 0.25 % trypsin-0.05 % ethylenediaminetetraacetic acid solution with phenol red.
2. HSC standard growth medium. Dulbecco's Modified Eagle Medium with high glucose, GlutaMAX and pyruvate (Gibco, United States of America) containing 10 % fetal bovine serum (Gibco, United States of America) and supplemented with 1 % penicillin-streptomycin solution. Store for maximum 1 month at 4 °C and protect from light.
3. Incubator (37 °C ± 1 °C, 90 % ± 5 % humidity, 5 % ± 1 % CO_2).
4. Laminar air flow cabinet.
5. Thermostated bath.
6. Tissue culture flasks (Techno Plastic Products AG, Switzerland).

2.3 Evaluation of the Response of Cultured HSCs to Profibrogenic Compounds

1. HSC low-serum and insulin-free medium. Dulbecco's Modified Eagle Medium containing 0.01 % fetal bovine serum and supplemented with 1 % penicillin-streptomycin solution (*see* **Note 2**). Store for maximum 1 month at 4 °C and protect from light.
2. Human hepatocellular liver carcinoma HepG2 cells (ATCC, United States of America).
3. HepG2 complete growth medium. Dulbecco's Modified Eagle Medium containing 10 % fetal bovine serum and supplemented

with 2 % 200 mM L-glutamine solution (Lonza, United Stated of America) and 1 % penicillin-streptomycin solution. Store for maximum 1 month at 4 °C and protect from light.

4. Poly(2-hydroxyethylmethacrylate) (poly-HEMA) (Sigma, United States of America).
5. Human recombinant platelet-derived growth factor (PDGF) expressed in *Escherichia coli*, lyophilized powder suitable for cell culture (Sigma, United States of America).
6. Tissue culture test plates (Techno Plastic Products AG, Switzerland).

3 Methods

3.1 Isolation of Human Quiescent HSCs

3.1.1 Mechanical and Enzymatic Digestion of the Human Liver Sample

1. Cut the liver sample in small pieces. It is important to carefully remove any fibrotic bands and the Glisson's capsule.
2. Filter all the enzymatic solutions using a 0.22 μm pore filter. Put the processed sample into the enzymatic solution 1.
3. Incubate the enzymatic solution 1 with the liver tissue during 30 min at 37 °C while gentle shaking.
4. Collect the product of enzymatic digestion 1 and filter using a 100 μm pore cell strainer. Distribute the product of filtration over four sterile tubes (i.e. 25 mL of digestion in each tube) and bring them up to 50 mL with Gey's balanced salt solution in order to dilute the enzymes.
5. Put the remaining solid product from the first digestion in the enzymatic solution 2 and incubate during 30 min at 37 °C while gentle shaking.
6. Collect the product of enzymatic digestion 2 and filter using a 100 μm pore cell strainer. Distribute the product of filtration over four sterile tubes (i.e. 25 mL of digestion in each tube) and bring them up to 50 mL with Gey's balanced salt solution in order to dilute the enzymes.
7. Centrifuge tubes from enzymatic digestion 1 and 2 at 1,200 × *g* and 4 °C for 10 min.
8. Discard the supernatants and resuspend the pellets in Gey's balanced salt solution.
9. Collect the pellets in two new tubes and centrifuge at 1,200 × *g* and 4 °C for 10 min in order to put all the cells together in two tubes (*see* **Note 3**).

3.1.2 Isolation of HSCs with a Nycodenz Gradient

1. Discard the supernatant of the two tubes and resuspend the pellets in 13 mL of Gey's balanced salt solution in each tube (*see* **Note 4**). Collect the resuspended cells in a single tube.

2. Distribute 6.5 mL of the cell suspension on each aliquot containing 18 % Nycodenz.
3. Mix gently in order to form the correct gradient concentration. For quiescent HSCs, a final concentration of 9 % Nycodenz will be required to obtain a pure HSC population.
4. Put 1 mL Gey's balanced salt solution carefully above the Nycodenz cell suspension to generate a 2-phase gradient. Carefully transfer the tube into the centrifuge and spin at 1,200 × *g* and 4 °C for 15 min with low acceleration and break-off.
5. Collect in a new sterile tube the upper phase of the gradient, including the whitish band at the interface of the gradient. This band contains mainly HSCs and some debris (Fig. 2).
6. Wash the cells with Gey's balanced salt solution and centrifuge at 1,200 × *g* and 4 °C for 5 min.
7. Discard the supernatant and resuspend the cells in complete culture medium.
8. Seed the cells in a flask or well (*see* **Note 5**).
9. Maintain the cells in 5 % CO_2 at 37 °C in complete culture medium. 24–48 h after the isolation, wash the cells with Dulbecco's phosphate-buffered saline without calcium and magnesium. Add complete culture medium and let the cells grow until the cells reach confluency. The cells are now ready to be passaged (Fig. 2) (*see* **Note 6**).

3.2 HSC Cultivation

Primary HSCs are cultured in 5 % CO_2 at 37 °C in complete growth medium.

1. Check cell confluency under the microscope. Cells will need to be passaged when reaching 95 % of confluency (Fig. 2).
2. Set the waterbath at 37 °C and warm the medium and the trypsin-ethylenediaminetetraacetic acid solution.
3. Remove the medium from the flask and carefully wash with Dulbecco's phosphate-buffered saline without calcium and magnesium.
4. Add 1.5 mL trypsin in a T75 flask and incubate 3–5 min at 37 °C and 5 % CO_2.
5. Check the cells under microscope to confirm that most cells are in suspension.
6. Add 3–4 mL medium containing fetal bovine serum to block the trypsin reaction.
7. Harvest the cells, put them in a 50 mL centrifuge tube and centrifuge cells for 5 min at 650 × *g* and 4 °C.
8. Resuspend the cells in standard growth medium and add the cell suspension into a new flask or well.

9. Incubate the cells in 5 % CO_2 at 37 °C and let them grow. Change the medium every other day.
10. In order to study the cross-talk between liver cell types, co-cultures that incorporate hepatocytes as well as HSCs can be used (*see* **Note 7**).

3.3 Evaluation of the Response of Cultured HSCs to Profibrogenic Compounds

3.3.1 Preparation of Cultured HSCs

1. Seed 10,000 cells/cm^2 in a 6-well plate or desired configuration to perform the assay (*see* **Note 8**). Check the cells under the microscope to confirm if they are distributed homogeneously in the well.
2. Incubate the cells in 5 % CO_2 at 37 °C for 24 h.
3. Remove the medium and wash the cells twice with Dulbecco's phosphate-buffered saline without calcium and magnesium.
4. Add low-serum and insulin-free medium to each well and incubate overnight in 5 % CO_2 at 37 °C to stabilize the cells.
5. Dilute in low-serum and insulin-free medium the stock solution of the profibrogenic factors to be tested, taking into consideration the size and number of wells.
6. Aspirate the low-serum and insulin-free medium and carefully add the pre-warmed low-serum and insulin-free medium with the diluted profibrogenic factors (*see* **Note 9**).
7. Incubate the cells in 5 % CO_2 at 37 °C for the desired time.
8. Harvest the cells in order to isolate RNA or proteins and to assess the HSC response to the stimulus as listed in Table 1 (*see* **Note 10**).

3.3.2 Wound Healing Assay

1. Culture the cells in a 6-well plate until they are confluent.
2. Remove the medium and wash the cells carefully with Dulbecco's phosphate-buffered saline without calcium and magnesium.
3. Add low-serum and insulin-free medium to each well and incubate overnight in 5 % CO_2 at 37 °C to stabilize the cells.
4. Using a sterile 100 μL pipette tip, scratch two parallel wounds across the cell monolayer in each well (*see* **Note 11**) (Fig. 3).
5. Carefully wash the wells with Dulbecco's phosphate-buffered saline without calcium and magnesium after the scratches to remove the detached cells.
6. Carefully add the pre-warmed low-serum and insulin-free medium with the diluted profibrogenic factors to be tested as inhibitors or inducers of cell motility.
7. To evaluate the HSC response to a desired compound, add to each well the compound at the desired final concentration and incubate the cells as indicated. It is important to include a negative control, typically the vehicle used to resuspend the

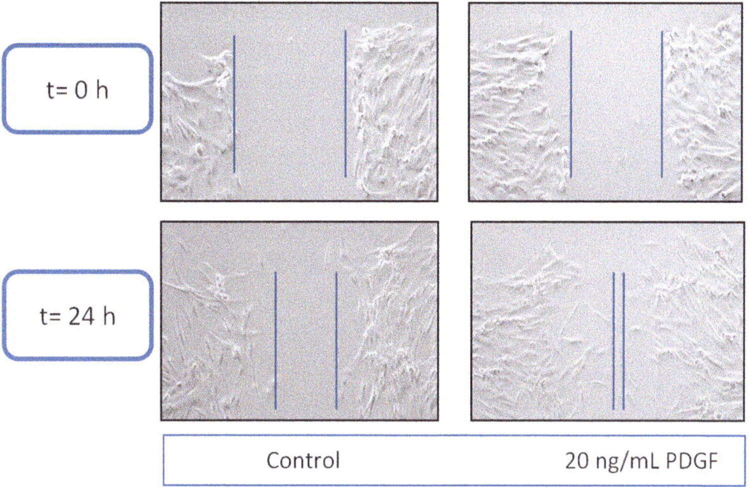

Fig. 3 Wound healing assay in activated HSCs. Representative pictures of a wound healing assay taken at 0 and 24 h after induction with 20 ng/mL PDGF and vehicle. Vertical lines show the closure or healing of the wound performed on the surface of the cultures

compound, as well as a positive control (i.e. 20 ng/mL PDGF or 20 % fetal bovine serum) (*see* **Note 12**).

8. Incubate the cells and take pictures of the same area 8, 16, 24, and 48 h after stimulation (Fig. 3) (*see* **Note 13**).

4 Notes

1. Filter the Nycodenz solution using a 0.22 μm pore filter and aliquot in 15 mL sterile tubes. Add 6.5 mL of the solution/tube. Tubes can be stored at −20 °C until needed. Usually, four aliquots are used for each isolation protocol.
2. The percentage of fetal bovine serum in HSC low-serum and insulin-free medium can be modified according to the requirements of the experiment.
3. Additional washes may be required if there is still debris in the first wash.
4. The volume used to resuspend the pellets will depend on the number of Nycodenz aliquots used. For each aliquot, 6.5 mL of cell suspension is needed. In a standard procedure with 25 g of liver tissue, cells are resuspended in 26 mL Dulbecco's phosphate-buffered saline to be distributed in four aliquots of Nycodenz.

5. It is advisable to plate the cells at a high density, since some of them will not attach after the isolation procedure. A reference seed concentration would be 30,000 cells/cm^2.
6. In order to have homogeneous and pure cultures of activated HSCs, cells from passage 3–7 should be used. If quiescent cells are required, HSCs will have to be used within the first 24 h after isolation.
7. To achieve a more physiological three-dimensional culture configuration, it is possible to generate aggregates of HSCs with HepG2 cells. A 2 % poly-HEMA coating in a 6-well plate prevents cells to attach and forms cellular aggregates. In order to properly form the aggregates, it is important to seed the HepG2 cells and HSCs in a 2:1 ratio (i.e. 200,000 HepG2 cells and 100,000 HSCs in a 6-well plate). Then, incubate in 5 % CO_2 at 37 °C over approximately 48 h to allow the formation of the spheroids prior to any assay.
8. Depending on the assay to be performed, cells will have to be seeded in a different configuration (Table 1).
9. It is important to add very carefully the pre-warmed low-serum and insulin-free medium, since shear stress may induce HSC activation of some signaling pathways. To evaluate the PDGF response, 20 ng/mL PDGF is added to each well and the cells are incubated for different periods in time, such as 4, 8, 12, and 24 h. The vehicle is used to resuspend PDGF [17].
10. Depending on the test to be performed after stimulation of the cells, it is required to incubate the cells for different periods of time. In order to investigate an early response to fibrogenic compounds, it is important to assess the induction of key cell signaling pathways. Phosphorylation of signaling molecules may be biphasic, with a short-term and a long-term response, thus it is advisable to collect proteins at 0, 5, 15, and 30 min, and 1 and 2 h after induction [17]. In order to see changes at the gene and protein expression level, cells are usually harvested 24 or 48 h after induction.
11. The gap width can be modified by using different pipette tip size.
12. Check the progression of the cultures under an inverted microscope and use image analysis software to quantify the gap closure. Make a mark under the well when you take the first image in order to ensure that you take the subsequent images in the same area.
13. Cell migration can also be assessed by using a Boyden chamber assay as an alternative to the wound healing assay.

Acknowledgements

This work was financially supported by grants from European Commission within its FP7 Cooperation Program and Cosmetics Europe (HeMiBio HEALTH F5 2010 number 266777). Pau Sancho-Bru and Mar Coll are funded by Instituto de Salud Carlos III and co-financed by Fondo Europeo de Desarrollo Regional (FEDER), Unión Europea, “Una manera de hacer Europa.”

References

1. Vinken M (2013) The adverse outcome pathway concept: a pragmatic tool in toxicology. Toxicology 4:158–165
2. Morales-Ibanez O, Dominguez M, Ki SH et al (2013) Human and experimental evidence supporting a role for osteopontin in alcoholic hepatitis. Hepatology 58:1742–1756
3. Mannaerts I, Eysackers N, Onyema OO et al (2013) Class II HDAC inhibition hampers hepatic stellate cell activation by induction of microRNA-29. PloS One 8:e55786
4. Friedman SL (2008) Hepatic stellate cells: protean, multifunctional, and enigmatic cells of the liver. Physiol Rev 88:125–172
5. Bataller R, Brenner DA (2005) Liver fibrosis. J Clin Invest 115:209–218
6. Mederacke I, Hsu CC, Troeger JS et al (2013) Fate tracing reveals hepatic stellate cells as dominant contributors to liver fibrosis independent of its aetiology. Nat Commun 4:2823
7. Troeger JS, Mederacke I, Gwak GY et al (2012) Deactivation of hepatic stellate cells during liver fibrosis resolution in mice. Gastroenterology 143:1073–1083
8. Friedman SL (2010) Evolving challenges in hepatic fibrosis. Nat Rev Gastroenterol Hepatol 7:425–436
9. Friedman SL (2004) Mechanisms of disease: mechanisms of hepatic fibrosis and therapeutic implications. Nat Clin Pract Gastroenterol Hepatol 1:98–105
10. Tsochatzis EA, Bosch J, Burroughs AK (2014) Liver cirrhosis. Lancet 383:1749–1761
11. Cardenas A, Gines P (2009) Portal hypertension. Curr Opin Gastroenterol 25: 195–201
12. Sancho-Bru P, Bataller R, Gasull X et al (2005) Genomic and functional characterization of stellate cells isolated from human cirrhotic livers. J Hepatol 43:272–282
13. Xu L, Hui AY, Albanis E et al (2005) Human hepatic stellate cell lines, LX-1 and LX-2: new tools for analysis of hepatic fibrosis. Gut 54:142–151
14. Thomas RJ, Bhandari R, Barrett DA et al (2005) The effect of three-dimensional co-culture of hepatocytes and hepatic stellate cells on key hepatocyte functions in vitro. Cells Tissues Organs 181:67–79
15. Krause P, Saghatolislam F, Koenig S et al (2009) Maintaining hepatocyte differentiation in vitro through co-culture with hepatic stellate cells. In Vitro Cell Dev Biol Anim 45: 205–212
16. Godoy P, Hewitt NJ, Albrecht U et al (2013) Recent advances in 2D and 3D in vitro systems using primary hepatocytes, alternative hepatocyte sources and non-parenchymal liver cells and their use in investigating mechanisms of hepatotoxicity, cell signaling and ADME. Arch Toxicol 87:1315–1530
17. Affo S, Morales-Ibanez O, Rodrigo-Torres D et al (2014) CCL20 mediates lipopolysaccharide induced liver injury and is a potential driver of inflammation and fibrosis in alcoholic hepatitis. Gut doi: 10.1136/gutjnl-2013-306098

INDEX

Mathieu Vinken and Vera Rogiers (eds.), *Protocols in In Vitro Hepatocyte Research*, Methods in Molecular Biology, vol. 1250, DOI 10.1007/978-1-4939-2074-7, © Springer Science+Business Media New York 2015

R

S

T

U

V

W

Y

MIX
Papier aus verantwortungsvollen Quellen
Paper from responsible sources
FSC® C105338

If you have any concerns about our products,
you can contact us on
ProductSafety@springernature.com

In case Publisher is established outside the EU,
the EU authorized representative is:
Springer Nature Customer Service Center GmbH
Europaplatz 3, 69115 Heidelberg, Germany

Printed by Libri Plureos GmbH
in Hamburg, Germany